THE LIBRARY
ST. MARY'S COLLEGE OF MARYLAND
ST. MARY'S CITY, MARYLAND 20686

BUSTARDS, HEMIPODES, AND SANDGROUSE

Bustards, Hemipodes, and Sandgrouse
Birds of Dry Places

PAUL A. JOHNSGARD

University of Nebraska–Lincoln

Colour plates of paintings by
MAJOR HENRY JONES
from the collection owned by
THE ZOOLOGICAL SOCIETY OF LONDON

Oxford   New York   Tokyo
OXFORD UNIVERSITY PRESS
1991

Oxford University Press, Walton Street, Oxford OX2 6DP

Oxford New York Toronto
Delhi Bombay Calcutta Madras Karachi
Petaling Jaya Singapore Hong Kong Tokyo
Nairobi Dar es Salaam Cape Town
Melbourne Auckland

and associated companies in
Berlin Ibadan

Oxford is a trade mark of Oxford University Press

Published in the United States
by Oxford University Press, New York

© (text) Paul A. Johnsgard, 1991
© (Plates 2–51) Zoological Society of London, 1991
© (Plate 1) Mark Marcuson, 1991

All rights reserved. No part of this publication may be reproduced,
stored in a retrieval system, or transmitted, in any form or by any means,
electronic, mechanical, photocopying, recording, or otherwise, without
the prior permission of Oxford University Press

This book is sold subject to the condition that it shall not, by way
of trade or otherwise, be lent, re-sold, hired out, or otherwise circulated
without the publisher's prior consent in any form of binding or cover
other than that in which it is published and without a similar condition
including this condition being imposed on the subsequent purchaser

A catalogue record for this book is available from the British Library

ISBN 0–19–857698–6

Phototypeset by Wyvern Typesetting Ltd, Bristol
Printed by Butler and Tanner Ltd, Frome, Somerset

Preface and acknowledgements

In the spring of 1986, when I was in England working at the Zoological Society of London on the manuscript for *The quails, partridges, and francolins of the world*, I decided that I should take the time to review all the remaining original watercolours by Major Henry Jones in the Society's library collections. These paintings, which total over 1200, are still mostly unpublished, although they had served as the nuclear illustrative basis for not only my recent book on the quails but also an earlier one on the pheasants of the world. As I looked through these magnificent leather-bound sets, I did so with the thought that perhaps one of the taxonomic groups illustrated by Major Jones might still be badly in need of monographic treatment, and could provide me with the incentive to undertake such a project.

While reviewing these paintings, it became apparent that the family of bustards, all currently recognized species of which had been fully illustrated by Major Jones, offered perhaps the greatest potential for a book. Indeed, Sir Peter Scott had suggested to me in the mid-1970s that a monograph on the bustards of the world would represent an excellent research project, inasmuch as several of the species were even at that time in a probable threatened or endangered status. I also knew that the mating systems of the bustards included some types distinctly similar to those of lek-forming grouse, and that probably some interesting examples of convergent or parallel evolution might exist with that group. It seemed to me, however, that perhaps the available ornithological literature on the bustard family was inadequate to build an entire book around, so I also reviewed and indexed Major Jones's plates of hemipodes. These birds represent a similarly unmonographed group that, although not in any measurable danger from a conservation standpoint, are nearly unique among avian families in that apparently all of the included species have uniformly reversed sexual dimorphism and associated polyandrous mating systems. Additionally, their uncertain taxonomic status and dubious position within the class Aves has been receiving increased attention as a result of recent DNA hybridization data. Major Jones had painted all of the commonly recognized species of hemipodes in the genus *Turnix*, as well as the aberrant Australian plains-wanderer *Pedionomus torquatus*, a species that is no longer included in the family Turnicidae. On the other hand, he had not illustrated the African lark-quail *Ortyxelos meiffrenii*, another aberrant desert-adapted bird that subsequent to Major Jones's painting activities had been taxonomically transferred into the Turnicidae.

In considering these two apparently disparate Old World groups of mostly steppe- and desert-adapted birds, it seemed to me that it might be possible to undertake an interesting behavioural and ecological comparison of the two groups, which are quite divergent anatomically, in body size, and in breeding systems. I thus set about surveying and reviewing the available technical literature on them. I also began contacting various ornithologists who I thought might perhaps already be working on similar monographs and on whose work I might possibly be intruding, or who at least might know of any such work in progress. As a result of these enquiries, I learned from Paul Goriup of the International Council for Bird Preservation that he and Nigel Collar had contemplated writing a book on the bustards of the world and their conservation, and had been assembling information on the group for several years with that possibility in mind. With this news I began to reassess my own position on the desirability of doing a book strongly emphasizing bustards, and tried to think of ways in which I might reduce potential conflicts of interest.

At almost precisely the same time, I learned from Dr Marcia Edwards of the Zoological Society of London that Dr David Thomas of the University of Wales, Cardiff, had reviewed the Jones paintings some years previously, and had been considering writing a book on the sandgrouse of the world for Oxford University Press. She wondered if it might be worth undertaking some kind of a collaborative project, and suggested I contact him. I realized the potential advantages of bringing the sandgrouse into the picture: it would allow for a broader comparative taxonomic treatment, and would be of particular interest because of convergent ecological similarities among these three groups with respect to their generally arid-adapted niche traits. Finally, it would allow for the first illustrative monographic treatment of the entire sandgrouse family; all the species of this family had been beautifully painted by Major Jones, but the paintings had never previously been published. Although Dr Thomas initially agreed to participate in the project, the burden of his administrative duties later made it impossible for him to write his proposed sections, and thus I have necessarily proceeded without him.

Inasmuch as all of these groups have been relatively neglected by taxonomists, there is no general agreement as to the appropriate taxonomic sequences or vernacular names that might be best applied to them. I have therefore diverged in some cases from the taxonomic treatments and vernacular names used in some recent regional references, such as *The birds of Africa* and *The birds of the Western Palearctic*. In particular, I have consistently used 'buttonquails' (rather than bustard-quails, hemipode-quails, or other possible variant choices) as a basis for the English vernacular names of all *Turnix* species, and have limited the term 'hemipodes' to a more inclusive familial designation (namely, including both *Ortyxelos* and *Turnix*, and thus encompassing all the turnicids). I have also avoided using the generally inappropriate Afrikaans-based English vernacular name 'korhaan' for any of the African bustards, and when alternative choices for English vernacular names are available for a species (such as Vigor's bustard vs. black-throated bustard) I have preferentially used the descriptive name rather than the patronymic one. I have not personally reviewed all the subspecies of these groups, and instead have generally accepted the most recently available comprehensive reviews. When such reviews have not been available, I have resorted to accepting those subspecies that were recognized by Peters (1934, 1937) in his *Check-list of birds of the world*, and have added all the more recently described forms known to me. Authors and dates of original descriptions, but not complete citations, are provided for all of these relatively recently described taxa.

Among the persons and institutions who must be individually thanked on my part are especially the Zoological Society of London, for allowing the reproduction of the splendid watercolours by Major Jones that initially stimulated me to consider writing this book. Dr Marcia Edwards and Reginald Fish of that institution were especially helpful in this regard. I also wish to thank Mark Marcuson, of the University of Nebraska State Museum, for rounding-out the colour section of species illustrations with his fine watercolour rendition of the African lark-quail, and to Drs Lester Short, Jun., and Allison Andors of the American Museum of Natural History, for the loan of a specimen from which to paint this rare hemipode. A special debt of gratitude is owed Paul Goriup of the ICBP Bustard Study Group, for sharing his wealth of information, providing photocopied materials, and offering a great deal of advice on bustards, particularly inasmuch as he was then preparing his own book on the bustards of the world. Loren Grueber, of Long Beach, California, also provided me with reprints, translations, and observations on and photos of captive sandgrouse. Ahmad Rahmani of the Bombay Natural History Society provided reprints of a substantial number of critical and relatively inaccessible publications on various Indian bustards. The library resources at the Edward Grey Institute (Oxford University), the Zoological Society of London, and the University of Nebraska–Lincoln were all helpfully placed at my disposal. Additionally, photocopies of otherwise unavailable library materials were provided by William Andelt, M. G. Boobyer, Scott Johnsgard, W. Tarboton, and the Wilson Ornithological Society Library at the University of Michigan. Weights or measurements from museum specimens were very helpfully provided by David Niles (Delaware Museum of Natural History), Belinda Gillies (Museum of Victoria, Australia), and Walter Boles (Australian Museum). Angie Bjorth kindly helped me translate some German publications, and E. M. Mwangli provided me with unpublished data on some Kenyan bustards. Mr Ken Fink read almost my entire manuscript, and various parts of it were also critically read by David Allan (African bustards), Paul Goriup (bustards), Loren Grueber (various sandgrouse), Ahmad Rahmani (Indian bustards), Holger Schulz (little and black-bellied bustards, black-bellied sandgrouse) and Warwick Tarboton (several African bustards).

Lincoln, Nebraska   P. A. J.
1990

Contents

List of colour plates ix
List of figures x
List of distribution maps xi

I COMPARATIVE BIOLOGY

1 Taxonomic history and phyletic relationships 1
2 Zoogeography and evolutionary trends 15
3 Behaviour 30
4 Breeding biology 45
5 Exploitation and conservation 56

II SPECIES ACCOUNTS

A. Hemipodes (Family Turnicidae)

Key to the genera and species of hemipodes 67
Lark-quail 68
Madagascan buttonquail 70
Spotted buttonquail 72
Yellow-legged buttonquail 74
Black-rumped buttonquail 77
Striped buttonquail 79
Red-backed buttonquail 84
Little buttonquail 87
Red-chested buttonquail 89
Painted buttonquail 94
Chestnut-backed buttonquail 97
Black-breasted buttonquail 98
Barred buttonquail 100

B. Bustards (Family Otididae)

Key to the genera and species of bustards 105
Houbara bustard 106
Little bustard 116
Great bustard 125
Australian bustard 137
Great Indian bustard 142
Arabian bustard 150
Kori bustard 153
Denham's (Stanley's) bustard 158
Nubian bustard 163
Heuglin's bustard 165
Ludwig's bustard 167
Black-bellied bustard 169
Hartlaub's bustard 173
White-bellied bustard 176
Little Brown bustard 180
Black-throated (Vigor's) bustard 182

Rüppell's bustard 184
Little black bustard 185
Rufous-crested bustard 189
Blue bustard 194
Bengal florican 196
Lesser florican 202

C. Sandgrouse (Family Pteroclidae)

Key to genera and the species of sandgrouse 209
Pallas's sandgrouse 210
Tibetan sandgrouse 216
Spotted sandgrouse 217
Pin-tailed sandgrouse 221
Chestnut-bellied sandgrouse 227
Namaqua sandgrouse 231
Black-bellied sandgrouse 236
Crowned sandgrouse 239
Black-faced sandgrouse 242
Madagascan sandgrouse 244
Lichtenstein's sandgrouse 246
Double-banded sandgrouse 249
Four-banded sandgrouse 251
Painted sandgrouse 253
Yellow-throated sandgrouse 255
Burchell's sandgrouse 261

References 263
Index 274

Colour Plates

Plates 1–23 appear between pp. 112 and 113; Plates 24–51 appear between pp. 208 and 209

1. Lark-quail
2. Madagascan buttonquail
3. Spotted buttonquail
4. Yellow-legged buttonquail
5. Black-rumped buttonquail
6. Striped buttonquail
7. Red-backed buttonquail
8. Little buttonquail
9. Red-chested buttonquail
10. Chestnut-backed buttonquail
11. Painted buttonquail
12. Barred buttonquail
13. Black-breasted buttonquail
14. Houbara bustard
15. Little bustard
16. Great bustard
17. Australian bustard
18. Great Indian bustard
19. Arabian bustard
20. Kori bustard
21. Nubian bustard
22. Denham's bustard
23. Heuglin's bustard
24. Ludwig's bustard
25. Black-bellied bustard
26. Hartlaub's bustard
27. White-bellied bustard
28. Little brown bustard
29. Black-throated bustard
30. Ruppell's bustard
31. Little black bustard
32. Rufous-crested bustard
33. Blue bustard
34. Bengal florican
35. Lesser florican
36. Tibetan sandgrouse
37. Pallas' sandgrouse
38. Spotted sandgrouse
39. Pin-tailed sandgrouse
40. Chestnut-bellied sandgrouse
41. Namaqua sandgrouse
42. Black-bellied sandgrouse
43. Crowned sandgrouse
44. Madagascan sandgrouse
45. Black-faced sandgrouse
46. Lichtenstein's sandgrouse
47. Double-banded sandgrouse
48. Four-banded sandgrouse
49. Painted sandgrouse
50. Yellow-throated sandgrouse
51. Burchell's sandgrouse

Figures

1. Hypothetical phylogeny of the extant species of hemipodes
2. Hypothetical phylogeny of the extant species of bustards
3. Hypothetical phylogeny of the extant species of sandgrouse
4. Arid and sub-arid regions of the Old World
5. Major desert and steppe areas of the Old World
6. Species–density distribution map of the hemipodes
7. Species–density distribution map of the bustards
8. Species–density distribution map of the sandgrouse
9. Interspecific variation in adult female hemipodes
10. Representative natal plumages of hemipodes, bustards and sandgrouse
11. Interspecific variation in adult male bustard plumages
12. Interspecific variation in wing patterns of bustards
13. Interspecific variation in sandgrouse rectrices
14. Interspecific variation in adult plumages of sandgrouse
15. Interspecific variation in the ventral plumage patterns of male sandgrouse
16. Egocentric behaviour of the great bustard
17. Water-related behaviour of sandgrouse
18. Egocentric behaviour of sandgrouse
19. Defensive crouching in sandgrouse and bustards
20. Crouching incubation posture and defensive 'shock' display of bustards
21. Growth rates of captive-raised barred buttonquails
22. Growth rates of captive-raised great bustards
23. Appearance of great bustard chicks at various ages
24. Social behaviour and tracheal and oesophageal anatomy of the striped buttonquail
25. Heads and bills of *T. sylvatica whiteheadi, T. p. pyrrhothorax, T. p. worcesteri* and *T. p. everetti*
26. Behaviour of the red-chested and barred buttonquails
27. Social behaviour of the painted buttonquail
28. Social behaviour of the houbara bustard
29. Wing plumage variation in the little bustard
30. Social behaviour of the little bustard
31. Aerial displays of male little bustard
32. Development of plumage in the great bustard, and comparison of tracheal and oesophageal anatomy of adult male great bustard and Australian bustard
33. Social behaviour of the great bustard
34. Balloon-display sequence of male great bustard
35. Balloon-display of Australian bustard
36. Balloon-display sequence of male Australian bustard
37. Balloon-display of the great Indian and Arabian bustards
38. Balloon-display sequence of male great Indian bustard
39. Behaviour of the great Indian bustard
40. Stages of neck-inflation and balloon display by male kori bustard
41. Display sequence of male kori bustard, and fighting behaviour
42. Social behaviour of male Denham's bustard
43. Social behaviour of male black-bellied bustard
44. Social behaviour of male Hartlaub's bustard
45. Social behaviour of male little black bustard
46. Social behaviour of rufous-crested bustard
47. Social behaviour of male Bengal florican
48. Social behaviour of male lesser florican
49. Social behaviour of Pallas's sandgrouse
50. Behaviour of Pallas's sandgrouse
51. Parental behaviour of spotted sandgrouse
52. Social behaviour of Namaqua sandgrouse
53. Social behaviour of yellow-throated sandgrouse

Distribution maps

1. Distribution of the lark-quail
2. Distribution of the Madagascan and black-rumped buttonquails
3. Distribution of the spotted and barred buttonquails
4. Distribution of the yellow-legged buttonquail
5. Distribution of the striped buttonquail
6. Distribution of the red-backed buttonquail
7. Distribution of the little buttonquail
8. Distribution of the red-chested buttonquail
9. Distribution of the painted and chestnut-backed buttonquails
10. Distribution of the black-breasted buttonquail
11. Distribution of the houbara bustard
12. Distribution of the little bustard
13. European distribution of the little bustard
14. Distribution of the great bustard
15. European breeding distribution of the great bustard
16. Distribution of the Australian bustard
17. Distribution of the great Indian bustard
18. Distribution of the Arabian and kori bustards
19. Distribution of Denham's bustard
20. Distribution of Ludwig's, Heuglin's and Nubian bustards
21. Distribution of the black-bellied bustard and blue bustards
22. Distribution of Hartlaub's bustard
23. Distribution of the white-bellied bustard
24. Distribution of the little brown, black-throated and Rüppell's bustards
25. Distribution of the little black bustard
26. Distribution of the rufous-crested bustard
27. Distribution of the Bengal florican
28. Distribution of the lesser florican
29. Distribution of Pallas's and Tibetan sandgrouse
30. Distribution of the spotted sandgrouse
31. Distribution of the pin-tailed sandgrouse
32. Distribution of the chestnut-bellied sandgrouse
33. Distribution of the Namaqua, Madagascan and black-faced sandgrouse
34. Distribution of the black-bellied sandgrouse
35. Distribution of the crowned sandgrouse
36. Distribution of the painted and Lichtenstein's sandgrouse
37. Distribution of the four-banded and double-banded sandgrouse
38. Distribution of the yellow-throated sandgrouse
39. Distribution of Burchell's sandgrouse

*In memory of
Peter M. Scott (1909–1989)
who greatly helped me to
fulfil my life's goals*

I · COMPARATIVE BIOLOGY

1 · Taxonomic history and phyletic relationships

It is extremely difficult to write with any confidence on the evolutionary relationships of the bustard, hemipode, and sandgrouse groups; in almost every case they represent varying degrees of historic taxonomic controversy or mystery, which even today remain somewhat unsettled. Each of the three groups requires a separate discussion, for there is little if any indication that they are in any way closely related to one another.

HEMIPODES (family Turnicidae)

The taxonomic history and evolutionary relationships of the buttonquail assemblage are perhaps as confusing and unsettled as those of any group of birds. The taxon was unknown to Linnaeus, and when the most widely distributed species (*sylvatica*), was finally described in 1787 it was variously icluded by late eighteenth-century taxonomists in the inclusive grouse genus *Tetrao* or the correspondingly inclusive partridge genus *Perdix*. The buttonquail genus *Turnix* was erected in 1791 by Joseph Bonnaterre in the *Tableau encyclopédique et méthodique*. Interestingly, the aberrant African lark-quail was initially included within this same genus by Louis-J.-P. Vieillot when he first described this species in 1819. However, he transferred it to the monotypic genus *Ortyxelos* six years later, where it has since remained. Most if not all authorities of the early 1800s continued to retain these distinctively three-toed (or 'hemipode') quail-like species within the general assemblage of gallinaceous birds that includes the true quails, junglefowl, grouse, and similar forms having fowl-like bills and feet. For example, in his landmark review of avian genera, Gray (1844–9) included the hemipodes in the grouse family Tetraonidae.

Perhaps the first person to try to cope seriously with the unusual anatomical attributes of the hemipodes was T. H. Huxley. In 1867 he followed long-standing tradition and included them with the other more typical quail-like birds in his gallinaceous assemblage 'Alecteromorphae'. However, shortly afterwards (1868) he decided that the hemipodes ('Hemipodidae') differ more from the Alecteromorphae, the sandgrouse, and the dove-pigeon assemblage, than these groups do from one another, a decision apparently based mainly on osteological traits that had been documented a few years previously by Parker (1864). Parker had thereby judged the hemipodes to belong 'between the Tinamous and the true Gallinae', but Huxley argued that the chief relationships of the genus *Turnix* (= '*Hemipodius*') are on the one hand with the tinamous, on another with sandgrouse, and on a third with the plovers, *Pedionomus* being perhaps 'the connecting link' between the plovers and the buttonquails. Huxley concluded that it was 'impossible to include *Hemipodius* with either the Tinamorphae (= tinamous) or the Charadriomorphae (= plovers), and still less with the Pteroclomorphae (= sandgrouse); and I see no alternative but to make it the type of an independent group, which may be called the Turnicomorphae'.

The perspicacious advice went unheeded at the time, and biologists generally continued to affiliate the hemipodes with the gallinaceous birds. As late as the 1890s some authorities continued to argue in favour of a close galliform–hemipode relationship (e.g., Gadow 1892). However, in Ogilvie-Grant's (1893) catalogue of the birds of the British Museum, he elevated the hemipodes to the ordinal rank 'Hemipodii', a category that had initially been proposed two years earlier by R. Bowdler Sharpe. Ogilvie-Grant defined this group as containing the two genera *Turnix* and *Pedionomus*, and placed it in taxonomic sequence following his orders Gallinae (containing the families Tetraonidae, Phasianidae, Megapodiidae, and Cracidae) and Opisthocomi (containing only the aberrant South American hoatzin *Opisthocomus hoazin*). The monotypic lark-quail genus *Ortyxelos* was thus excluded from the hemipodes, and instead was tentatively incorporated a few years later into the courser group by Sharpe (1896).

This general state of affairs persisted until Lowe (1923) investigated the anatomy of *Ortyxelos*. Lowe judged that this genus, rather than closely related to such charadriiform birds as the pratincoles and coursers of the family Glareolidae, is indeed a close relative of *Turnix*. He then proceeded to discuss the possible affinities of the hemipodes as a group. He believed that, far from being a part of the gallinaceous assemblage ('with which they have only the very feeblest connections'), hemipodes are 'indeed so generalized and so vastly ancient in origin—being a medley of columbine, pteroclidine, passerine, and even charadriine factors ... that they might indeed be regarded as living relics of a group so ancient and generalized that they cannot be associated with any of the recognized bird-groups of the present day'. Lowe concluded that the hemipodes are the most 'composite' of all avian groups, and so generalized as to 'resist any effort to definitely identify them with any present-day group'. He thought that, together with the sandgrouse and pigeons, they probably drew their ancestral characters 'from some far-distant early bird-group which was so composite as to contain also coracomorphine (= crow-like) and even charadriine factors'.

Unfortunately, hardly anyone has seriously evaluated these arguments, nor has anyone independently examined the possible relationships of the hemipodes since Lowe's time. Thus, when Peters (1934) published the second volume of his world check-list of birds, he considered the hemipodes as simply comprising a suborder (Turnices) in the Gruiformes, presumably following the earlier classification of Wetmore (1930). Based on his interpretation of electrophoretic data from egg-white proteins, Hendrickson (1969) concluded that the hemipodes are apparently closely related to the Rallidae, and that both are part of a larger natural group of gruiform birds that includes the Psophiidae, Eurypygidae, and Heliornithidae. However, in his recent proposed revision of the world's birds, Cracraft (1981) considered the turnicids as only dubious (*incertae sedis*) members of the Gruiformes, suggesting that some of their more primitive traits might serve to unite them with the columbiform birds.

When Sibley and Ahlquist (1985) initially reported the results of their DNA hybridization studies on African birds, they simply stated that the genus *Turnix* appeared to have no close relatives. Later, Sibley *et al.* (1988) gave the hemipodes a problematic (*incertae sedis*) taxonomic status, pointing out that their 27 available comparative data points for this group exhibited an unusually large range of numerical values. The smaller delta values (indicating closer phyletic relationships) occurred between *Turnix* and members of the typical gruiform birds on the one hand, and on the other with their recognized shorebird assemblage (Charadriides'). Relatively large delta values occurred between *Turnix* and the ratites, galliforms and anseriforms. These authors thus regarded the hemipodes as comprising at least a monotypic order (Turniciformes), and perhaps even constituting a unique but unnamed infraclass.

However, they cautioned against accepting these somewhat heretical conclusions prematurely, noting that their data were still not sufficiently convincing to rule out the possibility that the hemipodes might actually belong taxonomically among the gruiform birds, where they have regularly been placed by most taxonomists in recent decades.

Unfortunately, these authors were apparently unable to include *Ortyxelos* in their within- and between-group comparisons. They did, however, confirm a relationship of the Australian plains-wanderer *Pedionomus* with the charadriiform shorebirds, specifically with the seedsnipes (Thinocoridae). These results agree with the findings of Bock and McEvey (1969), who in analysing the osteology of *Pedionomus* concluded it to be quite distinct from that of *Turnix*, and of Olson and Steadman (1981), who argued for its seedsnipe affinities.

Beyond these considerations of general placement of the entire hemipode group, little can be said of their more detailed affinities, such as possible intrageneric relationships of *Turnix*. All the species of buttonquails are extremely similar in most aspects of their morphology, and they have not attracted the attention of taxonomists to any great degree. They are small and extremely unobtrusive birds that are notoriously difficult to collect, and thus are poorly represented in most museum collections. Furthermore, the fact that they often exhibit insular or disruptive distributions tends to cause local phenotypic variations that make the establishment of realistic species limits difficult at best. For example, Ogilive-Grant (1889) recognized 23 species of *Turnix*, nearly half of which are currently regarded either as subspecies or are considered as representing synonyms of some other valid form. He observed that 'to define each of the several species ... is by no means too easy a matter', and furthermore noted that 'most of the species pass through intricate changes of plumage, and every character seems to be subject to variation'. In 1894 the number of hemipode species recognized by him was reduced to 21, but basically his taxonomy otherwise remained essentially unchanged (Table 1).

With these kinds of confusing geographic evidence and complex morphological variations at hand, it is not surprising that the hemipodes have since remained largely ignored or avoided by taxonomists. Generally, most recent taxonomists have followed Peters's (1934) revision as to species-level taxonomy. He accepted a total of thirteen species of *Turnix*. This included the recognition of *worcesteri* as a full species, but Peters did not recognize *everetti* as specifically distinct from *sylvatica*. Peters's linear arrangement of species (Table 1) has also since been generally followed by most recent authorities.

Table 1. Some representative classifications of the hemipodes (Turnicidae)

Ogilvie-Grant (1893)	Peters (1934)	This study
Turnix taigoor[1]	*Turnix sylvatica*	*Ortyxelos meiffrenii*
T. fasciata[1]	*T. worcesteri*	*Turnix nigricollis*
T. rufilatus[1]	*T. nana*	*T. ocellata*
T. powelli[1]	*T. hottentotta*	*T. tanki*
T. sylvatica	*T. tanki*	*T. hottentotta*
T. dussumieri[1]	*T. suscitator*	*T. sylvatica*
T. nana	*T. nigricollis*	*T. maculosa*
T. hottentotta	*T. ocellata*	*T. velox*
T. blanfordi[2]	*T. melanogaster*	*T. pyrrhothorax*[4]
T. tanki	*T. varia*	*T. castanota*
T. albiventris[2]	*T. castanota*	*T. varia*
T. maculosa	*T. pyrrhothorax*	*T. suscitator*
T. saturata[3]	*T. velox*	*T. melanogaster*
T. rufescens[3]	*Ortyxelos meiffrenii*	
T. ocellata		
T. nigricollis		
T. melanogaster		
T. varia		
T. castanonota		
T. pyrrhothorax		
T. velox		
Pedionomus torquatus		

[1] Synonymous with *suscitator* (in part) of later authors.
[2] Synonymous with *tanki* (in part) of later authors.
[3] Synonymous with *maculosa* (in part) of later authors.
[4] Tentatively including *worcesteri* and *everetti*.

Subsequent to Peters's revision, Sutter (1955a) recommended the 'splitting' of Peters's Australasian species *sylvatica* into two mostly allopatric species. According to him, an assemblage of seventeen races should be included within his newly recognized species *maculosa*, whereas six were retained within the nominate assemblage *sylvatica*. Limited sympatry perhaps occurs between representatives of these two groups in the southern Philippines. Sutter also considered both of the distinctively large-billed and allopatric forms *everetti* and *worcesteri* to comprise full species in the *pyrrhothorax* species-group. He did not, however, undertake a complete revision of the genus *Turnix* as a whole. Additionally, some recent Australian authorities have advocated the splitting of *castanota* into two species, the more easterly and wholly allopatric Cape York population *olivei* being recognized as distinct.

Given the low level of morphological or other taxonomically relevant information so far available on the henipodes, any proposed classification of the genus *Turnix* must be considered a highly tentative one at best. Thus the one that I use in this book (Table 1), an eclectic and tentative sequence based mainly on plumage characteristics, differs somewhat in species sequence from previous ones. In terms of species that are recognized, it is nevertheless quite similar to Peters's (1934) classification, except that *maculosa* is accepted as a full species as recommended by Sutter (1955a), *nana* and *hottentotta* are considered conspecific as in most recent treatments of African birds, and *worcesteri* is tentatively considered as a race of *pyrrhothorax* rather than as a full species. The associated phylogram that I have provided (Figure 1) is similarly a very tentative one.

BUSTARDS (family Otididae)

The traditional order Gruiformes, as constituted by Peters (1934), is one that has been rather generally regarded as a collection of seemingly rather disparate and perhaps distantly related forms. For example, the same array of species had earlier been placed in no less than ten orders by Stresemann (1927–34). In addition to the bustards, which Peters placed in a monotypic suborder Otides, the hemipodes were also included by him in this same gruiform order, but were designated as a separate suborder (Turnices).

Bustards, Hemipodes, and Sandgrouse

Earlier, the bustards had been given a variety of taxonomic treatments, including an early placement by Linnaeus among the ratites. This treatment continued to receive attention until at least the mid-1800s, when for example Gray (1844–9) included the bustards as a subfamily of the ostrich family Struthionidae. Perhaps the first relatively modern assessment of the bustards was that of Huxley (1867), who erected an order (Geranomorphae) that included the crane-like and rail-like birds as 'typical members'. Huxley also regarded the bustards as members of this order, albeit ones having apparent evolutionary connections with his shorebird order Charadriomorphae.

A crane- and rail-like affinity of the bustards was also suggested by Garrod (1874a), based on his pelvic musculature studies. He included both the bustards and the rails as adjacent families within his large order Galliformes. However, he also included within his enlarged family Otididae such dissimilar birds as the South American cariamas (*Cariama* spp.), the African secretary-bird (*Sepentarius serpentarius*), and tentatively even the flamingos.

Somewhat later, Gadow (1892) proposed a classification for the world's birds, in which the bustards comprised one of five families (Otides) within his order Gruiformes. Shortly thereafter, Sharpe's (1894) landmark catalogue of the gruiform birds in the collection of the British Museum appeared. There the bustards comprise the last of seven families recognized within his order Alectorides (the others being the Aramidae, Eurypygidae, Mesitidae, Rhinochetidae, Gruidae, and Psophiidae). Sharpe thereby excluded the buttonquails from this group.

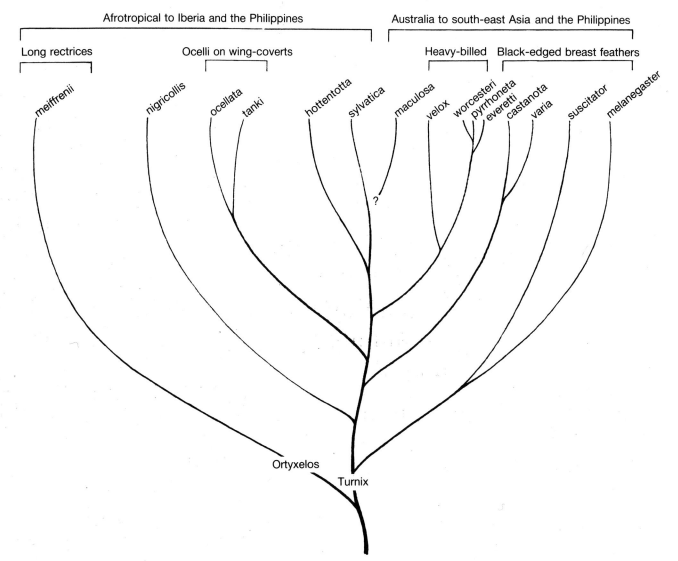

Fig. 1. Hypothetical phylogeny of the extant species of hemipodes. The forms *worcesteri* and *everetti* are of uncertain taxonomic status and, although sometimes regarded as full species, are here considered races of *pyrrhothorax*.

Sharpe recognized twelve genera within the bustard family Otididae, and a total of thirty species. Not surprisingly, this classification was the one used by Major Henry Jones as a systematic basis for painting the taxa of bustards in the British Museum, just as he used the corresponding taxonomy proposed at about the same time by Ogilvie-Grant (1893) for illustrating the hemipodes and sandgrouse.

Thus, by the start of the twentieth century, ornithological opinion had rather clearly placed the bustards fairly firmly inside the gruiform order, with only a few dissenting views. One such dissenter was Stresemann (1927–34), who questioned their gruiform affinities. Similarly, Hopkins (1942) believed on the basis of their mallophagan parasites that the bustards were not related at all to the gruiform birds, but instead might have their affinities with the galliform group. Nevertheless, Peters (1934) recognized the bustards as comprising a monotypic suborder of the Gruiformes, and furthermore constituted the group as a family consisting of eleven genera and twenty-four extant species (Table 2).

Since the appearance of Peters's *Check-list*, little additional taxonomic juggling of the bustards has occurred, other than a slight reduction in the numbers of species and genera recognized by more recent authorities (Table 2). There has seemingly been a nearly random juggling of the sequence of bustard genera by various authors, but no effort on the part of anyone to examine seriously any objective basis for defining these genera or establishing their most realistic taxonomic sequence. Similarly, the nearest relatives of bustards remain somewhat uncertain. Using electrophoretic data from egg-white proteins, Hendrickson (1969) suggested that the bustards should be removed from the order Gruiformes, and that their closest relatives are unknown. Recently, Cracraft (1981) similarly suggested that the bustards could not be easily placed within the typical Gruiformes (suborder Grues), nor with the South American seriamas (Cariamidae), and in his view their actual evolutionary relationships are still unresolved. However, a recent chromosomal study (Nishida *et al.* 1981) has noted some apparent similarities between the karotypes of *Otis tarda* and typical cranes (*Grus*).

Perhaps the most important single new source of evidence on bustard relationships comes from the recent molecular systematics work of Sibley and Ahlquist (1985). Based on data from DNA hybridization, these authors judged the bustards to have branched off from the other gruiform birds about 77 million years ago, as part of a lineage that also gave rise to the typical cranes (Gruidae). This same lineage also produced such New World endemics as the limpkin (*Aramus quarauna*), the sungrebe (*Heliornis fulica*), the trumpeters (*Psophia* spp.) and the seriamas, and the endemic New Caledonian kagu (*Rhynochetos jubatus*). More recently, these data from DNA hybridization have been refined and

Table 2. Some representative classifications of the bustards (Otididae) (number of accepted species per genus is indicated parenthetically)

	Peters (1934)	Osborne *et al.* (1984)	Urban *et al.* (1986)	This study
	Tetrax (1)	*Otis* (1)	*Tetrax* (1)	*Chlamydotis* (1)
	Otis (1)	*Ardeotis* (4)	*Neotis* (4)	*Tetrax* (1)
	Neotis (5)	*Neotis* (4)	*Chlamydotis* (1)	*Otis*[3] (1)
	Choriotis[1] (4)	*Eupodotis* (9)	*Ardeotis* (4)	*Ardeotis* (4)
	Chlamydotis (1)	*Sypheotides* (1)	*Otis* (1)	*Neotis* (4)
	Lophotis (2)	*Houbaropsis* (1)	*Eupodotis* (9)	*Eupodotis* (11)
	Afrotis (1)	*Tetrax* (1)	*Houbaropsis*[2] (1)	
	Tetrax (1)	*Chlamydotis* (1)	*Sypheotides*[2] (1)	
	Eupodotis (5)			
	Lissotis (2)			
	Houbaropsis (1)			
	Sypheotides (1)			
Total genera	11.0	8.0	8.0	6.0
Total species	24.0	22.0	22.0	22.0
Monotypic genera	6.0	5.0	5.0	3.0
Species/genus	2.2	2.5	2.5	3.7

[1] Synonymous with *Ardeotis* of later authors.
[2] Extralimital to book's geographic coverage, but recognition implied.
[3] *Otis* could easily be expanded to include both *Tetrax* and *Chlamydotis*, as suggested by Dementiev and Gladkov (1951).

re-evaluated, and a formal proposed taxonomy of the birds of the world has been set forward (Sibley *et al.* 1988). In their taxonomy the bustard group has been elevated to the status of an infraorder (Otidides), and is sequentially placed between the neotropical sunbittern (*Eurypyga helias*, infraorder Eurypygides) and the cranes and just-mentioned crane-like birds (infraorder Gruides).

At the generic level, the bustards have traditionally been classified in what might almost be called a chaotic manner, as is apparent from the lack of evident consistency shown in Table 2. As with various other largely polygynous bird groups, many of these supposed genera have been largely or entirely based on male display structures such as crests and ornamental plumes that probably function as interspecific isolating mechanisms or even as intraspecific sexual signals and have no significance at higher taxonomic levels. Snow (1978) commented that many of the smaller African species traditionally placed in small or monotypic genera might easily be placed in the single *Eupodotis*, and his suggested classification was later used by Urban *et al.* (1986). Classifications relying entirely on bustard traits other than those associated with adult males are, unfortunately, still lacking. Although only an incomplete DNA survey of the world's bustard taxa has been made, Sibley and Ahlquist tentatively judged that perhaps some of the bustard genera that have been recognized as distinct by Peters (1934) and most other previous authors (*Lissotis, Lophotis,* and *Afrotis*) could probably be encompassed within the genus *Eupodotis*.

A summary of the species-level classification of the bustard family that has been used in this book is presented in Table 2, and an associated highly tentative phylogram is provided in Figure 2. The suggested classification attempts to avoid placing much associative value on traits typical of adult males, and all of the most questionable bustard genera except perhaps *Tetrax* have thereby been eliminated. Superimposed on the phylogram are brackets that associate apparent clusters of species according to their general zoogeographic relationships and to available information on male display features or behavioural categories. It should be emphasized that this is a highly tentative classification and associated phylogeny, and that both might well have been constructed in other ways, given the still highly limited array of information available on the group.

SANDGROUSE (family Pteroclidae)

The taxonomic status and phyletic relationships of the sandgrouse have been one of the most persistently knotty problems of avian taxonomy, and a full historical survey of this question would require much more space than is warranted here. Nearly all the significant aspects of the controversy have been admirably summarized (with over fifty relevant references) by Sibley and Ahlquist (1972), and a somewhat more recent summary has been provided by Fjeldså (1976).

The sandgrouse were placed in the grouse genus *Tetrao* by Linnaeus (who in the 1766 edition of his *Systema Naturae* recognized only the single species *alchata*). Similar treatments of the sandgrouse were typical of other early post-Linnaean writers, such as P. S. Pallas and J. F. Gmelin. This gallinacious association was presumably based on the birds' grouse-like feathered tarsi, precocial young, feathered nostrils, and seed-eating anatomical adaptations such as a well-developed crop and a relatively short, blunt bill. However, in 1815 the sandgrouse were first separated from the grouse by Coenraad Temminck, who in his *Histoire Naturelle Générale des Pigeons et Gallinacés* devised the generic term *Pterocles* for the group.

A few decades later, Gray (1844–9) recognized the sandgrouse family Pteroclidae. However, he placed it within his order Gallinae, a large taxonomic category that encompasses the present-day Galliformes. Similarly, Huxley (1867) included the sandgrouse in his taxonomic category 'Alecteromorphae', which similarly includes the present-day Galliformes, as well as such non-galliform birds as the Turnicidae. However, Huxley noted that in some respects the sandgrouse approached the pigeons (his 'Peristeromorphae'), and a year later (1868) he said that sandgrouse might better be called the 'pigeon-grouse'. He further stated that in many ways the sandgrouse 'are completely intermediate' between pigeon-like and grouse-like birds and provide 'an absolute transition between these groups'. This being the case, he suggested that, rather than including them in either group and thus destroying its definition, they should be placed in a group by themselves, and of equal taxonomic rank to the other two. He thus proposed a new avian group 'Pteroclomorphae', to provide this degree of recognition.

The pigeon-like similarities of the sandgrouse that had been noted by Huxley were by no means a novel idea, for at least as early as 1840 C. L. Nitzsch had included the sandgrouse in his 'Columbinae' assemblage on the basis of their feather-tract similarities. Parker (1864) had similarly noted their pigeon-like syringeal structures. Huxley had based his conclusions as to the intermediacy of sandgrouse between grouse and pigeons primarily on five groups of osteological traits, which are essentially grouse-like in one case (vertebrae number), pigeon-like in

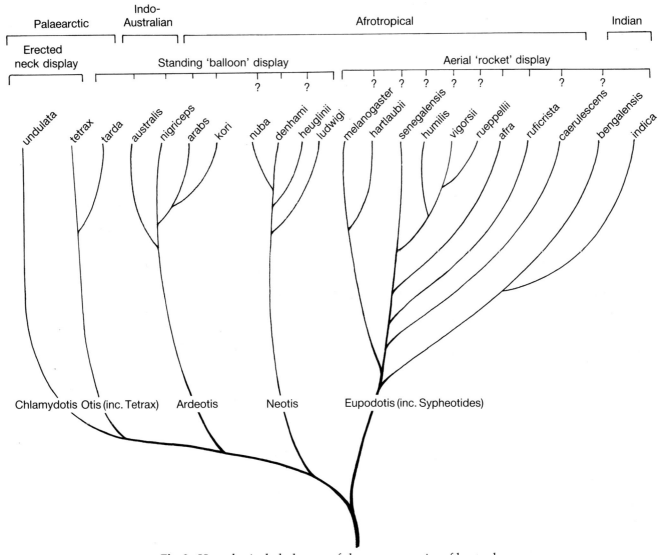

Fig. 2. Hypothetical phylogeny of the extant species of bustards.

another (sternal and pectoral bones), and variously intermediate in the other three (skull, pelvic, and foot structures). Soon thereafter, Garrod (1874b) concluded that the grouse-like skeletal features of the sandgrouse were simply the result of convergence, and that the majority of their features placed them clearly in the pigeon group. He thus included the sandgrouse family Pteroclidae within his larger group (Columbae) or pigeon-like birds. However, D. G. Elliot (1878, 1885) continued to argue for an intermediate position of the sandgrouse between the gallinaceous birds and the pigeon group, based mainly on the evidence assembled by Huxley, Parker, and others. Gadow (1882) similarly reviewed the anatomical and biological evidence then available as to the relationships of the sandgrouse, and believed them to be more closely related to the galliform birds ('Rassores') than are the pigeons, and in turn more closely related to the pigeons than to the shorebirds.

This argument for an intermediate position of the sandgrouse between the gallinaceous and columbiform birds persisted in the literature through the remainder of the nineteenth century with little variation; Ogilvie-Grant (1893) placed the sandgrouse in his order Pterocletes, immediately prior sequentially to his galliform order Gallinae, and considered them 'a well-marked order between the Columbae and the Gallinae'. Virtually the same position was taken by Shufeldt (1901), whose osteological analysis of the skull suggested to him that an intermediate taxonomic position between these two groups was most appropriate.

Although adopting this argument in part, Beddard (1898) emphasized that the sandgrouse occupied a 'lower' phyletic position than do the pigeons, and whilst they might have given rise to the pigeons, the reverse was not true. Further, he stated that arguments that had been advanced, as for example by Huxley (1868), favouring a relationship with the gallinaceous birds, might be applied with equal strength as indicating an affinity with the shorebird assemblage. He concluded that it 'seems reasonable to look upon the Ptercletes as not far from the stock which produced the Limicolae, which itself was possibly not far again from the primitive gallinaceous stock'.

In an examination of microscopic feather structure, Chandler (1916) judged that sandgrouse were more like the gallinaceous birds than any other group, and that at least in this regard there was no indication of any close relationship with the pigeons. Reinforcing this view, Waterston (1928) concluded from the evidence provided by mallophagan parasites that sandgrouse and pigeons were apparently not closely related, but rather the taxonomic position of the sandgrouse 'would appear to be between the grouse and the pheasants'.

Nevertheless, both Wetmore (1930) and Peters (1934) included the sandgrouse as a distinct suborder in the pigeon assemblage Columbiformes. This position was upheld by Niethammer (1934), who found the crop structure of sandgrouse to be more like that of pigeons than of galliform birds, warranting in his opinion the status of a separate family within the columbiform group. In a general survey of the biology and natural history of the sandgrouse, Hüe and Etchécopar (1957) accepted the idea that sandgrouse belonged taxonomically within the Columbiformes. During the same year Verheyen presented a quite different sandgrouse taxonomy on the basis of his somewhat questionable osteological analyses. In his dubious classification the sandgrouse were included in a new order 'Turniciformes' that also included not only the hemipodes but three additional aberrant families represented by the Madagascan mesites (Mesiornithidae), the Australian plains-wanderer (Pedionomidae) and the South American seed-snipes (Thinocoridae). However, only a few years later (1961) he realigned the sandgrouse with the pigeons, placing them again in their traditional place, as a suborder of that group.

Adding to a long-standing controversy as to the method by which sandgrouse drink (which apparently began with D. G. Elliot's assertion in 1885 that sandgrouse drink in a pigeon-like manner), Wickler (1961) stated that sandgrouse drink in a different manner from that of pigeons, which do not have to raise the head periodically to swallow. Instead, Wickler affirmed that the birds do indeed occasionally interrupt their drinking to make swallowing movements with the head tilted up. Goodwin (1965) confirmed this point, saying that their drinking methods did not indicate any relationship betweeen sandgrouse and pigeons. Rather, he believed that 'sandgrouse are probably most closely related to the true plovers to which, in spite of the differences involved in their adaptation to living in arid regions and feeding on seeds, they show many similarities, especially of behaviour'. He also believed that drinking methods of closely related birds may vary considerably, and may not even be useful for taxonomic purposes as 'a specific character'. For example, hemipodes also sometimes drink continuously without raising the head, presumably sucking or pumping up water (Cramp and Simmons 1980). The drinking behaviour of sandgrouse was also documented by Cade et al. (1966), who noted that the birds typically drink by taking a draught of water into the mouth by sucking once or twice, and then raising the head to swallow, thus differing both from the pure sucking method of pigeons and from the typical 'dip-and-tilt' method of most other birds.

Maclean (1967) shortly thereafter provided a thorough discussion of the phylogenetic relationships of the sandgrouse. In this paper he reviewed the previously available evidence from morphology (mainly using data of Beddard 1898), behaviour (drinking, flying, calling, stretching, scratching, etc.) and breeding biology (nest location, clutch size, egg type and downy pattern). Based on this previously available evidence, which was supplemented by new electrophoretic information of egg-white proteins, Maclean firmly rejected any relationship between sandgrouse and pigeons, and instead argued for a relationship between the sandgrouse and shorebirds. He concluded by recommending that the sandgrouse be considered a suborder of the Charadriiformes, placed in taxonomic sequence adjacent to the sub order Charadrii.

Maclean's conclusions were immediately attacked by Stegmann (1968, 1969). Stegmann claimed on the basis of his earlier anatomical studies that, rather than being more primitive than pigeons and giving rise to them (as most earlier writers had asserted), the sandgrouse had instead been derived from them and in the process of becoming ground-adapted had become more specialized than pigeons. However, both groups in Stegmann's view were part of the same general phyletic assemblage (Pteroclo-Columbae), and of more ancient lineage than either the galliform or charadriiform birds, which in his opinion were only convergently similar to the sandgrouse.

Stegmann was in turn responded to by Maclean (1969), who not only chided him for considering only morphological traits and ignoring Maclean's data, but also criticized his position that sandgrouse were derived from arboreal pigeon-like ancestors. Instead, Maclean argued that virtually all their traits indicated that sandgrouse had never been arboreal, and indeed pointed out that the essentially arboreal traits of pigeons (such as nest-building retention and the lack of development of cryptic egg colouration or tendency toward precocity in the young) were still strongly evident in those species that had become ground nesters.

Supplementing this controversy was Timmerman's (1969) observations that the mallophagan lice of sandgrouse indicated a close relationship with neither the pigeons nor the shorebirds. However, Stegmann's position was later defended by Olson (1970), who likewise relied entirely on skeletal similarities between sandgrouse and pigeons to argue for this relationship. After summarizing all these diverse opinions and adding their own interpretations of the egg-white electrophoresis data, Sibley and Ahlquist (1972) noted that the sandgrouse were more similar to the shorebirds than to any other group, but because of conflicting data from other sources they recommended a continued inclusion of the sandgrouse in the order Columbiformes. They suggested that perhaps the Columbiformes had an evolutionary connection with the shorebirds through the sandgrouse, but on the other hand the sandgrouse might represent a separate evolutionary trend. Somewhat similarly, Wolters (1975) regarded the sandgrouse, pigeons and parrots as comprising a single phyletic group, based on cladistic principles.

Only a few years later, Fjeldså (1976) added additional fuel to the controversy with his extensive study of the sandgrouse. He applied cladistic techniques to data associated with internal anatomy, feathers, pterylosis, egg traits, breeding biology, drinking behaviour, water-transport behaviour, mallophagan parasites, and zoogeography. He concluded that the sandgrouse 'stand below' the pigeons in their cladistic branching sequence, and that coursers (Glareolidae), sandgrouse and pigeons shared a common ancestor that existed subsequent to an ancestor that had been shared by shorebirds in general. He proposed a phylogeny in which the sandgrouse diverged from Old World ancestral coursers that had probably abandoned insect-eating for seed- and bud-eating on low ground vegetation. This perhaps occurred during the very early Tertiary era, when xeric conditions were spreading over much of the Old World. This line of vegetarian courser-like birds produced the sandgrouse, from which doves and pigeons eventually diverged, perhaps as early as during the Eocene. Fjeldså believed that this phylogeny explained the resemblances between sandgrouse and both waders and pigeons, and avoided presupposing any major reversals in evolutionary processes.

In order to accommodate his cladistic findings to taxonomic requirements, Fjeldså recommended moving both the pigeons and the sandgrouse into the order Charadriiformes, as a suborder Columbae, and suggested placing this group sequentially between the Charadrii and the larid–alcid group. He further recommended dividing the Columbae into two super families, the pigeon-like Columboidea and the Glareoloidea, the latter group containing the two families Glareolidae and Pteroclididae.

Taking such conflicting views into consideration, Voous (1973) placed the sandgrouse between the charadriiform and columbiform birds as a separate order Pteroclidiformes, a procedure that was later followed by Cramp (1985). In his proposed classification of the birds of the world, Cracraft (1981) stated that in his opinion the sandgrouse were a sister-group with the columbids, but that this collective columbiform group might either be a sister-group of the Charadriiformes in general, or a sister-group with some portion of the Charadriiformes, possibly the Glareolidae as suggested by Fjeldså. Cracraft believed that these and other alternatives had not yet been satisfactorily evaluated.

Perhaps the most significant recent contribution to this question has been that provided by the DNA hybridization evidence of Sibley and Ahlquist (1985), who not only clearly placed the sandgrouse 'within the charadriiform cluster', but also estimated its point of branching away from the more typical shorebirds at about 77 million years. This time estimate is not very different from Fjeldså's estimate of a Tertiary separation of proto-pteroclidid stock, but, in contrast to Fjeldså's view, Sibley and Ahlquist placed the glareolids (Fjeldså's proposed ancestral stock for the sandgrouse) well away from the sandgrouse and adjacent to the alcid–larid assemblage. These authors concluded that their infraorder 'Pterocletes' represented the sister group of the plovers and their near allies. This classification was further refined and modified in Sibley et al. (1988), who retained an infraorder status (now called the 'Pteroclides') for the sandgrouse within the Charadriiformes. However, at this time they modified their shorebird taxonomy so as to include the scolopacid sandpipers and their relatives within their proposed infraorder Charadriides, the suggested shorebird sister-group to the sandgrouse. The pigeon and dove family was retained in its own monotypic order Columbiformes, and this order was sequentially grouped more closely with the gruiform birds than with the charadriiform assemblage.

It thus seems likely that, after about a century and a half of controversy, the real phyletic relationships of the sandgrouse can be stated with some degree of confidence as being with the typical shorebirds, much in the way that Goodwin and Maclean have proposed.

Beyond the position of the sandgrouse in the broader taxonomic scheme, there still remains the question as to the number of genera and species of sandgrouse that should be recognized, and their most appropriate linear sequence. In his landmark study of the family, D. G. Elliot (1878) suggested that two genera were 'amply sufficient' to account for the variation existing among the species of the group, and that these two taxa exhibited several differences in skull structure, as well as differences in the structure of the sternum and scapula. He recognized a total of 14 species of *Pterocles* and two of *Syrrhaptes*, which collectively included nearly all of the currently recognized species excepting *burchellii*, which had not yet been discovered. However, Ogilvie-Grant (1893) recognized three sandgrouse genera, by separating out the four pin-tailed species of *Pterocles* as a separate genus *Pteroclurus*.

Another important contribution to the generic-level taxonomy of the Pteroclidae was that of Bowen (1927), who used relative diurnal activity and plumage traits (number and shape of rectrices, shape of wingtips and male plumage patterns) to reclassify the group (Table 3). On the basis of these data he recommended the recognition of three genera, *Syrrhaptes*, *Pterocles*, and *Dilophilus*, the last-named newly proposed to encompass five apparently closely related species (*lichtensteinii*, *bicinctus*, *fasciatus*, *quadricintus* and *decoratus*). However, in his world *Check-list*, Peters (1937) did not follow the advice of Bowen. Instead, Peters continued to accept only two genera (Table 3), although the sequence and number of species that he adopted was identical to that proposed by Bowen.

Nearly all subsequent authors have followed Peters's classification, although Maclean (1984) suggested on the basis of plumage and zoogeographic considerations that 'six or seven more or less well defined groups, each of which might qualify as a subgenus, although not all of equal rank' might be recognized (Table 3).

Finally, for purposes of overall summary and comparison, a tentative phyletic dendrogram has been drawn for the sandgrouse (Figure 3), which also provides bracketing information on zoogeography and plumage characteristics that seem to be useful in assessing possible relationships within the group. There is a generally greater amount of such seemingly useful information on the sandgrouse than on the bustards, and as a result a somewhat higher level of potential confidence in the affinities that have been suggested here. There seems to be little doubt

Table 3. Some representative classifications of the sandgrouse (Pteroclidae)

Bowen (1927)	Peters (1937)	Maclean (1984)
Syrrhaptes tibetanus	*Syrrhaptes tibetanus*	*Syrrhaptes* (Group 2)
S. paradoxus	*S. paradoxus*	*S. paradoxus*
Pterocles alchata	*Pterocles alchata*	*S. tibetanus*
P. namaqua	*P. namaqua*	*Pterocles*
P. senegalensis[1]	*P. exustus*	Group 1
P. senegallus	*P. senegallus*	*P. alchata*
P. arenarius[2]	*P. orientalis*	*P. exustus*
P. coronatus	*P. coronatus*	*P. senegallus*
P. gutturalis	*P. gutturalis*	*P. namaqua*
P. burchelli	*P. burchelli*	Group 3
P. personatus	*P. personatus*	*P. orientalis*
P. decoratus	*P. decoratus*	Group 4
Dilophilus lichtensteinii	*P. lichtensteinii*	*P. coronatus*
D. bicinctus	*P. bicinctus*	*P. decoratus*
D. indicus	*P. fasciatus*[3]	*P. personatus*
D. quadricinctus	*P. quadricinctus*	Group 5
		P. lichtensteini
		P. quadricinctus
		P. bicinctus
		P. indicus
		Group 6
		P. gutturalis
		Group 7
		P. burchelli

[1] Synonymous with *exustus* of later authors.
[2] Synonymous with *orientalis* of later authors.
[3] Synonymous with *indicus* of other authors.

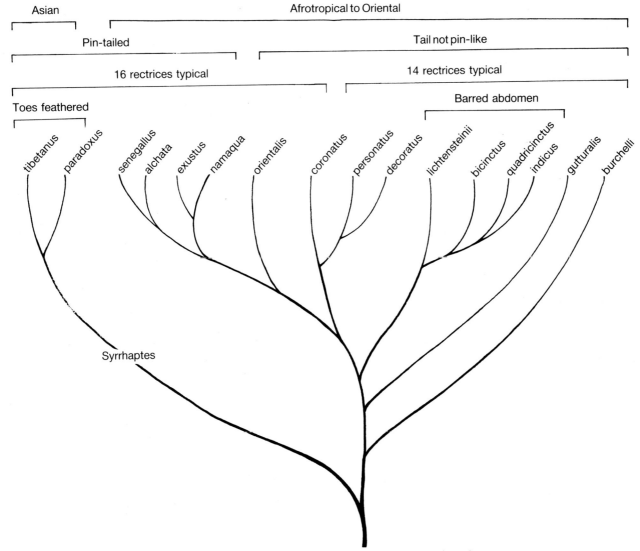

Fig. 3. Hypothetical phylogeny of the extant species of sandgrouse.

as to the desirability of accepting only two sandgrouse genera, and there is a reasonable level of agreement as to probable affinities within the large genus *Pterocles*. The sequence of species that I have adopted for use in my species accounts differs only slightly from that proposed by Maclean (1984).

SUMMARY

It is clear from this review that, although the details are still to be determined, the relationships of hemipodes, bustards and sandgrouse to one another are remote at best, and the features distinguishing them from one another are generally far more apparent than those tending to associate them. A summary of some of the major attributes of each of the three groups (Table 4) helps to point out some of these distinguishing features, although the meaning and significance of some of the more technical points may be lost on the average reader. However, many of them should be made clearer upon reading the later chapters of this book.

In summary, traditional recent classifications such as that of Wetmore (1960) would provide a higher-level taxonomy of these three groups as follows:

Order Gruiformes
    Suborder Turnices
        Family Turnicidae (Hemipodes)
    Suborder Otides
        Family Otididae (Bustards)
Order Columbiformes
    Suborder Pterocletes
        Family Pteroclidae (Sandgrouse)

If, however, one accepts the admittedly rather rad-

Table 4. Some comparative traits of hemipodes, bustards, and sandgrouse

	Hemipodes	Bustards	Sandgrouse
Centre of distribution	Australia (6/13)	Africa (18/22)	Africa (14/16)
Longest primaries	9–10	6–9	10
Total secondaries	8–10	16–24	17–18
Secondary organization	Eutaxic	Diastataxic	Diastataxic
Total rectrices	8–12	16–20	14–18
Powder downs	Absent	Present	Present
Porphyrin pigmentation	Absent	In some spp.	Absent
Oil gland	Present	Absent	Present
Tarsal surface	Scutellate	Reticulate	Feathered
Nares	Impervious	Pervious	Impervious
Nasal bones	Schizorhinal	Holorhinal	Holorhinal
Basipterygoid processes	Present	Absent	Present
Sternal notching	Single	Double	Double
Penis	Absent	Rudimentary	Absent
Oesophageal crop	Absent	Present	Present
Enlarged trachea	Females only	Not enlarged	Not enlarged
Inflatable oesophagus	Females only	Males (some)	Neither sex
Inflatable gular pouch	Neither sex	Males (some)	Neither sex
Sexual maturity attained	4–5 months	2–6 years	First year
Typical pair-bonding	Polyandrous	Polygynous	Monogamous
More colourful sex	Female	Male	Male
Heavier sex	Female	Usually male	Male
Typical adult weight	30–125 g	0.5–10 kg	150–400 g
Sex defending territory	Female	Male	Both?
Communal male display	Absent	In some spp.	Absent
Male display flight	Not reported	In some spp.	Present (?)
Courtship feeding	Present (all?)	Rare (1 sp.?)	Not reported
Allopreening	Present (all?)	Not reported	Not reported
Sex constructing nest	Both	Female	Both
Usual nest type	Lined, hidden	Nest scrape	Nest scrape
Usual clutch size	2–7 eggs	2–4 eggs	3 eggs
Egg pigment spotting	Heavy	Slight	Moderate
Egg-laying interval	1 day (or less)	2–3 days	1–2 days
Incubation period	12–16 days	20–28 days	23–28 days
Incubating sex	Male	Female	Both
Multiple-brooding	Regular	Absent	Possible
Parental feeding of young	By male	By female	Lacking
Parental watering of young	Lacking	Lacking	By male
Parental care of young	By male	By female	By both
Dorsal down pattern	Striped	Spotted	Spotted
Fledging period	c. 7–15 days	30–35 days	14–28 days
Juvenal plumage	Distinctive	Female-like	Female-like
Seasonal plumage variation	In some spp.	None in most	None in most
Typical breeding cycle	Opportunistic	Seasonal	Opportunistic
Tpical mobility pattern	Mostly nomadic	Some migratory	Often nomadic
Typical mobility pattern	Solitary	Semigregarious	Gregarious

ical taxonomy of Sibley *et al.* (1988), these three groups would be classified as follows:

    Infraclass?
        Order Turniciformes (*incertae sedis*)
            Family Turnicidae (Hemipodes)
    Infraclass Neoaves
        Order Gruiformes
            Infraorder Otidides
                Family Otididae (Bustards)
        Order Ciconiiformes
            Suborder Charadrii
                Infraorder Pteroclides
                      Family Pteroclidae (Sandgrouse)

From the standpoint of the present book, either classification serves the basic purpose of emphasizing a rather distant relationship among these groups, regardless of their nearest phyletic relatives.

2 · Zoogeography and evolutionary trends

It seems likely that all three groups of these birds evolved in the Old World, for there is no fossil trace of any of them from the New World, nor any other reason to suggest New World origins or invasions. They all additionally have broad present-day distributions within the Old World, occupying most or all of their seemingly suitable habitats throughout this hemisphere.

THE DISTRIBUTION OF ARID LANDS IN THE OLD WORLD

To a considerably greater extent than is true of the land masses in the New World, substantial portions of Eurasia, Africa and Australia are presently characterized by semi-arid to arid climates, and the largest of the world's deserts are to be found there. To a considerable degree, this reflects the fact that much of the land mass of the Old World is situated between the latitudes 15° and 35°N and S, which are regions having climatic zones dominated by permanent high-pressure systems and associated relatively cloudless weather persisting throughout the year (Hills 1966, McGinnes et al. 1968). Such enormous Old World deserts as the Sahara (having an area of some 8–9 million square kilometres, or more than half the total world's desert surface) and the Arabian desert (2·4 million square kilometres), as well as the somewhat smaller but still immense Iranian and Indian ('Thar') deserts, occur within this general latitudinal zone in the northern hemisphere, whereas the Australian desert lies in the comparable southern hemisphere zone.

The other largest deserts of the Old World, those of interior Asia (Turkestan, Takla Makan and Gobi), extend substantially further north, and thus well beyond the drying effects of such permanent high-pressure zones as just mentioned. However, these more temperate regions are situated too far from the sea to receive significant amounts of the cyclonic precipitation that is carried inland on westerly winds from the Atlantic. Furthermore, these areas are effectively isolated by topographic barriers such as the Himalayas, which produce a 'rain-shadow' effect that prevents monsoon rains originating in the Indian Ocean from reaching these regions.

A final major cause of deserts is cold ocean currents passing along extended coastlines. This causes deepwater upwelling of nutrient-rich waters and associated coastal fogs but reduces the likelihood of significant precipitation inland from the coast itself, the air's moisture having effectively been discharged as a result of the chilling received from the cold coastal waters. Such is probably the primary source of the Namib desert of south-western Africa, which merges southwardly and interiorly with the somewhat less arid Kalahari desert of southern Africa, forming what is called the Kalahari–Namib desert by McGinnes et al. (1968). To the south, the climate of the Kalahari is ameliorated progressively by increasing winter precipitation, and this desert gradually merges with the arid steppes of the Karoo subdesert. Another coastal area under the influence of cold upwelling effects is the Somali–Chalbi desert area of Africa's eastern horn, where the low annual rainfall is related to a combination of orographic rain-shadow effects, air pressure patterns and coastal upwelling (McGinnes et al. 1968).

Although it may be impossible to define deserts or arid lands meteorologically by using such a simple statistic as annual precipitation (owing to the marked effects of temperature and wind on photosynthesis and transpiration rates, as well as the potential for marked seasonal or annual precipitation variations, for example), Hills (1966) suggested that an annual average precipitation of 250–300 mm could in most areas serve as a convenient if crude boundary mark for separating arid from sub-arid climatic zones. With this in mind, a map of the Old World showing regions of comparable average annual precipitations is presented, as mapped by the *American Oxford Atlas* (Figure 4). It may be compared with a similar map showing the distribution of the major Old World deserts and steppe (arid grassland) vegetation types, as mapped in *The Times Atlas of the World* (Figure 5). This comparison suggests a reasonable degree of congruence between these meteorological and botanical criteria of arid lands, given the necessary mapping simplifications and the potentially subjective aspects of defining the vegetational boundaries of particular deserts. Further, each of these maps provides a convenient basis for zoogeographic comparisons by using the individual or collective species distribution maps presented later in this book.

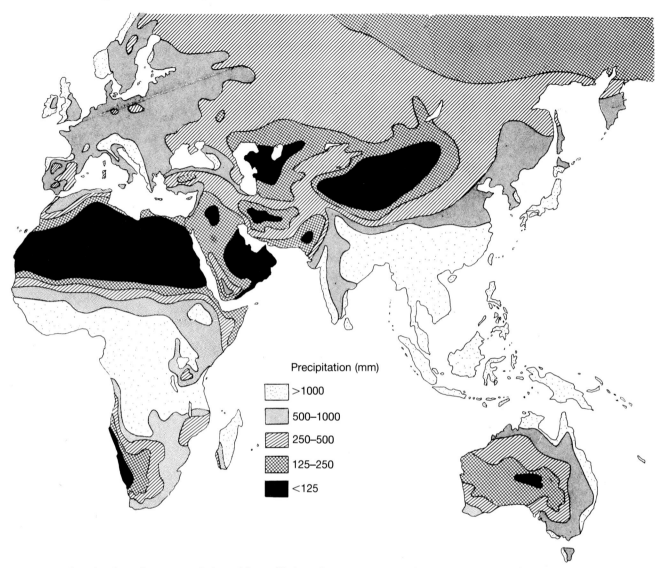

Fig. 4. Arid and sub-arid regions of the Old World, based on mean annual precipitation, as mapped in the *American Oxford Atlas* (1951).

ZOOGEOGRAPHY OF HEMIPODES, BUSTARDS AND SANDGROUSE

It is generally obvious that all three of the groups of birds included in this book are to varying degrees arid-adapted. Goriup (1988b) identified these three families (as well as the stone-curlew family Burhinidae, represented in the western Palearctic by a single species) as being highly characteristic of, and the only avian families whose representative species are entirely limited to, the treeless and relatively arid ('steppic') habitats of Western Europe, North Africa and the Middle East. By superimposing individual range maps of all the extant species of hemipodes, bustards and sandgrouse comprised by these larger groups categories, and using the distribution maps prepared for this book's species accounts, collective taxon-density maps have been generated (Figures 6–8). Such maps might be expected to provide more information of potential zoogeographic interest than would one of any single species, and offer potentially interesting comparisons with climatic, vegetational or land-form maps of the same general region.

Hemipodes

Taking the hemipodes for initial consideration, it is apparent that no clear centre of ancestral origin of the Turnicidae is immediately evident, even though Australia collectively holds the largest number (six) of contemporaneously extant species. From there

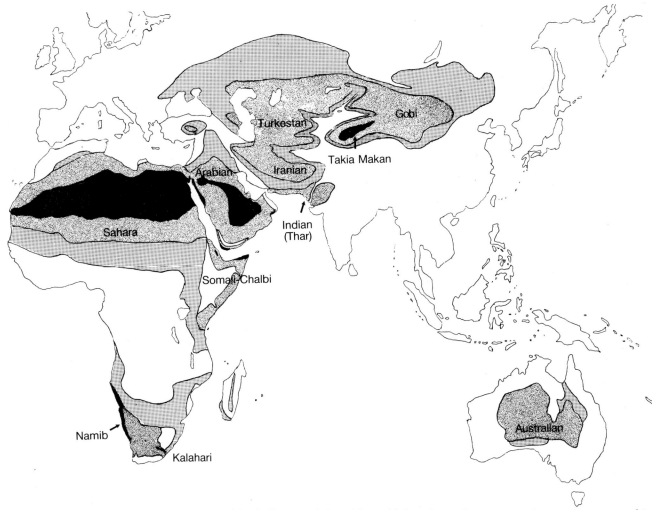

Fig. 5. Major desert (stippled) and steppe (shaded) areas of the Old World, based mainly on vegetation types as mapped in *The Times Atlas of the World* (1983). Areas of extreme desert (inked) are generally based on a map in Gore (1976).

they extend northwardly in a rather unpredictable pattern across the Melanesian islands of the south-western Pacific, occurring on all the larger East Indian islands with the exception of Borneo, and extending well beyond the Sunda Platform to the Philippines.

On the mainland of south-eastern Asia the group extends northward to the Himalayas and to eastern China, where perhaps they become limited by colder temperatures from further northward expansion. They also do not occupy the more arid interior of China (Figure 6). In Africa the group collectively extends across nearly all of the sub-Saharan interior with the exception of the Somali Peninsula, the Namib and Kalahari deserts, and parts of the Congo basin. Only one species, the aberrant lark-quail, penetrates into the southern portions of the Sahara desert.

By comparison with the other two groups, it is apparent that the hemipodes are certainly the least arid-adapted of the three. Indeed some species have ranges that climatically encompass monsoon or even tropical rain-forests and perhaps locally receive over 250 cm of annual precipitation. In such areas the birds occupy jungle-edge or early successional habitats following forest clearing or fires, where rank, weedy vegetation exists, or where low, tangled thickets are present. Ridley (1983) reviewed the ecological affinities of the buttonquail genus *Turnix*, and judged six species to be associated with grassland habitats, four more with combinations of grasslands with scrub, semidesert or savanna, two with dry savanna, two with the edges of tropical rain-forest, and one with woodland and grassy scrub. The only species that he classified as specifically associated with semidesert is the yellow-legged buttonquail, whilst the two species that he associated with rain-forest are the black-breasted buttonquail of northern Australia and the spotted buttonquail of Luzon Island in the Philippines.

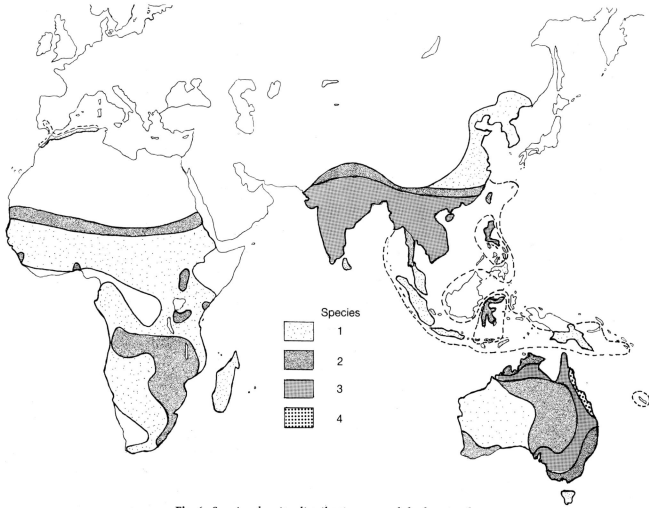

Fig. 6. Species-density distribution map of the hemipodes.

The hemipodes are generally non-migratory and of relatively low mobility as compared with bustards and sandgrouse, which perhaps helps to account for their complete absence from the extensive steppe habitats of Turkey and the Middle East. Their present absence from these regions also makes it difficult to hypothesize how the group arrived in Africa from a presumed Asian or Australian point of origin, for the Arabian peninsula and the Indian desert now effectively separate these two components of the family's collective range. In contrast to the bustards, it seems unlikely that any major changes in the group's collective range has occurred in historical times inasmuch as the birds seem to be little affected by human activities.

Bustards

The collective distribution of the bustard family (Figure 7) has many clear differences from that of the hemipodes. Most obviously, the group is concentrated in sub-Saharan Africa, which presumably represents the bustards' area of ancestral origin. However, bustards also occur in all the other major arid lands of the Old World, crossing temperate Eurasia from Europe to eastern Siberia, and occurring discontinuously in India and again in Australia. The Eurasian ranges of the little and great bustards have been so seriously influenced by human activities, in terms of both habitat changes and direct exploitation, that little can be learned from present-day patterns, but presumably the heart of the group's range in Africa is still only slightly influenced by these factors. There, it would appear that three focal points of bustard distribution exist. One of these extends along the southern boundary of the Sahara desert (the Sahel zone) in a broad east-west band of steppe or sub-desert vegetation. The second East African locus is centred in dry scrub in the vicinity of present-day Lake Rudolf, and extends more generally southward along the grasslands of the Rift Valley. The third is situated in the desert-like northern Karoo and Orange River region of South Africa, and extends north through the Kalahari desert and along the southern limits of the even drier Namib desert.

In general, the individual bustard species of Africa tend to have rather limited geographic distributions, at least compared with the hemipodes and sandgrouse, perhaps at least in part as a reflection of their generally heavy bodies and correspondingly reduced flying abilities. Thus no single species ranges into all three of these 'nuclear' areas just described, although the collective ranges of the Arabian and kori bustards rather neatly delineate these three regions. In that regard, it is interesting that ranges of the four species in the genus *Ardeotis* as a whole similarly rather effectively circumscribe the basic bustard family distribution, with the exception of the temperate-zone steppes and grassland of Europe and central Asia that are now occupied by *Otis* and *Tetrax*.

Whether or not this broad distribution pattern indicates that *Ardeotis* approaches a 'primitive' or generalized type of ancestral bustard is questionable, although if one imagines the earliest bustard type to be a large, omnivorous and steppe-adapted seriama-like or crane-like form, such a conclusion might be easily reached. *Ardeotis* is furthermore the only bustard genus to occur on all three of the major continental focal points of the bustard family (Africa, India, Australia), although it has not speciated nearly to the extent as has *Eupodotis*, which, depending upon how inclusively the genus is defined, is essentially limited to Africa or to Africa plus India.

With these thoughts in mind, one might hypothesize that the bustards originated in Africa, perhaps in the general sub-Saharan region, and possibly separating from proto-seriama stock at the time that Africa and South America began to drift apart more than 70 million years ago. Early *Ardeotis* stock presumably moved into what is now Eurasia, and eventually also spread into India and Australia,

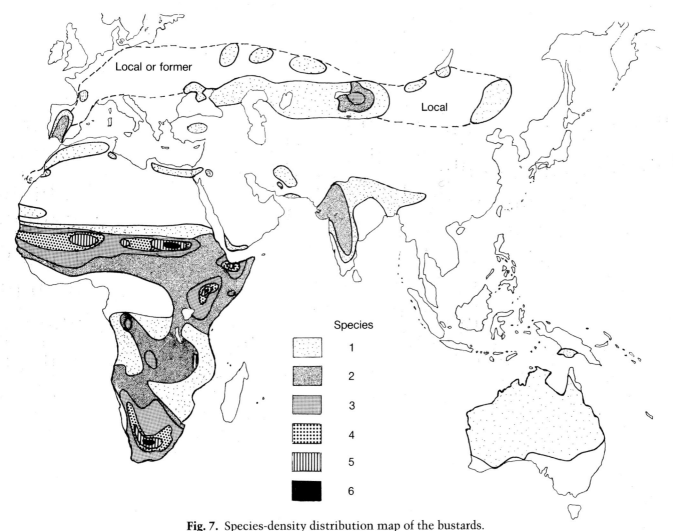

Fig. 7. Species-density distribution map of the bustards.

necessarily crossing substantial oceanic barriers to achieve the latter. Stock that gave rise to *Otis, Chlamydotis* and *Tetraix* also moved into the Mediterranean basin and beyond into the developing arid lands of central Eurasia. Similarly, in Africa a variety of generally small to moderate-sized species of bustards began to speciate locally in areas of steppes and semidesert across southern and eastern Africa.

Sandgrouse

Of the three groups of birds under consideration, the sandgrouse show the highest level of correspondence with the mapped zones of deserts and arid lands shown in Figures 4 and 5, although they are curiously lacking from the Australian deserts. With this exception, they show a high degree of conformity to desert distributions (Figure 8), and are the only group with species occupying the driest parts of the Namib, Sahara, and other major Old World deserts.

Like the bustards, the sandgrouse have a high degree of species diversity in Africa, and indeed have similar loci of maximum diversity in the Sahel zone immediately south of the Sahara, in the Rift Valley of east-central Africa, and in the general vicinity of the Kalahari desert of south-western Africa. However, they have even higher levels of species diversity in the south-western parts of the Arabian peninsula and in the Indian desert, both consisting of generally low-altitude permanently hot deserts with extensive sandy areas or salt flats present. Interestingly, although three species of sandgrouse (*paradoxus, orientalis* and *exustus*) have been introduced into grassland or desert habitats of Washington, Nevada and perhaps elsewhere in North America at various times, these introductions have not proved successful. Similarly, several other introductions into less likely climates and habitats elsewhere, such as in New Zealand and southern Australia, have generally been failures, although a very limited degree of success has occurred with *exustus* in Hawaii (Long 1981, Paton *et al.* 1982).

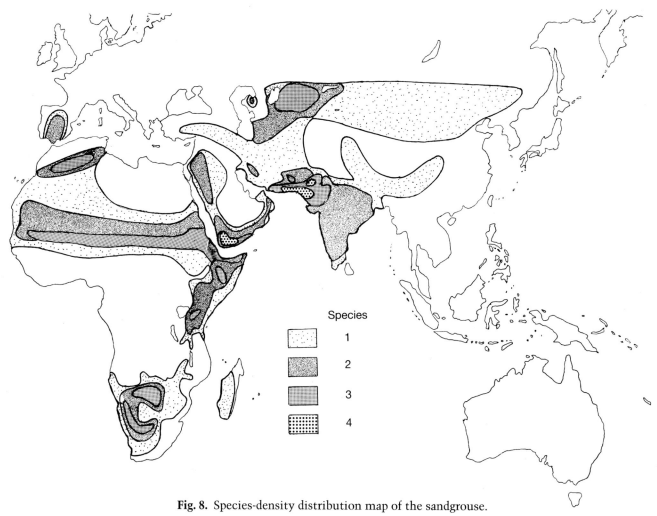

Fig. 8. Species-density distribution map of the sandgrouse.

Maclean (1976) earlier mapped the zoogeographic distribution of the sandgrouse, using somewhat different base maps for his data but finding somewhat similar overall patterns. He indicated a maximum possible species density of six species in the Iran–Afghanistan region, and suggested that the family probably had its origins in North Africa or the Middle East. From these areas the birds radiated out towards Asia, India, Madagascar and southern Africa, the last-named area perhaps having been invaded as many as five times. Specific centres of adaptive radiation were judged by him to have been located in the Thar (Indian) desert, Baluchistan, the north-western Sahara, and the Kalahari. According to his analysis each of these areas might support up to as many as five coexisting species. However, Thomas (1984a) noted that within each of these broad geographic regions, ecological differences among the species tended to reduce the actual numbers of ecologically interacting species to no more than about two, or at most three, even in areas having a relatively rich sandgrouse fauna. He observed that ecological distinction was perhaps most complete when there were also marked size differences between sympatric species, and seasonal migration or nomadism might further reduce actual contacts among species.

Maclean (1984) later hypothesized that the sandgrouse probably originated somewhere between the south-western Sahara and north-western India, in the broad band of low-altitude desert that extends almost continuously along this latitudinal zone. He suggested that *P. orientalis* exhibited the nearest approach to a 'basic' sandgrouse plumage pattern or hypothetical ancestral type that presumably arose in North Africa or the Middle East. The high mobility of sandgrouse makes it extremely dangerous to speculate much on their areas of origin and speciation based on current distribution patterns, but Maclean's conclusions about the group's approximate geographic origins are not difficult to accept. He considered *Syrrhaptes* to be a derived sandgrouse type, although not far removed from his hypothetical ancestral type, its differences primarily resulting from adaptations to the low environmental temperatures characteristic of arid central Asia.

POSSIBLE EVOLUTIONARY TRENDS IN STRUCTURE AND BEHAVIOUR

Hemipodes

There is very little that can be said about structural and behavioural variations in the hemipodes, inasmuch as the group is still only very poorly documented in most biological respects. The major morphological variations in the group are to be found at the generic level between *Ortyxelos* and *Turnix*, the former genus apparently lacking the specialized oesophageal condition of *Turnix*, and having longer and stiffer rectrices than occur in *Turnix*. The apparent absence of an inflatable oesophagus in *Ortyxelos* is a conclusion based in part on the species' lack of low-frequency calls, and in part of Lowe's (1923) statement that there is 'no indication of any dilatation of its oesophagus'. However, Lowe did not indicate whether specimens representing both sexes were examined with respect to this trait. These conditions might constitute evidence that *Ortyxelos* is somewhat more generalized in structure than is *Turnix*. Nevertheless, Lowe found very few osteological differences between these two genera, the primary one being that the sternum's episternal spine points downwards in *Ortyxelos* and upwards in *Turnix*, a point of obscure and questionable phyletic significance. Although the adult female is slightly larger and brighter in *Ortyxelos* than is the male, there is still 'no evidence of polyandry' in that little-studied form, and in any case even in *Turnix* polyandry may be varied with monogamy under varied circumstances or local conditions (Urban *et al.* 1986).

Mainly to show their geographic affinities, and to provide an idea of the overall similarities and differences among them, a drawing of the thirteen species of hemipodes is provided (Figure 9) in which geographically allied species are grouped, but with no other suggestion of relationship thereby implied.

Bustards

Among the bustards we have a larger array of taxa to consider than is the case with either hemipodes or sandgrouse, and a substantially greater variation in morphology and behaviour exists. Rather strangely, nobody has yet tried to assemble the available information on these variations and to place them in any sort of evolutionary perspective, in the same way that nobody has seriously examined the species- and genus-level taxonomy of the group.

It is possible that some indications of phyletic relationships may be apparent from a study of the patterns of bustard downy young (as may also be true for sandgrouse and hemipodes), but few illustrations of these have been published, and very few specimens exist in museum collections. Fjeldså (1977) illustrated the downy patterns of *Otis*, *Tetrax* and *Chlamydotis*, and noted several similarities between the last two, while *Otis* had several differences from each of these. Of course, all downy bustard patterns represent prime examples of effective concealing colouration, and rather little vari-

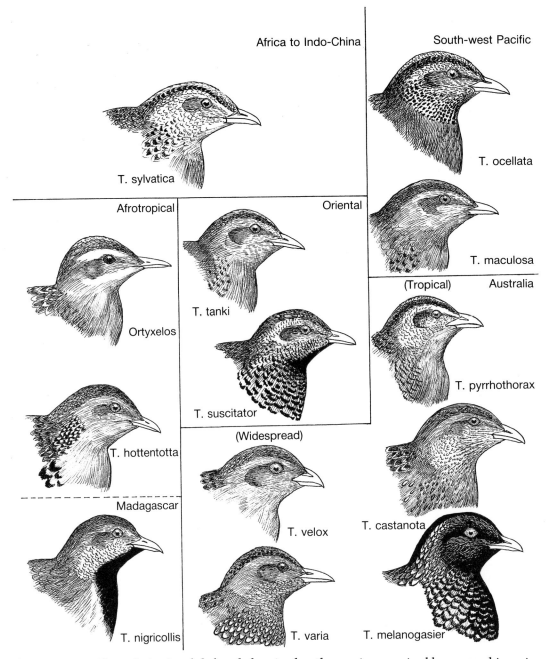

Fig. 9. Interspecific variation in adult female hemipodes, the species organized by geographic regions.

ation is to be expected throughout the entire group. Indeed, the downy young of bustards, hemipodes, and sandgrouse show comparable degrees of disruptive patterning that probably help to provide maximum concealment in a sandy, stony, or grassy environment, depending upon the species (Figure 10).

Among bustards, one major difference that is apparent is that several of the smaller species (the red-crested, black, black-bellied, and Hartlaub's of Africa, and the Bengal and lesser floricans of India) have black underparts in the adult males or both sexes. The other species have white underparts or, as in the case of the blue bustard, bluish grey underparts. This is quite a surprising situation, as all bustards live in environments characterized by intense sunshine, where white-bellied countershading would seem to be highly advantageous for maximum concealment. According to Kemp and Tarboton (1976), at least most of these black-underpart African species are savanna-adapted forms, whereas some closely related but white-bellied

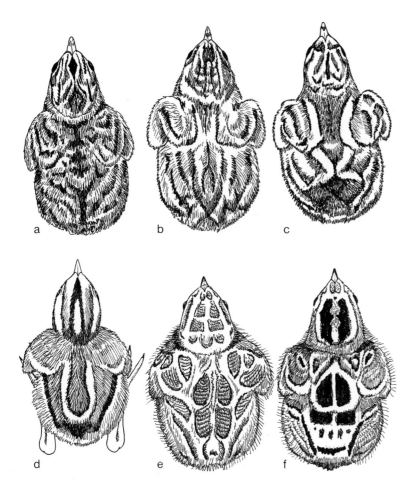

Fig. 10. Representative natal plumages of bustards, hemipodes and sandgrouse: (a) great, (b) little and (c) houbara bustards; (d) striped buttonquail; (e) Pallas's sandgrouse; and (f) four-banded sandgrouse. After sketches in Fjeldså (1977).

forms are more open-country species of grasslands and semi-arid steppes. Although the females and immatures of a few of these black-underpart species do have white underparts, in some species such as the black and red-crested bustards even these birds have black underparts. It seems most likely that these black underparts are associated with male territorial display, perhaps especially flight display, when the black underparts would be especially conspicuous, as Schulz and Schulz (1986) have already suggested for the black-bellied bustard.

Although at least some of these species just mentioned are known to have aerial displays, and others are known to expose their black underparts during terrestrial display, correlations between these conditions must for the present remain tentative at best. Similarly, a few of the sandgrouse species have black or at least blackish underpart colouration in males or in both sexes, for equally obscure reasons that seem more likely to be related to species-specific signalling functions than to any apparent ecological adaptations. The case is perhaps somewhat similar to that of the grey and golden plovers (*Pluvialis* spp.), in which the black underparts of the breeding adults are contrary to expectations based on principles of effective counter-shading during nesting in open-country environments but may be significant visual signals during aerial advertisement displays over tundra breeding habitats.

In any case, the distribution of black, or black contrasted with white, is a conspicuous plumage feature of many adult bustards, and these traits can be used for a convenient method of classifying male plumage variation within this family. This can be illustrated by the sketches in Figure 11, by which scheme the genus *Otis* becomes the most generalized, and *Sypheotides* plus two species of *Eupodotis* become the most specialized (i.e., having the greatest degree of black in the plumage). At least in part by chance, the arrangement of species within this figure generally parallels the highly tentative phylogeny of the family that was presented earlier in Figure 2.

As further evidence of the probable signalling importance of black-and-white plumage patterning in the bustard for possible species-recognition or other display purposes, it is of interest to look at the upper wing surface plumage patterns (Figure 12). This trait is usually completely hidden while the birds are in normal resting posture, but a strikingly

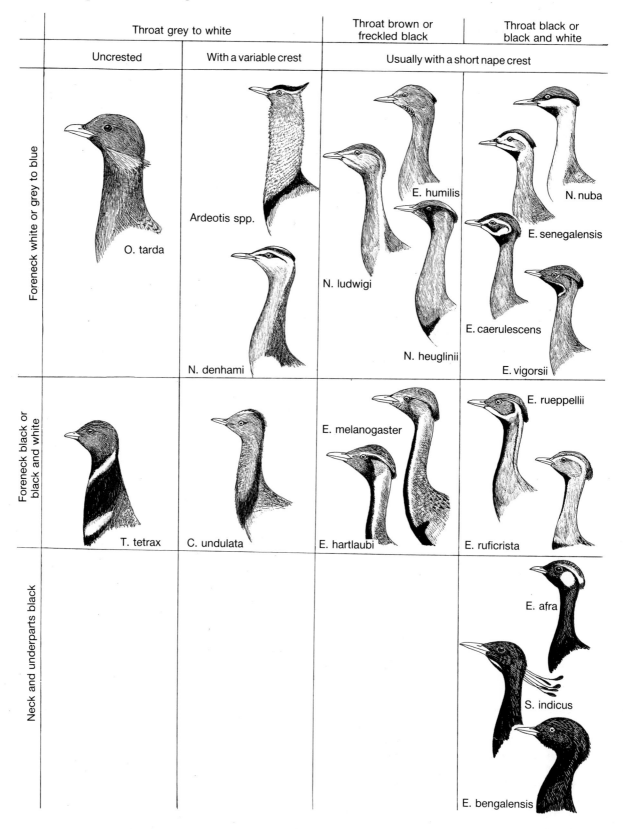

Fig. 11. Interspecific variation in bustards, the species organized on the basis of adult male plumage characteristics of the head and neck.

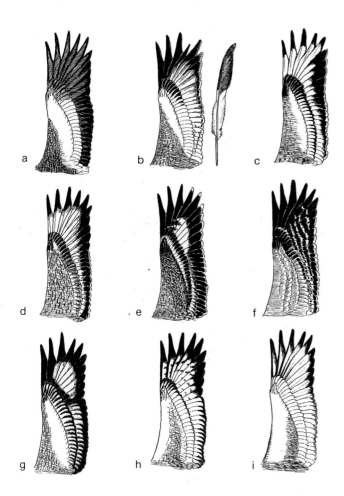

Fig. 12. Interspecific variation in wing patterns of representative bustard species, showing typical dorsal plumage patterns as well as the specialized seventh primary of the little bustard: (a) great bustard; (b) little bustard; (c) houbara bustard; (d) Nubian bustard; (e) Denham's bustard; (f) Arabian bustard; (g) little black bustard; (h) black-bellied bustard (*E. m. melanogaster*); and (i) Hartlaub's bustard. After various sources.

conspicuous and often species-specific patterning becomes evident as soon as the birds stretch their wings or take flight. Any correlation between the amount of black on both the upper wing surface and the underparts is weak at best, and indeed the relationship often tends to be somewhat inverse. For example, a generally dark upper wing surface is typical of all four of the white-underparted species of *Ardeotis*, whereas the black-bellied male Hartlaub's bustard has a nearly white upper wing surface that perhaps helps to accentuate the male's black underpart colouration, as does the essentially entirely black-bodied male of the Bengal florican.

Further evidence that upper wing plumage patterning may be important in male display lies in the fact that male little bustards have specialized seventh primaries (counting from the inside) that are both distinctly shorter and more pointed than are either of the immediately adjacent primaries (see Figure 11b). In flight, the vibrations of these feathers result in a distinctive whistling noise that calls attention to the bird. Thus during its territorial-advertisement display flights the male alternates periods of wing-beating and gliding whilst distinc-

tively erecting its black neck feathers (Schulz 1986a). The lesser florican has a similar attenuation of all of its primaries that produces a humming sound, and calling by territorial males during flight display is also quite widespread in bustards.

Sandgrouse

Like the hemipodes, the sandgrouse family consists of only two rather well-defined genera, including one small genus ecologically adapted to extreme conditions (extreme aridity in the case of the hemipode genus *Ortyxelos* and extreme cold in the case of the sandgrouse genus *Syrrhaptes*), and one large genus consisting of about a dozen species that are extremely similar in morphology. The typical sandgrouse are indeed so similar that most taxonomists have had to rely on features such as male plumage patterns for purposes of classification. Thus D. G. Elliot (1878) used the presence or absence of a conspicuous breast-band in males as a basis for a major division of the genus *Pterocles*, and Bowen (1927) used traits such as relative diurnality, number of rectrices, wing shape, and colour patterns of adult males for this purpose.

26 *Bustards, Hemipodes, and Sandgrouse*

Although Bowen did not consider the genus *Syrrhaptes* in detail or comment on relative ancestral patterns of speciation, he did offer some interesting comparisons within the genus *Pterocles*. Thus he noted that two distinct tail-shape types occur among sandgrouse. One group, having wedge-shaped tails, includes all the long-tailed species of *Syrrhaptes* and *Pterocles*, as well as five relatively short-tailed species (*orientalis, coronatus, gutturalis, burchelli* and *personatus*) that are sometimes placed within a separate genus. The remaining five species (*decoratus, lichtensteinii, bicinctus, indicus* and *quadricinctus*) all have rounded tails in which the central pair of rectrices is scarcely longer than the adjoining pair (Figure 13). In all but two species of the wedge-tailed group the tail normally consists of sixteen rectrices (although some individual variation may exist in this trait), with the central pair unlike the others in colour pattern. The two exceptional species (*burchelli* and *personatus*) typically have only fourteen rectrices, with the central pair similar to the adjoining feathers.

Bowen believed the relative degree of elongation of the central tail feathers to be of doubtful taxonomic significance, and the number of tail feathers to be of 'indefinite' value, but proposed that all of the five species having fourteen strongly barred tail feathers, broadly rounded inner primaries, and male forehead patterns of black and white be separated generically (*Dilophilus*). Most of these five species

Fig. 13. Interspecific variation in the rectrices of representative adult sandgrouse: (a) black-faced (female); (b) Lichtenstein's; (c) double-banded; (d) crowned; (e) spotted; (f) Madagascan (from incomplete specimen); (g) yellow-throated; and (h) black-bellied. Redrawn after Bowen (1927).

also exhibit crepuscular drinking behaviour and barred abdomen patterns, and have a double or triple pectoral band in the male. Bowen did not suggest which of these might represent primitive traits and which are apparently derived ones.

Maclean (1984) provided a hypothetical pattern of evolutionary trends in the sandgrouse with respect to their external morphology. He judged that the black-bellied sandgrouse most closely approached the ancestral plumage type, differing from it only in lacking elongated central rectrices, which Maclean considered to be an ancestral trait. A second evolutionary modification, leading to the pin-tailed sandgrouse, involved some minor pattern changes, such as the loss of the black underparts and reduced sexual dichromatism, whilst another change (perhaps actually the oldest) involved plumage as well as foot-structure modifications (feathered toes, fusion of the front toes, and loss of the hallux) that led to *Syrrhaptes*. A fourth hypothetical evolutionary line moved southward in Africa, producing the chestnut-bellied and Namaqua sandgrouse, both of which retained several ancestral plumage traits such as elongated central rectrices.

Maclean hypothesized that a fifth evolutionary line involved the loss of the elongated central rectrices and the development of a black-and-white forehead pattern, producing in Africa the spotted, crowned, and black-faced sandgrouse, as well as the Madagascan sandgrouse. An offshoot of this fifth line produced a fairly well-defined subgroup of four species consisting of the Lichtenstein's, double-banded, four-banded, and painted sandgrouse. The last-named of this group is the most geographically removed from Africa and is the most differentiated in plumage.

Two final hypothetical evolutionary lines that lead to single species consist of the yellow-throated sandgrouse, a somewhat aberrant species that has retained a number of ancestral features, and the highly aberrant Burchell's sandgrouse, which is the most divergent of all sandgrouse in its plumage and morphology, according to Maclean.

In a general review of the group's natural history, Hüe and Etchécopar (1957) provided a useful graphic summary of male plumage characteristics in sandgrouse in the form of a dichotomous key, which with minor changes has been redrawn as Figure 14 here. Although the groupings used by Hüe and Etchécopar do not directly correspond with Maclean's proposed patterns of evolutionary trends, these drawings nevertheless allow for a convenient visualization of their ideas.

As with the bustards, there is a seemingly unpredictable appearance and disappearance of black underpart patterning in adult males, so much so that, for example, it is lacking in one species of *Syrrhaptes* but well developed in the other, in spite of the ecological similarities in the two. It is similarly well developed in the black-bellied sandgrouse but wholly lacking in the crowned sandgrouse, which on the basis of other considerations would seem to be fairly closely related. As in the case of bustards, it may be necessary for males (perhaps during flight displays or erect frontal posturing) to exploit their underpart patterning as species-specific signalling devices, inasmuch as their dorsal plumage patterning is largely or entirely given over to colour patterning that maximizes their concealing colouration from overhead. Such male signalling requirements evidently outweigh any undesirable visual effects that dark underparts produce in terms of nullifying the concealing benefits of effective countershading in the form of lighter underparts. Apparently much of the visual signalling of sandgrouse is done through frontal displays, during which the face, foreneck and chest are visible to the other bird, and it is in these areas that most of the distinctive and contrasting plumage patterns occur. Tail-fanning with an erected tail is apparently also a common sandgrouse display, and this trait may help to account for the distinct colours, patterns and shapes of sandgrouse tails.

Unlike the bustards, the sandgrouse have not significantly modified their upper wing surfaces for the development of species-specific plumage patterning, but instead typically exhibit distinctive underwing patterns that range from almost entirely immaculate white to nearly black, as well as various intermediates of grey or brown. Some of these variations are illustrated in Figure 15, which further shows that although black underparts are sometimes accompanied by white underwings or vice versa, there is no clear pattern to this, and in several species there is a general similarity of overall underpart and underwing colouration. Interestingly, one species of sandgrouse, Pallas's, has evolved highly specialized outermost primaries with tips that are highly elongated and attenuated in adult birds, and generate loud whistling noises in flight. This feature is somewhat analogous to that found in the little bustard, although different primary feathers are involved, and in that species only adult males exhibit the specialized feather trait. Maclean (1976) commented that the flying abilities of sandgrouse species might perhaps be related to the notching of their inner three or four primaries, which occurs in the yellow-throated, Namaqua and Burchell's sandgrouse, and which he suggested might be a characteristic of the faster-flying and more open-country species. Maclean also noted that adult sandgrouse were generally concealingly coloured dorsally so as

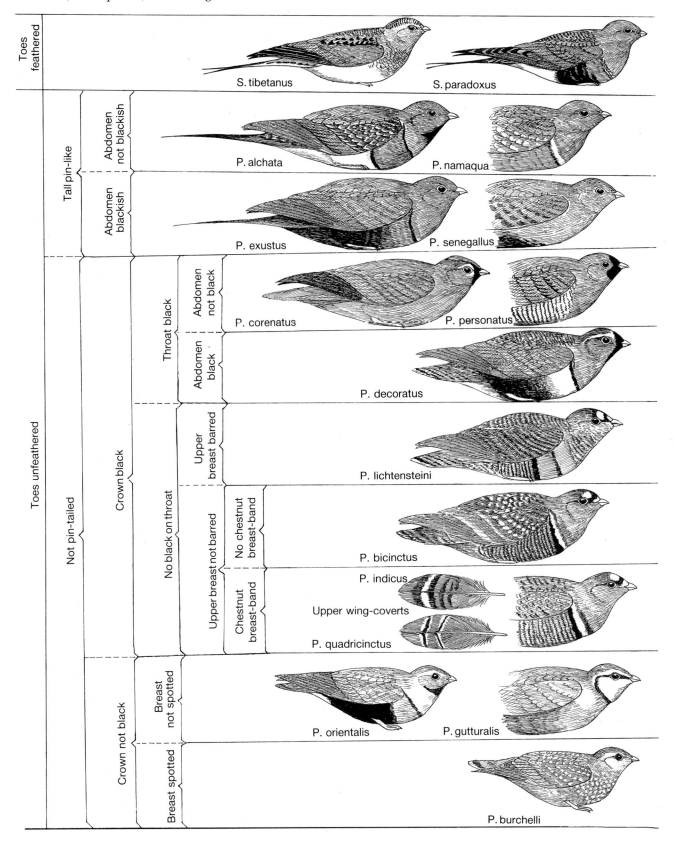

Fig. 14. Interspecific variation in adult male plumages of the sandgrouse. Redrawn from Hüe and Etchécopar (1957).

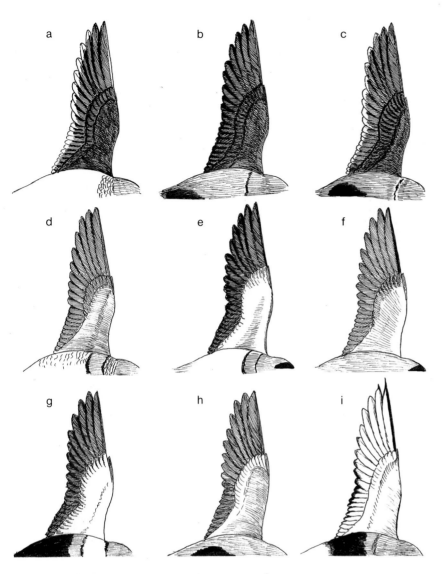

Fig. 15. Interspecific variation in adult wing and underside plumages of representative male sandgrouse: (a) Tibetan; (b) painted; (c) chestnut-bellied; (d) Lichtenstein's; (e) pin-tailed; (f) crowned; (g) black-bellied; (h) spotted; and (i) Pallas's. After various sources.

to match their usual backgrounds, which range from reddish sands to olive-coloured stony pavements and lichen-covered stones. A similar situation applies to the downy young of sandgrouse, which show an extremely high degree of concealment in their normal environment.

3 · Behaviour

Like all birds, bustards, hemipodes and sandgrouse spend a great deal of time in day-to-day individual survival and maintenance activities. As they serve similar functions in all birds, these behaviours tend to be rather similar in all avian groups and as such have limited potential for inter-group comparisons. Although these activities certainly grade almost imperceptibly into social interactions such as intraspecific fighting behaviour, it may be convenient to consider these separately before a formal discussion of strictly social interactions.

No complete behavioural inventory of 'ethogram' seems to have been performed on any species of the groups considered here, although Collins (1984) provided a useful description of the general behavioural activities of the houbara bustard, and his classification of these activities has to some degree been adopted here. In modified form it may be summarized as follows:

I. Egocentric (self-orientated) behaviour
   A. Integument-care activities
      1. Preening and oiling
      2. Feather-drying and sunbathing
      3. Dusting and shaking
      4. Self-scratching

   B. Comfort and maintenance activities
      1. Feather-ruffling and sunbathing
      2. Wing- and leg-stretching
      3. Head-shaking and beak-wiping
      4. Throat-fluttering or panting
      5. Wing-drooping
      6. Drinking
      7. Yawning
      8. Defecating

   C. Roosting and related behaviour
      1. Diurnal roosting
      2. Rain-roosting or huddling
      3. Nocturnal roosting
      4. Incubation behaviour

II. Protective and escape behaviour
    1. Alert posture
    2. Squatting or 'freezing'
    3. Distraction behaviour (injury-feigning)
    4. Shock-display (of bustards)
    5. Creeping
    6. Bush-circling
    7. Running and hiding
    8. Explosive take-off
    9. Evasive flight and landing

EGOCENTRIC BEHAVIOUR

One of the few studies of the egocentric behaviour of hemipodes is that of Fulgenhauer (1980) on the red-chested buttonquail. Apart from sitting, standing, walking and feeding activities, the most commonly observed egocentric activity was self-preening. This occurs in both standing and sitting positions, and is sometimes also combined with sunbathing, when one or both wings are spread out on the ground and preening of the wing feathers is performed. At far lower frequencies dust-bathing, wing-and-leg stretching, wing-arching (overhead wing-stretching), head-scratching, body-fluffing, and drinking were observed. During wing-and-leg stretching one leg is passed through the wing feathers whilst the bird is standing on the other. Wing-arching typically follows this activity. Body-fluffing consists of the shaking of the body whilst all the feathers are ruffled. Yawning was observed only rarely. Roosting is done by huddling on the ground, with the head resting on the shoulders. Mated birds often roost side by side, their flank feathers in contact, and the birds usually facing in the same direction.

Dust-bathing in hemipodes is a fairly complex behaviour, having leg-scratching, beak-wiping, pecking, head-wiping, dust-tossing, and head-scratching components. It occurs in two stages, the first being site preparation. This consists of short or prolonged ground-scratching movements, whilst the body is rotated. Pecking at the soil and beak-wiping accompanies this phase. The second stage consists of squatting, dust-tossing, head-rubbing, ground-pecking, and beak-wiping. Although dust-bathing is seemingly a purely egocentric behaviour, Fulgenhauer noted that it reached its peak frequency during pair-bonding, so perhaps it may serve some social function during pairing.

As in many other birds, stretching in bustards takes two common forms, including wing-and-leg stretching (Figure 16a) and overhead wing-stretching (Figure 16b). Like stretching, periods of feather-preening (Figure 16c) often occur during seemingly idle periods of resting or roosting. These activities

Fig. 16. Egocentric behaviour of the great bustard: (a) wing- and leg-stretching; (b) two-wing stretch; (c) preening; and (d) alert posture (following sunbathing). After photos in Heinroth and Heinroth (1927–8).

may be periodically interspersed with an alert scanning of the environment (Figure 16d). Bustards typically rest with the head drawn back on the shoulders, rather than with the beak tucked into the scapulars. This same general posture is maintained during incubation (see Figure 20).

Frisch (1976) and especially Schulz (1985c) have provided detailed descriptions of egocentric behaviour patterns of the little bustard. Sand-bathing in this species occurs in birds as young as only five or six days old. Bathing and dusting are performed by lying prostrate on the ground, with all the feathers fluffed. During sunbathing one wing is often extended on the ground. This may be combined with dust-bathing, when the birds shuffle their tarsi, writhe the neck on the soil, and toss soil into their plumage with scooping movements of the neck (Cramp and Simmons 1980) (See Figure 39). Sand-bathing is especially common just prior to the moulting period, when it perhaps helps in the shedding of loose feathers. Panting is performed with the bill wide open, but to avoid extremely high temperatures the birds often seek out shady locations. Yawning is done rather quickly, and is especially frequently performed by incubating birds. Daytime and nocturnal roost sites are not fixed, but although daytime roosts are often in thick grass or shady sites, nocturnal roosts are typically on bare ground or at least very open areas, apparently to facilitate predator protection.

Egocentric behaviour of sandgrouse is still only rather poorly documented. Like bustards they apparently do not bathe in water, and lack a well-developed oil-gland. They thus perform little if any oiling of the plumage, and probably rely primarily on dust-bathing for feather conditioning and maintenance. Most of the written observations on the egocentric behaviour of sandgrouse revolve around their pattern of drinking (Figure 17c), and whether or not they drink in a pigeon-like manner, without raising the beak. This subject was discussed in terms of its possible taxonomic significance in Chapter 1, and need not be repeated here. Interestingly, however, typical behaviour in drinking from a dish

Fig. 17. Water-related behaviour of sandgrouse: (a) water-soaking by male Namaqua sandgrouse; and (b) drinking by Namaqua sandgrouse chicks (both after photos in Cade and Maclean 1967). Also drinking by adults (c) and chicks (d) of spotted sandgrouse (after photos by George 1969, and in Grzimek 1972).

of water has been observed in sandgrouse chicks that were only a day old (Frisch 1970). Far more remarkable than the drinking behaviour of sandgrouse is their water-transport behaviour, during which water is soaked up in the belly feathers by adult males (Figure 17a), and then returned to the nest site to be extracted and drunk by the chicks (Figure 17b, d). Inasmuch as abdominal water transport by sandgrouse represents parental rather than egocentric behaviour, it is not appropriate to discuss it here,

and in any case this unique behaviour is dealt with later in the text.

Like bustards and hemipodes, sandgrouse rest and roost huddled down with the head resting lightly on the shoulders. They appear to enjoy standing in the rain, letting water droplets accumulate on the head and back feathers (Figure 18a). During diurnal or nocturnal roosting, members of a pair often sit side by side in contact with one another (Figure 18b). Grouped clumping or huddling behaviour is also

Fig. 18. Egocentric behaviour of sandgrouse: (a) rain-bathing; (b) roosting; and (c) dust-bathing postures of yellow-throated sandgrouse (all after photos by L. Grueber). Also (d) dust-bathing of spotted sandgrouse (after sketch in Urbán *et al.* 1986).

common during both diurnal and nocturnal roosting in sandgrouse flocks, the birds variously organized in either a back-to-front orientation or facing in the same direction. Unlike some quail species such as bobwhites (*Colinus* spp.), sandgrouse evidently do not use circular clumped roosting arrangements. Wing-drooping and gular fluttering are often performed by heat-stressed sandgrouse; the latter is often performed by incubating birds and the former is more commonly observed in dehydrated individuals. Variable feather erection or ruffling may also play a role in thermoregulation by sandgrouse (Thomas *et al.* 1981).

Dust-bathing is also frequently performed by sandgrouse, and probably serves several different functions. For example, this activity frequently occurs prior to drinking, and also prior to abdominal feather-wetting and subsequent water transport. Dust-bathing has also been observed following wading and drinking, apparently for the purpose of drying the body feathers. During dust-bathing the birds regularly turn on their sides, with both legs raised in the air (Figure 18c, d), and they may even roll over on their backs while rubbing sand or dust into their feathers.

PROTECTIVE AND EVASIVE BEHAVIOUR

One of the most widely used protective devices of hemipodes, bustards and sandgrouse for avoiding detection is that of 'freezing', or becoming immobile in a crouched or huddled posture. Downy young of all groups regularly perform this behaviour, probably instinctively, and in all cases their natal plumage patterns are sufficiently cryptic as to make visual detection extremely difficult (Figure 19a, b). Sandgrouse chicks not only crouch, but may even dig themselves partially into the sand, further obscuring them from easy view (George 1977). Similarly, adults of all groups often crouch when danger threatens, perhaps especially danger from above by visually hunting avian predators. Even those contrastingly patterned species of bustards such as the little black bustard (Figure 19c) become relatively inconspicuous by crouching, thereby hiding most of their black underparts and primarily exposing the disruptively patterned upperparts to overhead view.

When an incubating or brooding bird has been discovered, injury-feigning is a common protective response. In sandgrouse this injury-feigning often takes the form of a 'broken-wing act' (George 1977). At least in the spotted sandgrouse, parents have been observed flying at about half-speed in front of a falcon, apparently as a distraction-lure deviced (Cramp 1985). In bustards the female may similarly distract the intruder's attention by running about with drooping wings and raised tail or, in the case of the largest species, may even attack intruders. Injury-feigning may sometimes also be performed by hemipode parents, but more commonly the families simply hide or steal away (Fjeldså 1977).

Among bustards, one of the interesting anti-predator behaviours that seems to be unique to the group is 'shock display' (Figure 20). This defensive display is performed towards various terrestrial and aerial threats, or even to any sudden and unexpected stimulus, but especially towards aerial predators. It somewhat resembles the sexual displays of some adult male bustards but more closely resembles the defensive or threat display of the painted snipe (*Rostratula benghalensis*). The tail is raised, fanned and vibrated, the wings are extended, vibrated and

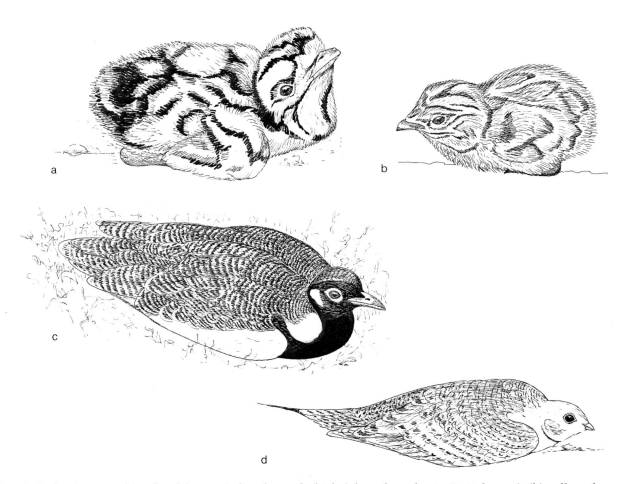

Fig. 19. Defensive crouching by: (a) great Indian bustard chick (after photo by A. R. Rahmani); (b) yellow-throated sandgrouse chick (ater photo by L. Grueber); (c) adult little black bustard (after photo by the author); and (d) adult Pallas's sandgrouse (after photo by L. Grueber).

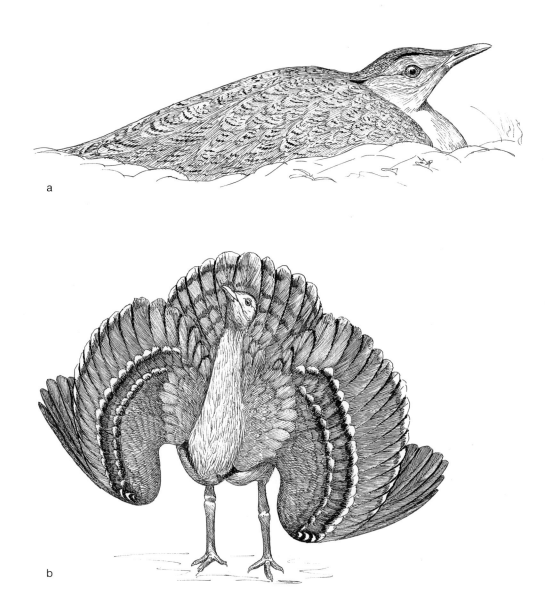

Fig. 20. Crouching posture of an incubating Nubian bustard (a) (after photo by P. Ward, in P. Osborne *et al*. 1984), and defensive 'shock' display of female Australian bustard (b) (ater photo by P. Goriup, in P. Osborne *et al*, 1984).

tilted upwards, exposing their entire upper surfaces, and the bill is directed toward the intruder whilst the head is drawn back. Frisch (1976) observed the shock display performed by little bustards when they were only 7 to 9 days old.

More commonly, evasive behaviour such as creeping, zigzag runs, running and hiding, or bush-circling may be used as defensive or anti-predator ruses by bustards. Similar devices are probably used by hemopodes, although there are few descriptions of these. Often buttonquail will flush suddenly when disturbed, fly only a few yards, and then pitch quickly into heavy cover or grass. Thereafter they refuse to take flight again, no matter how carefully one searches for them. Such actions are fairly straightforward responses that require little or no detailed explanations or descriptions. Because of their very short legs, sandgrouse perform little or no running, and instead are much more inclined to take flight when danger approaches. Often the birds will crouch in whatever cover is available until the last possible moment, and then take flight in a noisy eruption. The take-off may perhaps be followed by a quick landing and hiding, or by subsequent evasive

FLOCKING, SPACING AND TERRITORIAL BEHAVIOUR

Birds that live in almost any environment imaginable are likely to be caught between conflicting social tendencies. Group-living may normally be favoured to facilitate breeding, to obtain possible protection from predators through the advantage of collective awareness and defensive mechanisms, and to enjoy the potential benefits of sharing of important ecological information, including possible locations of food and water. On the other hand, extreme environments such as deserts are often able to support only limited animal populations, and thus competition for such restricted resources may effectively outbalance any or all of these advantages, and make it more advantageous to spread out the total population as widely as possible. It is these two conflicting tendencies, namely those that favour integrated flocking behaviour and positive intraspecific social interactions such as heterosexual attraction, versus those resulting in spacing-out tendencies, such as territoriality, individual distance maintenance and other aspects of negative intraspecific interactions, that will concern us here.

Hemipodes

There is no evidence that the African lark-quail occurs in groups larger than pairs outside the breeding season. However, the two indigenous African species of *Turnix* sometimes occur in 'loose aggregations' or in 'small groups or scattered flocks', the latter perhaps most often developing during nomadic wandering outside the breeding season (Urban *et al.* 1986). Similarly, the striped buttonquail reportedly occurs in India as single birds or pairs, and is only rarely found in larger but still small numbers (Ali and Ripley 1983), these last-named assemblages perhaps being family groups (Cramp and Simmons 1980). Likewise, Frith (1969) stated that in Australia the painted buttonquail usually occurs singly or in pairs, and that the little buttonquail similarly is usually seen singly or in small groups. Assuming that these are indeed family units, it is unlikely that such groups would ever consist of more than six birds, this number representing a pair and a full-sized brood of four dependent young.

These and similar observations by naturalists suggest that hemipodes generally occur in a highly dispersed social pattern. Thus in Namibia, nests of the striped buttonquail have been found in unusually favourable habitat to be dispersed at a density of ten nests in about ten hectares, or averaging about one nest per hectare (Hoesch and Niethammer 1940). This represents a breeding density that is only about 40 per cent of those typical of several species of the North American odontophorine quail such as the northern bobwhite (*Colinus virginianus*) under comparable ideal habitat conditions (Johnsgard 1973). Furthermore, the bobwhite has a body mass approximately three times that of *Turnix*, and thus had considerably greater food resource requirements.

This dispersion pattern of hemipodes is still only very poorly documented but is probably one that is established and maintained by the territorial and/or sexual calls of the female and the associated extreme aggressiveness of females towards one another, including her own chicks (Hoesch 1960, Wintle 1975). Although breeding birds are relatively sedentary, and presumably territorial, the females perhaps wander about rather freely in search of males, and may not remain within a fixed territorial location (Wintle 1975). In any case, their calling not only attracts males but also and perhaps even more regularly attracts other females, especially breeding females, who challenge the intruding bird. Jerdon (quoted by Baker 1930) noted that when a caged decoy female barred buttonquail was placed out and began to call, any females within earshot 'run rapidly to the spot and commence fighting with the caged bird', so that twelve to twenty birds could thus be captured in a day within a single patch of good habitat. All the birds thus trapped were females and often included birds that were so advanced in their egg-laying cycle as to deposit eggs in the bag in which they were placed, before the bird-catcher had returned to his house. Baker reported that males were also sometimes caught by essentially the same method, but by using nooses placed at distances of about 5 ft around a tethered female. However, males were apparently much less prone to approach the calling female closely than were other females, and were thus perhaps unlikely to be trapped in the decoy's cage, or even by nooses set out immediately around a tethered female.

Bustards

In contrast to the hemipodes, the bustards are distinctly gregarious, at least outside the breeding season. Although bustard populations are rarely large enough currently to support large flocks, such existed historically in several species. Thus the great bustard once formed winter flocks in Europe that collectively numbered several thousands, although in recent years these influxes have numbered only up to one or two hundred birds (Glutz von Blotzheim 1973). During the winters of 1969–70 and

1978–79, Western Europe was invaded by large numbers of great bustards (up to 300–400 birds, with individual flock sizes up to 20), the invasions apparently being triggered by unusually early cold weather (Hummel 1985).

Relatively large bustard flocks in Europe are still possible only in the case of the little bustard, and even these flocks are but tiny remnants of what once must have been typical. Historical records from the USSR suggest that migrating flocks of little bustards may have numbered as many as 20 000 birds per flock, or groups so large that 20 to 30 minutes may have been required for a single flock to pass (Grote 1936; Glutz von Blotzheim 1973). In recent years winter mixed-sex flocks of up to about 2000 birds have occurred in Portugal, which begin to separate into single-sex flocks in February and March, prior to the onset of male territorial display (Schulz 1985b,c).

Flock number and composition of bustards are probably highly variable by season and perhaps locality, but Ena *et al.* (1985) provided an analysis of great bustard flocks in north-eastern Spain (Leon province). During the spring, two general types of flock were found to occur. One of these was made up of single-sex flocks, of either sex, that often contained good numbers of individuals. These comprised adult birds that later dispersed to mating display grounds, but which for the rest of the year were sexually segregated. The second type was of mixed flocks of varied sex and/or age. One of these comprised flocks of adult males and females, which represented the reproductively active groups found during periods of courtship display. In these flocks the sex ratio favoured females over males in a ratio of about 4:1, supporting the idea that polygny represents the typical great bustard mating system. Flocks comprising adult males and other birds of undetermined sex were evidently actually made up of either all-male flocks or mixed-sex adult flocks of the type just mentioned. The remaining flocks comprised birds of mixed ages, such as adult and young males, or females and young males, suggesting that immature birds may remain with adult of either sex even during the spring display period, the latter meanwhile apparently not taking any active part in the spring display activities. Newly hatched chicks apparently remain in the company of their mothers throughout the summer and into the autumn, when they become integrated into the larger groups typical of late autumn and winter.

In species that are as rare as most bustards it is extremely difficult to judge maximum densities resulting from territoriality requirements, inasmuch as the birds probably only rarely reach an ecological or behavioural 'saturation point' in their breeding numbers. Fairly good observations do exist, however, for the little bustard. In that species, the male's territory is never less than a hectare, although it is perhaps normally about 4–6 ha (Frisch 1976, N. J. Collar, cited in Cramp and Simmons 1980). The density of territorial males has been estimated to be as high as thirteen to sixteen per square kilometre (or one male per 6·25–7·7 ha) in Portugal (Ferguson-Lees 1967). Schulz (1984a) estimated a similar density of nine to thirteen males per square kilometre (one male per 7·7–11 ha) in the same general area, his estimate including locations where birds were clustered as well as neighbouring areas with lower densities. André (1985) estimated densities of up to as many as six to eight males per square kilometre (one male per 12·5–16·6 ha) during spring in sample quadrats of an area of 3900 ha of the Crau, southern France, although most regions had far lower densities. Schenk and Aresu (1985) provided data for Sardinia suggesting local post-breeding densities as high as twenty little bustards (the sample including both sexes and all ages) per square kilometre, although the median densities estimated for five sample areas averaged about ten birds per square kilometre.

Sandgrouse

Perhaps more than any of the other two previous groups, the sandgrouse are highly social. This is presumably a reflection of the fact that, as with shorebirds for example, flocking is perhaps an effective anti-predation response for these birds, which necessarily must spend much of their time in open environments where concealment possibilities are few. One of the basic attributes of the sandgrouse is the clock-like regularity with which single birds, family parties or, more commonly during the dry season, large flocks gather at watering places. Bowen (1927) reiewed some of this literature, and cited a variety of accounts of such consistencies of sandgrouse drinking behaviour in terms of its timing, four species (*lichtensteinii, bicinctus, indicus* and *quadricinctus*) being crepuscular and the remaining sandgrouse species drinking in a diurnal rhythm.

Meinertzhagen (1954) stated that in Arabia the chestnut-bellied sandgrouse gathers in flocks of 200–300 birds to drink daily during morning hours, the birds typically arriving in small parties that unite as they fly in aggregations of up to as many as 50 000 individuals of the chestnut-bellied sandgrouse have been reported in extralimital areas of that species' range in Mali, where watering areas are very limited (Lamarche 1980). In India, groups of 2000–3000 chestnut-bellied sandgrouse are not uncommon during the summer period when surface water is limited, and as many as 50 000 pin-tailed sandgrouse have been in a single flock in Turkey (Christensen *et al.* 1964). Meinertzhagen similarly said that flocks of pin-tailed sandgrouse in Iraq and sometimes in Syria occur in incredible numbers, which darken the sky and produce deafening sounds as they pass overhead.

Watering flocks of 300–400 individuals of Namaqua sandgrouse often arrive together at watering holes, but most of the other African species of sandgrouse apparently gather in rather smaller flocks (Urban *et al.* 1986). Clancey (1967) described the double-banded sandgrouse as occurring in scattered pairs during the breeding season, but at other seasons living often 'in communities' of as many as 50 birds. He also stated that large flocks of *namaqua* and *burchelli* regularly gathered at early morning to mid-morning watering points of the Kalahari desert. It seems very likely that the size of such sandgrouse flocks not only in the Kalahari but elsewhere is largely a reflection of the overall local abundance of the species relative to the distribution of available watering sites.

Spacing or territorial behaviour in sandgrouse is little studied, but there is little or no evidence of territoriality so far available. Based on the number of adults frequenting an isolated waterhole in eastern Turkey, Christensen *et al.* (1964) judged that in the black-bellied sandgrouse the breeding density might be about one or two pairs per 40 ha, or 2·5–5 pairs per square kilometre (100 ha). This compares with a range of densities in various habitats of from 0·5 (on *Stipa* steppe) to 16 (on dry clay pans) pairs of this species per 100 ha in the USSR (Dementiev and Gladkov 1951), and a similar density of about 3 pairs per 100 ha in central Turkey (Lehmann 1971).

Such low breeding densities of sandgrouse are more likely to be reflections of a low carrying capacity of the habitat rather than the direct result of spacing behaviour; indeed there is evidence of semi-colonial nesting by sandgrouse in some areas. Thus nests of Pallas's sandgrouse in the USSR have been found as close as 4–6 m apart, the nests situated in a manner resembling colonies (Dementiev and Gladkov 1951). Although the Namaqua sandgrouse is not colonial, its nests are sometimes placed only 25 m apart (Urban *et al.* 1986). Pin-tailed sandgrouse nests in the USSR have likewise been found only about 20 m apart and also apparently clustered in colonies, according to Dementiev and Gladkov. Ferguson-Lees (1969) similarly noted that several pairs of pin-tailed sandgrouse often nested quite close together in loose groups, but on the other hand many of this species' nests are more widely isolated. In the Crau area in southern France this species occurs as a relatively small and disjunctive breeding population. There it is restricted to an isolated stony area of limited habitat, with a breeding denstiy of only about one pair per square kilometre (Guichard, 1961) or roughly one pair per linear kilometre (Cheylan 1975).

PAIR-BONDING BEHAVIOUR AND MATING SYSTEMS

Hemipodes

Although hemipodes are extremely secretive birds, and many of the species remain essentially unstudied, it is a reasonable assumption that at least most if not all of them are facultatively polyandrous. Monogamy may be a possible alternative mating system when populations are low or the length of the available breeding period is perhaps too short to allow for multiple matings to occur. In the species so far studied, the females are highly aggressive and are probably territorial, the hens being so aggressive toward one another that in captivity they are sometimes allowed to fight with one another as a form of spectator sport analogous to cock-fighting (Baker 1921, Ridley 1983).

Polyandrous mating seems to have been established for at least four *Turnix* species, namely the yellow-legged (Seth-Smith 1903), striped (Flieg 1973, Bell and Bruning 1974, Wintle 1975), painted (Seth-Smith 1905) and barred (Winkel 1969) buttonquails. Additionally, male incubation of the eggs, a necessary prerequisite to and concomitant of polyandry, has been established for the Madagascan (Butler 1905a) and little buttonquails (Lendon 1938), as well as reportedly occurring in all the other Australian species of *Turnix*, including the red-backed, red-chested, chestnut-backed, and black-breasted buttonquails (Frith 1976). The black-rumped button-quail is 'possibly polyandrous', incubation being performed by the male alone. Although there is similarly no direct evidence of polyandry in the African lark-quail, male incubation is likewise known to occur in that species (Urban *et al.* 1986).

On the other hand, there is some indication that fairly strong monogamous pair-bonding sometimes occurs in hemipodes, at least in captivity. This may perhaps also occur in nature when, for example, local ecological conditions might require that the female remains to help feed the pair's young (Trollope 1970). Strong, apparently preferentially monogamous, pair-bonds have sometimes been reported among hemipodes in captivity, the same male incubating successive clutches of the same female's eggs (Hoesch 1959, 1960). However, the conditions of captivity may often place certain restraints on the female's potential choice of mates. Alternatively, captive situations might produce unnatural strains on established pair-bonds before they would normally be broken in a breeding cycle. Thus when Trollope (1970) introduced an extra male striped buttonquail into a cage with an established pair present, it resulted in attacks on the new male by both pair members. When he introduced a female barred buttonquail into a cage in which he had previously successfully placed two males, the female chased both of the males, and fighting occurred between these previously mutually tolerant males. Typically, females will initially chase and attempt to fight with newly introduced males, and if these birds are not in breeding condition they may be promptly killed by the female. A male that is able to 'stand his ground' to the female may soon be accepted by her, and within a few minutes the pair may be feeding peacefully together. Shortly thereafter the pair will roost together and will be inseparable unless another male appears and is able both to evict the first male and to attract the attention of the female. This may help explain why, if a strange male is introduced into an aviary, the resident breeding male may neglect and even desert his eggs or chicks in order to attack the newly introduced male (Wintle 1975).

Female advertisement behaviour is apparently much the same in all the typical buttonquails of the genus *Turnix*, and consists in uttering extremely low-pitched notes (variously described as cooing, moaning, drumming, booming, and droning) which carry great distances and probably serve as effective acoustic territorial signals for these ecologically almost invisible birds (Ridley 1983). In at least several and probably all of the *Turnix* species, the unusual low-frequency calls that are generated in the female's relatively large trachea (see Figure 24) are effectively resonated by an inflation of the anterior oesophagus (Niethammer 1961). However, the African lark-quail apparently lacks both low-pitched calls and an inflatable oesophagus.

Bustards

The pair-bonding behaviour of the bustards is known only slightly better than that of the hemipodes, in spite of their greater size and increased conspicuousness. Cramp and Simmons (1980), in summarizing characteristics of the entire family, stated that monogamous pair-bonds are 'perhaps usual', but that at least one species (the great bustard) is promiscuous, and that many other species may form 'very tenuous' pair-bonds. Urban *et al.* (1986) stated that the mating systems of bustards are 'largely unelucidated', many species apparently having dispersed collective male display ground ('leks'), in which successful males may fertilize several females without any true pair-bonds forming at all. Limited data on sex ratios in bustards, as summarized by Schulz (1986b), support the idea that females typically outnumber males in the adult population, which also suggests that a non-monogamous mating system is probably typical.

Osborne *et al.* (1984) probably came closest to describing the true condition by commenting that 'it now seems that polygamy [= polygyny as used in this text] ... is the rule' for bustards, and suggesting that this condition evidently exists at least for the houbara, great, little, Australian and great Indian bustards, and the lesser and Bengal floricans. They suggested that although some of the smaller South African bustards may perhaps indeed be monogamous, that possibility still remains to be established. They proposed that five general categories of pair-forming behaviour exist in bustards. In the first group (white-bellied, black-throated, Rüppell's and blue bustards) elaborate displays are apparently lacking, and a monogamous mating system is a possibility. In the second group, made up of large species such as the great, kori, great Indian, Australian and Denham's bustards, 'balloon-type' displays are performed on the ground. In the third group, represented only by the houbara, the male performs an erratic run with its decorative neck feathers erected, although the houbara also sometimes jumps into the air while performing its ground displays. Such jumping-type displays are typical of the fourth group of bustards, such as the little bustard and lesser florican. Finally, in the fifth group, comprising such species as the Bengal florican and the red-crested, black-bellied, and little black bustards, a higher display-flight is typical, which usually ends with a gradual glide or very steep descent back to the ground.

It seems likely that these behavioural groups are more likely to be correlated with body size, agility, and possible variations in habitat height or density than with variations in strengths of pair-bonding *per se*. I further suspect that, with the possible exception of the first group, all are basically minor ecological variations on what is essentially a polygynous or promiscuous mating system. I would thus suggest that three intergrading groups might perhaps be recognized. The first would include those species in which males occupy large, mutually exclusive territories in which one or perhaps more females are attracted to mate and possibly remain nearby to nest and rear their young, albeit with little or no help from the male. It is possible that some only slightly sexually dimorphic species such as the little brown bustard fall into this category, although this is purely hypothetical. Widely separated or large but contiguous territories may also be typical of some of the *Ardeotis* bustards as well, such as the Australian and great Indian bustards, although male kori bustards have been described as gathering on low hilltops where direct physical contests to determine social dominance can occur. These gatherings may meet the minimal definition of a lek given by Bradbury and Gibson (1983), namely an assemblage of adult males that females visit solely for the purpose of fertilization.

Certainly there are several species of bustards in which displaying males form certainly interacting lek-like clusters, representing a second behavioural grade of organization. These males may occupy territories within sight or sound of one another and perhaps 100–1000 metres apart, where females can more or less simultaneously judge their relative attractiveness. Such may be the case with the Denham's bustard, in which displaying males may be spaced at about 700-m intervals (Wilson 1972), whilst in the great Indian bustard subordinate males may be tolerated at distances of 100–150 m from a dominant male (Ali and Rahmani 1984). This condition either represents or in any case probably leads to a 'dispersed lek' situation comparable with that of some grouse such as, for example, the blue grouse (*Dendragapus obscurus*) (Johnsgard 1973). In dispersed leks, as may occur in the little bustard (Schulz 1986a), males hold clustered display territories of varied sizes, the males copulating opportunistically with any females attracted to their own territories. A somewhat similar situation probably exists in the lesser florican, where the displaying birds form 'loose aggregations' and several displaying males are sometimes visible simultaneously at a single site (Osborne *et al.* 1984).

Finally, in highly sexually dimorphic species such as the great bustard, the males apparently exhibit a clear dominance-hierarchy on their common display grounds, forming a third-level degree of behavioural organization. In such species, reproductive opportunities are most clearly limited to the fittest (most dominant) individuals, and sexual selection is thereby most intense. The displaying males of the great bustard are typically spaced about 50 m or more apart, a minority of the males (the more dominant and presumably more experienced ones) holding mating territories that may attract up to nine females in a 'harem'. In such harem groups the females may also exhibit dominance ranking, the more dominant females chasing the less dominant ones, especially those receiving attention from the dominant male. However, females may also respond sexually to more than one male (N. J. Collar, in Cramp and Simmons 1980). This situation is a close parallel to that shown by sage grouse (*Centrocercus urophasianus*), which is the largest North American grouse, has the highest degree of sexual dimorphism, and perhaps exhibits the highest level of complexity and dominance-stratification in its lek-forming behaviour (Johnsgard 1973). However, unlike lek-forming grouse, the great bustard apparently does not hold strict territorial boundaries, and displaying males may shift about in their display sites with respect to one another. Additionally, greater age-related individual variation in male weights probably occurs in the great bustard and other large bustards than in any grouse species, and this factor may play an important role in mate-choice by bustard females.

Sandgrouse

Probably all the species of sandgrouse are obligatorily monogamous, based on the observation that in the species so far studied, the male assists in incubation and also assists in the care and protection of the young, and especially their watering. Urban *et al.* (1986) characterized all twelve of the African species of sandgrouse as being monogamous, solitary nesters. Indeed, among the six species so far studied in some detail as to their breeding behaviour, only the male is known normally to gather water in his belly feathers and return with it to the brood for them to drink. Although in the Namaqua sandgrouse the female may undertake this role, apparently this occurs only when its mate has been lost. Not only is seasonally persisting monogamy the apparently typical sandgrouse mating pattern, it is quite possible that, at least in the pin-tailed sandgrouse, the pair-bond may persist indefinitely and perhaps is lifelong (Meade-Waldo 1906, Cheylan 1975). At least in captivity, males will apparently copulate with any female that allows it, even when pair-bonded (Grueber personal communication).

Given the possibility that relatively permanent pair-bonds exist in sandgrouse, it is not surprising that, except in the pin-tailed sandgrouse, essentially no seasonal or age variations in plumage or general appearance occur. The only such variations occurring after the adult plumage has been attained are possibly in soft-part (eye-ring and bill) colouration. Certainly in some species such as Pallas's sandgrouse it is known that sexual maturity and initial egg-laying occur within a year of hatching, at least in captivity (Grumm 1985, Wilkinson and Manning 1986). The laying of two or even three clutches in fairly rapid succession within a single season is also well known in captive birds (Frisch 1970, Grummt

1985). Double-brooding in a single year by wild birds has been suspected for the black-bellied sandgrouse in Israel, and the same perhaps occurs at times with the pin-tailed sandgrouse in Africa (Urban *et al.* 1986). Retention of pair-bonds beyond a single breeding cycle would certainly facilitate prompt re-nesting following nest failure, as well as double-brooding within a single breeding season. Both of these breeding strategies might be distinctly advantageous for birds such as the sandgrouse which, judging from limited data on the Namaqua sandgrouse (Urban *et al.* 1986), seem to have relatively low (i.e. probably under 23 per cent) reproductive success rates.

PATTERNS AND TRENDS IN PAIR-BONDING BEHAVIOUR

It is apparent that the three groups of birds included in this book could hardly be more different from one another in their mating strategies, in spite of sharing some general ecological similarities such as all being ground-nesting, having small clutch-sizes and synchronized hatching, and all having precocial, well-camouflaged offspring. In recent years, various hypotheses have been advanced to try explain the diversity of mating systems found in birds and other organisms, such as those of Emlen and Oring (1977) and Murray (1984). The view of Emlen and Oring is that where ecological situations occur that allow one parent to raise the young alone, the other is freed to attempt to mate with other individuals, producing an 'environmental potential for polygamy'. In most cases among birds, this more often leads to polygyny (the acquisition of supplemental female mates by breeding males) rather than to polyandry (the acquisition of supplemental male mates by breeding females), largely because of the female's generally greater investment of energy in her fertilized eggs than the male's. According to Murray (1984), and as reviewed and updated by Jehl and Murray (1986), variations in the numerical ratio of breeding males to total males and of breeding males to breeding females are likely to cause major variations in optimum breeding strategies and enable one to predict the occurrence of several different combinations of mating systems and associated variations in sexual size dimorphism:

1. When most males can establish territories but few can obtain more than one mate, intrasexual competition is not intense, and sexual size dimorphism is slight, as in most monogamous mating systems. This situation probably applies generally in sandgrouse, and may perhaps occur in a few species of bustards.

2. When breeding females outnumber breeding males, and most males are able to establish territories, competition among breeding males is not intense, and sexual dimorphism remains slight even if polygyny is frequent. This situation probably does not occur in any of the groups considered in this book, but might perhaps apply to some bustards in which males perform conspicuous displays that place them at substantial risk to predation.

3. When only a few of the breeding males can establish territories of dominant breeding positions, the rewards to breeding males are great, and sexual selection for large males is intense. This situation occurs in at least some of the larger species of bustards.

4. When breeding males outnumber breeding females so that territorial or dominant males are still unsuccessful in obtaining males, a male can increase his chances for breeding by forgoing aggressive tendencies and accepting polyandrous relationships. This condition leads to selection for reversed sexual size dimorphism, as occurs in hemipodes.

Jehl and Murray (1986) later added an additional criterion to these four points, namely as to whether or not sexual display occurs mainly on the ground or in the air. This factor relates to the relative strength and manoeuvrability of birds, larger birds probably having physical dominance advantages during ground displays but smaller ones being relatively more manoeuvrable or more acrobatic during aerial activities. This idea might have some significance when considering the aerially displaying species of bustards, as will be noted later. Jehl and Murray proposed that in situations where females are scarce and become the aggressive sex, sexual dimorphism will not occur when females display acrobatically, but will occur when they display normally, as is the case in the hemipodes, for example.

Hemipodes

The polyandrous condition of hemipodes is easily the rarest of avian mating strategies. In an ecological sense, avian polyandry includes 'resource defense polyandry', in which females defend multipurpose territories, the quality of which varies and affects their abilities to attract or monopolize males, and 'female access polyandry', in which females compete among themselves for access to males (Emlen and Oring 1977). Hemipodes seemingly fall into the first of these two types. In a review of avian polyandry, Oring (1986) divided polyandry into various types, including classical (in which individual males breed solitarily with a female, whilst females divide their time among the males), and cooperative (in which groups of males share a single breeding effort with a female).

Classical polyandry can be further subdivided into simultaneous and sequential types; hemipodes are evidently of the sequential polyandry type. Oring listed sixteen bird species in which 'incidental' observations of polyandry have been observed, and eleven more species exhibiting what he regarded as representing 'classical' polyandry. This latter group included documented polyandry in two species of jacanas, but Oring further suggested, without specifically reviewing the literature, that probably all the species of this family are classically polyandrous. However, Johnsgard (1981) judged that one of the African jacanas (*Microparra capensis*) is more probably monogamous than polyandrous, a conclusion that was later also reached by Urban *et al.* (1986). Oring also did not review the situation in hemipodes, other than to note that more than a century ago Charles Darwin had correctly listed at least some species of *Turnix* as being among those birds exhibiting the polyandrous type of mating system. Jehl and Murray (1986) suggested that the hemipodes represent a classic example of reverse size and plumage dimorphism associated with sex-role reversal, and noted that a polyandrous mating system has been suggested by observations in the field as well as in captivity.

According to Oring, polyandry is most likely to evolve in avian populations having several ancestral attributes:

1. Males were already contributing significantly to parental care.

2. Food did not limit the potential for uniparental care and multiple egg production.

3. Uniparental care was adequate to cope with predation.

4. Males were more efficient than females in providing uniparental care.

5. Females were better able than males to exploit alternative reproductive opportunities.

6. Clutch-size was determinate, requiring multiple-clutching for increasing effective annual reproductive output.

7. Year-to-year variations in breeding conditions helped to enhance these previous preadaptations.

It is hard to say to what degree the hemipodes may have fulfilled these ancestral conditions, but with the possible exception of having determinate clutch-sizes, none of them obviously prevents the group from being potentially polyandrous. If we imagine the ancestors of the hemipodes to be monogamous with biparental care and a small clutch-size of three or four eggs, it is easy also to imagine the development of a polyandrous breeding system gradually evolving in tropical grassland habitats having short or highly variable periods available that are favourable for breeding.

Ridley (1983) suggested that the evolution of successive polyandry in hemipodes and their very rapid embryonic and post-hatching development periods have allowed them to take advantage of temporary flushes of vegetative growth and associated food supplies, as is typical of tropical grasslands. This allows hemipodes to respond quickly to sudden rains by moving into favourable areas and completing their breeding cycles before ecological conditions again deteriorate. Or, if the period of favourable breeding conditions persists, up to as many as seven broods might be reared in a single year (Wintle 1975). Ridley (1983) also suggested that, although several species of true quails are similar in their ecology to hemipodes, the quails achieve comparable rates of reproductive increase by laying larger but fewer clutches. Further, in this group all the species are strictly monogamous, both parents apparently being needed to rear their relatively large broods successfully (Johnsgard 1988).

It is apparent that reversed sexual dichromatism is present in all the species of hemipodes to varying degrees, and the same is apparently the case with sexual dimorphism, based on the still often rather limited available weight and measurement data (Table 5). With but a few exceptions that probably result from small sample sizes, the mensural ratios rather consistently indicate that the females of all species of hemipodes are on average significantly larger than males, the males generally having dimensions 90–95 per cent of those of females. The general consistency in these data suggest that all species of hemipodes are either regularly polyandrous or potentially polyandrous. The collective unweighted averages for the male–female ratios in all species range from 0·8 to 0·96 for the four dimensions shown, thus supporting a general reversed sexual dimorphism trend.

Bustards

As noted earlier, it seems likely that among the bustards there may be a diversity of mating systems present, although present evidence suggests that at least the majority of species are polygynous. Urban

Table 5. Male–female dimorphism ratios in hemipodes[1]

Species	Wing	Tail	Tarsus	Culmen
Lark-quail	0.94	0.87	0.94	–
Madagascan	0.94	–	–	–
Spotted	0.91[2]	–	–	0.93[2]
Yellow-legged	0.90[2]	0.88[2]	0.94[2]	–
Black-rumped	0.91	0.83	0.97	0.89
Striped	0.93	0.92	0.98	0.88
Red-backed	0.91[2]	0.84[2]	0.92[2]	–
Little	0.92	0.87	1.01	1.00
Red-chested	0.94	1.01	0.98	0.93
Chestnut-backed	1.00	0.84	0.92	0.89
Painted	0.88	0.88	0.91	0.93
Barred	0.96[2]	0.97[2]	0.96[2]	–
Black-breasted	0.98	0.91	1.01[2]	0.88
Mean	0.93	0.89	0.96	0.92

[1] Based on average dimensions of entire sample for sex unless otherwise indicated.
[2] Based on average of available range of measurements.

et al. (1986), in reviewing the eighteen African bustard species, stated that only one (Rüppell's bustard) was definitely monogamous, whilst the white-bellied bustard was 'probably monogamous, with group territorialism'. 'Group territoriality' (perhaps more accurately described as clustered male territories) was also attributed to the blue bustard. The little bustard exhibits male-dominance polygyny, with male territories more or less clustered in display-centres (Schulz 1985b). A comparable dispersed-lek mating system of aggregated displaying males perhaps also exists in the Denham's, rufous-crested and black-bellied bustards. Association of displaying males has also been reported for the kori bustard. The great bustard was classified by Urban et al. as apparently polygamous but sometimes exhibiting lek-like behaviour. However, monogamous matings are reported to occur occasionally in this species, and the leks are atypical in that displaying males do not defend specific territorial boundaries; rather, they simply aggregate on common display grounds. The mating system of the little black bustard was described by Urban et al. as unknown, but with no evidence of pair-bonding. The houbara is probably polygynous or promiscuous, based on recent observations by Collins (1984). The remaining African species (Ludwig's, black-throated, little brown, Nubian, Heuglin's, Hartlaub's and Arabian) were all reported by Urban et al. as having unknown or unspecified mating systems, although as noted earlier, the little brown bustard 'might be monogamous' (Osborne et al. 1984). All four of the non-African bustards (lesser and Bengal florican, great Indian bustard, and Australian bustard) are now known to have polygynous or promiscuous mating systems. Thus only one bustard species, Rüppell's, is probably monogamous, and it has been speculated that four others (little brown, black-throated, blue and white-bellied) are so.

It is painfully apparent that inadequate information exists on mating systems in the bustards, and that much of what is believed about them is based on anecdotal or highly limited information at best. In the belief that the degree of sexual dimorphism may be an indicator of relative polygyny tendencies, a summary of such data is shown in Table 6 for the bustard species, based on average measurements provided in the species accounts. Although inclusion of weight data would have been valuable as well, too little is available to provide meaningful comparisons for most species.

These results show a much wider range of variation in sexual dimorphism than do those for the hemipodes, although the collective unweighted averages of male–female ratio for all species consistently approximate to 1·10:1 in the three sets of dimensions shown. The greatest sexual dimorphism (males generally averaging 1·15–1·4 times as large in linear dimensions as females) occurs in the three genera that represent the largest and heaviest species (*Otis*, *Ardeotis*, and *Neotis*). In three potentially monogamous and relatively small species the males are generally 1·0–1·05 times as large as females. Species reportedly having dispersed male territories or dispersed leks typically have slightly less dimorphism evident (males generally averaging 1·0–1·15 times the size of females).

Interestingly, the two small Indian species of floricans seemingly exhibit slightly reversed sexual dimorphism, although the mensural data for these species are quite limited. Males of both of these spe-

Table 6. Male–Female dimorphism ratios in bustards[1]

Species	Wing	Tarsus	Culmen	Probable mating system
Houbara	1.09	1.10	1.08	Dispersed male territories?
Little	1.01	1.02	1.08	Clustered male territories
Great	1.27	1.26	1.19	Usually lek-like, promiscuous
Australian	1.13	–	1.14	Dispersed lek?
Great Indian	1.37[2]	1.25[2]	–	Dispersed lek?
Arabian	1.15	1.13	1.12	Unknown
Kori	1.21	1.27	1.28	Dispersed lek?
Nubian	1.18[2]	1.16	1.20	Unknown
Denham's	1.21	1.24	1.23	Dispersed lek?
Heuglin's	1.17	1.20	1.19	Unknown
Ludwig's	1.19	1.17	1.11	Unknown
Black-bellied	1.08	0.99	1.01	Dispersed lek
Hartlaub's	1.09	1.03	0.98	Unknown
White-bellied	1.02	1.04	1.04	Monogamous?
Little brown	1.02	1.03	1.00	Monogamous?
Black-throated	1.07	1.05	1.08	Monogamous?
Rüppell's[1]	1.05	1.06	1.06	Monogamous
Little black	1.04	1.04	1.06	Isolated male territories?
Rufous-crested	1.03	1.00	1.03	Dispersed lek?
Blue	1.01	1.05	1.00	Monogamous?
Bengal florican	0.94	1.01	0.79	Dispersed lek?
Lesser florican	0.84	–	–	Dispersed lek?
Mean	1.10	1.10	1.08	

[1,2] Footnotes as for Table 5.

cies are known to perform acrobatic aerial displays, and perhaps selection for small male size and associated improved aerial manoeuvrability has influenced body size in at least these species, just as it may have selected against large male body size in the several smaller African *Eupodotis* species that are also known to display aerially.

Sandgrouse

The apparently consistently monogamous sandgrouse offer the fewest opportunities for evolutionary and ecological speculation, among the three bird groups under consideration. Like most shorebirds (as well as all pigeons and doves), there is a consistent biparental care pattern, which is characterized by incubation participation by both sexes and the male continuing to help to care for the young, probably to fledging or even later. Weight and measurement dimorphism ratios shown in Table 7, based on data provided in the species accounts, indicate a rather consistent average male–female ratio of about 1·02–1·12. Above-average figures of up to 1·3:1 are associated with tail-length ratios in the long-tailed species (the first six species listed in the table),

Table 7. Male–female dimorphism ratios in sandgrouse[1]

Species	Wing	Tail	Tarsus	Weight
Tibetan	1.02	1.14	–	–
Pallas's	1.14	1.32	1.03	1.00
Namaqua	1.03	1.17	1.04	1.05
Pin-tailed	1.02	1.13	1.07	1.04
Chestnut-bellied	1.04	1.24	1.05	1.14
Spotted	1.05	1.21	1.06	–
Black-bellied	1.04	1.03	0.97	1.12
Crowned	1.05[2]	1.04	0.97	–
Madagascan	1.10[2]	1.14[2]	–	–
Black-faced	–	–	–	1.05
Lichtenstein's	1.06	1.08	1.08[2]	1.05
Double-banded	1.02	1.02	1.00	0.98
Four-banded	0.99	1.06[2]	–	–
Painted	Inadequate data			
Yellow-throated	1.02	1.08	1.02	1.02
Burchell's	1.01	1.06	1.02	1.12
Mean	1.04	1.12	1.03	1.02

[1,2] Footnotes as for Table 5.

where perhaps tail pattern characteristics and relative male tail-length may play a role in sexual display and consequent mate choice.

4 · Breeding biology

Although the breeding biologies of individual species are described later in this book, a comparative overview is helpful as a means of seeing similarities and differences within and between groups, and in searching for general patterns that might thus emerge. For the sake of descriptive continuity, the following discussion is organized according to approximate chronological sequence of breeding activities for each of the three groups under consideration.

HEMIPODES
Behaviour associated with fertilization

According to Flieg (1973), courtship in the striped buttonquail begins by the female swaying the body back and forth in a rocking motion, which attracts males into her territory. The next phase is performed in the presence of males and consists of uttering the booming call while bending the legs and standing at a 45-degree angle. The female then sits and makes scraping movements with her beak, which is followed by the male gently pecking at her head and nape, in turn followed by copulation. According to Flieg, the female may sometimes also mount the male. This reversed copulatory behaviour was observed in the same species by Hoesch (1959), especially during the egg-laying period, although Hoesch further noted that copulation appeared to be successful only when the male mounted his mate.

In his oberservations on this same species, Trollope (1979) stated that the booming call of the female presumably served to attract males rather than specifically serving as a challenge to other females, inasmuch as when a female uttered such a call it was not responded to by a hen in an adjoining aviary, or at least not before a marked time-lag had occurred. Trollope apparently saw the body-swaying display performed only by females of the barred buttonquail, and noted that in this species the female's major vocal signal was a more 'purring' note, uttered with much more distinctive head-lowering than occurred in the striped buttonquail. Wintle (1975) apparently observed the slow back-and-forth rocking or swaying behaviour in both sexes of the striped buttonquail, and thought that it might simply be an indication that the birds were nervous. It was also observed in the barred buttonquail by Dharmakumarsinjhi (1945), who suggested that it might mimic the movement of vegetation swaying in the wind and thus serve a visually protective purpose rather than an advertisement display function.

Mutual preening between members of the adult pair has been reported (Trollope 1970), and Flieg (1973) similarly saw the male preen ('gently peck') the female's neck and nape prior to copulation. Although courtship-feeding of the male by the female was not mentioned by Flieg, Trollope observed it in the barred buttonquail. It has also been reported to occur in the striped buttonquail by Wintle (1975), in the yellow-legged and painted buttonquails by Seth-Smith (1903, 1905), and in the Madagascan buttonquail as observed by German aviculturists and reported by Butler (1905a). It seems possible that the few cases in which females have been observed feeding young (Butler 1905a, Dharmakumarsinhji 1945) or preening young (Trollope 1970) may represent a relict of an earlier, presumably more typically monogamous, behaviour by buttonquails.

Nest-building and egg-laying

In contrast to other groups covered in this book, buttonquails perform a joint nest-building ceremony, which may have the function of showing the male where the eggs are to be laid (Ridley 1983). Such ceremonies have been observed in several species, including the yellow-legged (Seth-Smith, 1903), Madagascan (Butler 1905a,), striped (Butler 1905b, Wintle 1975) and barred (Trollope 1970) buttonquails. According to Wintle (1975), both sexes look for nesting sites and, although the male 'tries' each likely spot, it is the female that finally selects the nest location.

The nests of buttonquails are sometimes relatively elaborate structures, largely constructed by the female. Butler (1905a) has provided translations of the best descriptions (from a German account written by Dr Karl Russ) of nest-building in this group, based on observations of the Madagascan buttonquail:

The female prepared a clear space on the ground for the nesting-site, among the rush-stalks not far from the window. Standing on one foot and scratching with the other, also twisting herself in a circle she had with a few turns gained the desired width between the rush-stalks.

Meanwhile the resounding pairing note and the gentle call-note, sounding like *kru, kru*, had attracted the male; which now crept as far as possible in the space with her, and also began to twist and scratch. The hen now seized the stalks within reach with her bill, and with a rapid lateral movement flung them singly over herself, that is to say clear over her head on to her back; then the male dragged them farther toward himself.

When enough stalks lay upon the back of the bird which was in the nest, these were compacted by pressing upwards and at the sides, fresh ones being constantly added, until by degrees a complete enclosing arch was erected. Meanwhile, by turning at the nest and arranging softer stems, the male had modelled the nesting-hollow. The principal work fell to the lot of the female.

After she had driven the male out and slipped inside herself, she next drew almost all the stems at the entrance hole together, so that it was almost concealed; thereby the building from the outside appeared like a confused heap of stalks, though inside it was well-formed (whereby doubtless it affords protection to the little bird when in freedom). The latter crouched, stooping low among the loose-lying stalks before the entrance which then closed up on themselves behind her, so as to conceal the eggs and brooding cock from sight. It is noteworthy that the male showed himself extremely cautious every time he left the nest and slipped in again.

Later observations on the striped buttonquail (Butler 1905b) indicated that the process of nest-building was carried on in the same manner as in the Madagascan buttonquail, but the birds selected a different locational situation for it and performed somewhat different nest-building movements:

They do not stand or crouch like the latter, but sink the breast deep on the ground, and erect the hinder part of the body, so that the tail is directed straight upwards similarly to that of the doves; like the latter also they simultaneously make short tremulous movements with head and neck, whilst they peck immediately in front of them in the sand or their own breast-feathers; at times also shuffling a little with their wings and uttering soft notes. The female does her part more zealously and patiently; the building of the nest continues sometimes for days, sometimes only for hours. The fourth nest was not quite completed when the first egg rested in it . . .

Egg-laying normally occurs at approximate daily intervals in buttonquails (Butler 1905b, Wintle 1975), although Flieg (1973) made some divergent observations in an aviary containing three males and three laying females. These three females laid a total of at least 302 eggs in a seven-month period, or an average of one egg per female every other day during this entire period. These were laid in a total of eight different nests, some of which were perhaps used by different females. However, the shells of these eggs were described as being paper-thin, and at the time of maximum egg-laying some birds were laying several eggs per day, perhaps as many as five. (An editor's comment was inserted to the effect that 'some abnormal factor' must have been present to have produced these remarkable conditions, which are perhaps unique among birds.) Interestingly, Hoesch (1960) also observed several cases in which two eggs were laid in a single day by the same female, although the normal egg-laying interval was one day.

Incubation begins with the laying of the last (Butler 1905b) or penultimate (Hoesch 1970) egg in the clutch. Hoesch stated that the male usually begins to sit during the night before the last egg of the clutch is laid, regardless of the size of the clutch. The male's behaviour may, however, be dependent upon the behaviour of the female, which stays away from the nest on the day before the last egg is laid, whereas earlier in the laying cycle she may often sit on the nest during the day. According to Butler (1905b) the female may help to incubate the clutch for the first few days, but gradually abandons this to the male, so that eventually only the male is incubating.

According to Wintle (1975), the female will begin again to utter her advertisement call within a day of the male starting incubation, and one of the males that have been waiting for this 'invitation' will make his presence known. If accepted by the female, he will then chase off any other interested males, and by about 7 days later will be sitting on his own clutch of three eggs. Wintle stated that the interval between successive 'matings' (apparently meaning successive clutches) seldom exceeded 12 days, and was more often 7 days. This would indicate a resumption of egg-laying by the female within about 4 days, assuming an average clutch-size of three or four eggs. Hoesch (1960) stated that a female laid seven clutches during a nine-month period from early March until November, all of which were incubated and the young brooded by the same male, each succeeding clutch being started almost immediately following the rearing of the preceding brood. By introducing a second unpaired male, Hoesch was able to shorten the period between a female's laying of successive clutches (which averaged about 50 days when a single male was present) to only 11 days. This clearly indicates the advantage of utilizing polyandry as a means of maximizing reproductive efforts in buttonquails.

Hatching and brood-rearing

The incubation period of buttonquails is remarkably short. Flieg (1973) reported a 12–13 day period for the African race of the striped buttonquail, and

Hoesch (1959) found a 13-day incubation period for the same race. Winkel (1969) reported a 13-day incubation period for the barred buttonquail, and Butler (1905b) observed a 16-day incubation period for the Madagascan buttonquail. Incubation begins with the last egg or approximately so, hatching is synchronous, and the nest is deserted within about four hours after hatching has been completed (Wintle 1975). The hatched egg-shells are removed from the nest (Fjeldså 1977).

Newly hatched buttonquails are extremely small, averaging about 2·7 grams according to Flieg (1973) and as small as 1·5 grams according to Fjeldså (1977), and indeed are among the smallest of chicks of precocial birds, being comparable in size to those of the smallest species of true *Coturnix* (including *Excalfactoria*) quails. They are fed bill-to-bill by the male, who picks up food items and offers them to the young. He usually performs a 'titbitting' display, in which food is offered in the bill while the head is held stiffly in place, the wings are shivered, and soft calls are uttered. At two days of age, females can sometimes be distinguished from males, since they may begin to boom and display the courtship stance typical of adults. Within two or three days they are also starting to feed for themselves. However, the chicks are fed at least once per hour by the male, who frequently calls them back to him for brooding after feeding periods. When frightened, the young lie perfectly still, with their eyes closed (Hoesch 1959, Flieg 1973).

The primary feathers begin to emerge on the second day, but at varied growth-rates. Thus the inner primaries grow fastest but cease growth at a length much shorter than that of the final primaries and the outer first-generation primaries remain vestigial (Fjeldså 1977). Initial fluttering flights may occur as early as 7–11 days after hatching (Flieg 1973, Wintle 1975), but fledging in the form of spontaneous flights usually occurs at about 15 days after hatching (Hoesch 1959). During the second week, feathers appear on the back, wings, and sides of the crop (Fjeldså 1977). Full juvenal plumage is attained by the eighteenth day, and minimum adult proportions and measurements are attained by 27 days (Flieg 1973). Replacement of the mesoptile flight feathers begins in the third week, the feathers appearing in descending (outward) squence on the wings, while the outer first-generation or juvenal feathers are still growing (Sutter and Cornaz 1963, Fjeldså 1977). Initial copulations were observed by Flieg at 37 days, and by 39 days the sexes were distinguishable by overall size, weight, and measurements. Similarly, Winkel (1969) found that by three weeks the iris colouration had changed from dark grey to yellowish white, and by 40 days of age young barred buttonquails had reached adult weights and wing-lengths. Egg-laying was observed by Flieg to occur among female striped buttonquails as young as 4 months of age, whilst Hoesch (1960) stated that both sexes of this species are able to reproduce at the age of 5 months.

Winkel (1969) has provided some data on growth rates of the wing and tarsus in barred buttonquails and associated weight changes during the first 60 days after hatching, which are summarized in Figure 21.

Breeding success and productivity

Data on hemipode breeding success are so far limited to a few species bred under captive conditions, and as such are probably meaningless with respect to wild populations. However, even assuming a high rate of egg or chick loss, the reproductive potential of typical polyandrous buttonquails (the lark-quail may be monogamous) is little short of astonishing. According to Wintle (1975), some reproduction of the striped buttonquail occurs in South Africa throughout the year, with peak periods in March and September. Urban *et al.* (1986) additionally reported egg records for all months in Zimbabwe and Namibia. Monthly egg records spanning 9 or 10 months were also reported for Malawi and Zambia, whilst other regions of Africa had egg records ranging in collective length from 1 to 5 months. Clutch-sizes average 3·5 eggs in this species. The black-rumped buttonquail has a more restrictive span of egg dates that range up to as long as 6 months in Zimbabwe and 4 months in Zambia. Clutch-sizes average 3.0 eggs in this species. Fjeldså (1977) stated that a female buttonquail may lay as many as seven clutches during an 8 months breeding season, but said that the number of clutches tended per male in one season was unknown. In captivity as many as eight clutches have been laid in a single year by the Madagascan buttonquail (Butler 1905a), and Wintle (1975) suggested that a single female striped buttonquail should be able to keep five males busy rearing young the whole year round. However, he doubted whether natural populations have that many males available.

Assuming conservatively and for simplicity's sake that buttonquail clutch-sizes average 3·5 eggs, that there is an average interval of 2 weeks between successive clutches, and that the effective breeding season is 6 months, it is apparent that a single female buttonquail might possibly generate as many as twelve clutches. This represents about forty-two potential offspring in a single year, assuming that enough males are available to incubate all these clutches. Wintle (1975) judged that, on average, two

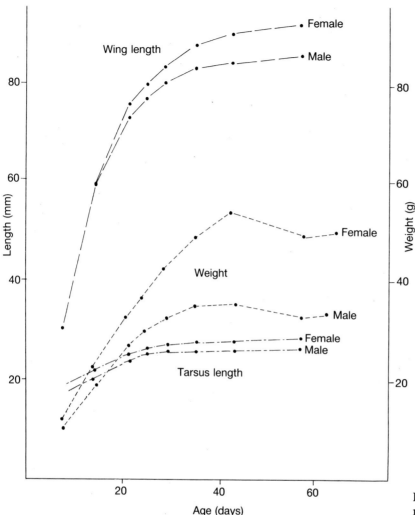

Fig. 21. Growth rates of captive-raised barred buttonquails. After Winkel (1969).

chicks per brood are raised as a rule, reducing the theoretical number of fledged offspring to twenty-two per female. Slightly more liberal estimates of clutch-size, hatching and rearing success and/or breeding season length could easily significantly increase this potential annual output. Even assuming a 75 per cent rate of egg plus chick mortality, the resulting adult female–young ratio at the end of the breeding season would be about 1:10 if these conditions prevailed, or greater than any observed reproductive rates in gamebirds such as true quails (Johnsgard 1973, 1988). If indeed young of the early broods begin to breed at 4 or 5 months of age, it is even possible that these in turn may begin to breed before the end of a single breeding season, enlarging the potential rate of annual increase still further. I can think of no other species of bird with precocial young having a comparably high reproductive potential, or even one that is remotely close to it. Unfortunately there appear to be few available data on adult–young ratios in wild buttonquail populations on which realistic productivity estimates might be based. For example, Morris and Kurz (1977) observed some sex and age ratios from wild-trapped red-chested and little buttonquails (using spotlighting techniques, which should have smaller age of sex trapping biases). Of 123 birds of both species captured, 41 per cent were juveniles or immatures, suggesting that a very high recruitment rate might indeed exist. Of the seventy-two adults, 66 per cent were males, or a 2:1 adult male–female ratio, which would be very desirable in a polyandrous breeding system. However, a small series of museum specimens did not exhibit a similarly distorted sex ratio.

Given these impressive potential reproductive statistics, one might wonder not why buttonquails are polyandrous, but rather why more species aren't! Advantages for successive polyandry obviously progressively diminish as the length of the available breeding season declines and the length of each breeding cycle increases, so it is not surprising that buttonquails have a pattern of geographical distribution that centres on the equator and barely extends into temperate climates. Hemipodes also have what are collectively among the shortest incubation (c.12–16 days) and rearing (c.18–27 days) cycles of any precocial bird species. It is also not surprising that females should be so extremely aggressive towards other hens which may be competing for the available males in a local population, since the number of available males probably places distinct restraints on any female's reproductive potential. A detailed analysis of adult sex ratios in buttonquail populations would clearly be of considerable interest, as would any data on the proportion of young birds in natural populations. Both of these ratios would represent highly significant statistics in terms of estimating actual or potential buttonquail productivity.

For these various reasons, Ridley's (1983) suggestion, that buttonquails utilize successive polyandry and rapid development of their young not only to breed prolifically but also to exploit quickly the temporary flushes of growth typical of tropical grasslands, is perhaps less convincing than an argument that they can instead thereby take best advantage of extremely prolonged breeding seasons through having a succession of broods lasting as long as favourable breeding conditions might prevail. The tiny size of the newly hatched young (under 2 grams in the smallest species) may also make them highly sensitive to chilling, and may prevent the birds from occupying areas that regularly have cold night temperatures. Such factors might help explain why they have not been able to colonize the vast steppes of Asia and the Middle East, where seemingly suitable grass-dominated habitats are almost unlimited but where the length of the summer breeding season is quite short and daily temperature fluctuations are likely to be considerable.

BUSTARDS

Behaviour associated with fertilization

Fertilization behaviour of bustards is in general only very poorly known, inasmuch as the birds are typically very shy during this period, copulations evidently occur extremely rapidly, and perhaps much of the activity associated with fertilization occurs around dawn and dusk. It has perhaps been best observed and described in the little bustard (Frisch 1976, Schulz 1985b, 1986a), but has also reportedly been seen and variably described in the houbara (Collins 1984a; summarized in Urban et al. 1986), the great bustard (Gewalt 1959, Cramp and Simmons 1980), the Australian bustard (Fitzherbert 1978), the red-crested bustard (Astley-Maberly 1967, Kemp and Tarboton 1976), the blue bustard (Kemp and Tarboton, 1976), the black-bellied bustard (Kemp and Tarboton 1976), the lesser florican (Dharmakumarsinhji 1950) and the Bengal florican (Narayan and Rosalind 1988).

The lesser florican can be regarded as a typical bustard in its copulatory behaviour. Copulation is preceded in this species by the male chasing the female, which apparently serves to stimulate and prepare females for copulation. In the little bustard the male rapidly runs behind the female with his head retracted into his black neck-collar, repeatedly stopping abruptly and calling whilst throwing his head and body sideways. This display apparently reduces the aggressive and defensive behavioural tendencies of the female long enough for the male to achieve copulation. Treading may only last for a few seconds, with the male initially pecking strongly at the defenceless female's neck and nape, Similar strong pecking behaviour occurs in the great bustard for a minute or more prior to actual insemination, the actions often drawing blood, and in both of these species such aggressive activity leads quickly to separation and escape by the female almost as soon as she has been inseminated (Schulz 1986a).

The frequency of bustard copulations in the wild is unknown, but considering the small clutch-size of these birds and the apparent great reluctance to mate on the part of the female, it is quite likely that single insemination is adequate for fertilizing a female's entire clutch. A second fertilization may well be required for any replacement clutches, as seems to be typical, for example, of various promiscuous species of grouse (Johnsgard 1973, 1983a).

Nest-building and egg-laying

Nest-building is essentially absent in bustards; the nest is a shallow scrape that is either unlined or may have the vegetation of the site trampled down by the female. The male is not known to play any role in nest site selection or nest-building in any species of bustard.

Most bustards have clutches of no more than two eggs, the major exceptions being the little bustard (usually three or four eggs, but five-egg clutches are reportedly common in Portugal) and the lesser florican ('nearly always' four eggs, according to Baker 1935). Egg-laying intervals are typically 1 or 2 days, and incubation begins approximately with the laying of the last egg of the clutch, or perhaps slightly earlier, so that hatching of all the young is essentially synchronous. At least in the little bustard, most laying apparently occurs during morning and evening hours. A laying interval of 2 days is typical of this species, but intervals of up to 3 days or more sometimes occur, at least in captivity (Frisch 1976).

Replacement clutch-sizes may be smaller on average than initial clutches; for example, two eggs have been observed as a replacement clutch in the little bustard (Labitte 1955). The incidence of replacement clutches is still unknown, but in addition to the little bustard thay have been reported for the houbara and great bustard, and quite possibly occur in several other species as well, judging from the recorded seasonal durations of their egg-laying periods.

Hatching and brood-rearing

Hatching and post-hatching stages of bustard biology remain virtually unstudied in the wild, and indeed bustards have been successfully hatched and raised so few times under captive conditions that little more than bare statements can be made about this phase of their life cycle. For example, Heinroth and Heinroth (1927–8) successfully raised great bustards in captivity from wild-taken eggs, very probably for the first time. They provided data on weight changes of a male and female great bustard for several months after hatching (Figure 22) and presented photographs of the appearance of the young birds at various ages (Figure 23).

Since the early efforts by the Heinroths to hatch and rear birds from wild-taken eggs, little success has been achieved in hatching and raising bustards in captivity. However, Frisch (1969b) succeeded in getting little bustards that had been captured as chicks and hand-raised to breed in captivity. Goriup (1985b) noted that, although for nearly a century great bustards had been maintained in captivity and hundreds had been raised from wild-taken eggs, these birds had been successfully bred in captivity only at five locations. Gewalt (1964) reported the first success in breeding great bustards in West Germany. This account was later supplemented by a detailed report by the Gewalts (1966), which provided information on the development of young great bustards. Captive breeding of the black (Gregson 1986), Australian (D. M. White 1985) and

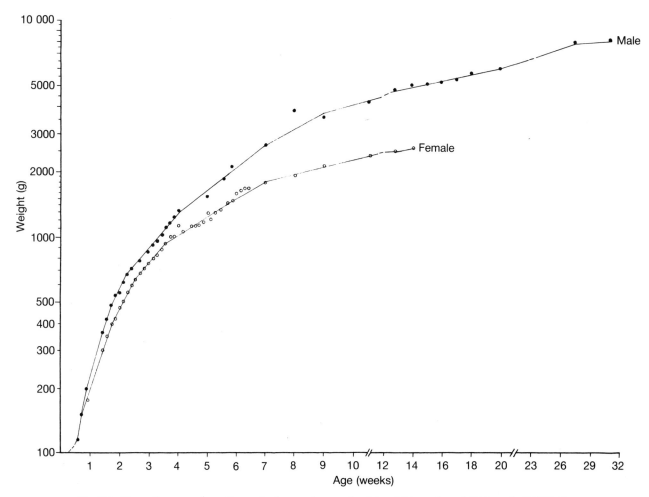

Fig. 22. Growth rates of captive-raised great bustards. After Heinroth and Heinroth (1927–8).

Fig. 23. Appearance of great bustard chicks at various ages: (a) 3 days; (b) male at 10 days; (c) male at 20 days; (d) female at 29 days; (e) male at 40 days; (f) female at 68 days; (g) male at 8 months. After Heinroth and Heinroth (1927-8).

houbara (Mendelssohn *et al.* 1979) bustards has also occurred in recent years.

Goriup (1985a) provided information on early post-hatching life of two newly hatched great bustard chicks, and gave comparative data on post-hatching weight changes during the first ten days following hatching, based on earlier published data by the Heinroths, the Gewalts, and Fodor (1966), this last being a Hungarian record of captive-rearing. L. Osborne (1985) provided similar information on the feeding, growth and development of six chicks hatched and raised from eggs taken in Portugal and Hungary, and illustrated weight-change curves for chicks up to 1 kg. She likewise provided comparative data for growth rates observed in other studies. The substantial differences in these curves suggest that differences in rearing conditions may have considerable impact on the rates of observed growth in young bustards, although the possibility of genetic differences influencing these growth-rate variations was not ruled out by Osborne.

Breeding success and productivity

Almost no real data exist on the population dynamics of any bustard species. Longhurst and

Silvert (1985) noted that with regard to the best-studied species of bustard, the great bustard, there were no available estimates of annual adult or juvenile survival rates, of fledging success, or even of average clutch-sizes. Their most useful statistic regarding probable annual population recruitment rate or overall productivity was an estimate of approximately 8·5 per cent annual population increase in Hungary over a span of 5 years, during a period immediately after hunting had been prohibited. They were able to simulate this approximate rate of population increase by assuming initial female breeding at 3 years, an average clutch-size (and apparently a 100 per cent hatching success rate) of 2·5 eggs, a 50 per cent survival of chicks to fledging, a 50 per cent survival of fledged juveniles to the end of their first year, a 75 per cent annual survival of second-year birds, a 90 per cent survival of third-year birds, and a 95 per cent survival of older age classes, with a sex ratio of unity among adults. However, the high implicit hatching success of this simulation is unlikely to be matched in wild populations, where pre-hatching nest losses are likely to exceed 50 per cent.

Schulz (1986b) attempted to analyse the population structure of little bustard flocks during spring in Portugal, when he counted flocks totalling 1537 birds. Using certain reasonable assumptions, including one that 1·3 juveniles per breeding female survived their first winter, he judged that 34 per cent of the flock total represented juvenile birds in their first spring, and that the adult sex ratio was 1·36 males per female. This is a substantially larger ratio of young birds in the population than was indicated in the computer simulation described above (in which first-year juveniles never exceed about 16 per cent of the total population), suggesting either that little bustards must have considerably higher productivity than great bustards, or that some assumptions in one or both of these estimations may be in error.

SANDGROUSE

Behaviour associated with fertilization

Observation on display behaviour associated with copulation in sandgrouse are still surprisingly limited, considering the fact that several species have been successfully bred in captivity in recent years. This suggests that such behaviour is relatively infrequent or unobtrusive.

Loren W. Grueber has observed (1987, and personal communication) that in his captive yellow-throated sandgrouse the male's courtship display is quite subdued, and consists of erecting the black feathers of the lower neck-band and jumping upwards while making a 90-degree turn with the tail erected and fanned. He has seen very few if any specific pre-copulatory displays, and has noted that the female simply indicates her invitation to copulate by assuming a submissive posture. A pair-bonding display consists of the two birds walking slowly together while making pecking movements. The post-copulatory display is similar to the pair-bonding display, but is performed by the male only. After dismounting he walks slowly in front of the female with his head held forward and slightly downwards.

One published account of pre-copulatory display in sandgrouse is by Frisch (1970), who stated that in the pin-tailed sandgrouse the male displays by following or chasing the female for a few metres with a stiff-legged gait and a downward-tilted and fanned-out tail, after which he jumps up on her and mates. Frisch also heard the male produce a very soft call, which he described as being similar to that of a black grouse (*Tetrao tetrix*); it was used to lead the female to a specific nest-site.

Nest-building and egg-laying

The nest-site of sandgrouse is a simple scrape on bare ground. Frisch (1970) stated that in the pin-tailed sandgrouse both members of the pair made a nest-scrape with their feet, in a manner very unlike that of a chicken scratching for food. However, they also picked at the depression with their beaks. Small stones, feathers and dung fragments were also apparently placed in the nest, in addition to other materials lying nearby. Marchant (1961) stated that four of five nests found of the spotted sandgrouse and fifteen of twenty-three nests of the pin-tailed sandgrouse were in depressions caused by hoof-marks of camels, which were up to about 8 cm deep. In the remainder of the sites the birds may have scratched the ground surface slightly, but in no nests was there a trace of any added materials.

In most species of sandgrouse the usual clutch-size is apparently of three eggs, with two-egg clutches next most frequent but sufficiently rare as to suggest that they are distinctly abnormal. Thus the mean size of sixty-two clutches of Namaqua sandgrouse was 2·9 eggs (Urban *et al.* 1986). Marchant (1961) noted that all five of the clutches he found of the spotted sandgrouse, and all but one of sixteen clutches of pin-tailed sandgrouse were of this number, the exception being of two eggs. He estimated that the eggs were either laid about 48 hours apart or were laid at irregular intervals. Grummt (1985) likewise observed in captive Pallas's sandgrouse that the eggs were laid at about 48-hour intervals, with one exceptional case of about 72 hours. Frisch (1970) judged that in his captive pin-tailed sandgrouse the eggs were consistently laid 2 days apart, and always in the later afternoon. Although egg-covering by both sexes and apparent incubation may begin with the first egg in species such as the pin-tailed and Namaqua sandgrouse, 'proper' incubation begins with the second or third egg (Urban *et al.* 1986). In Frisch's observations of the pin-tailed sandgrouse, 'firm-sitting' was started only after the third egg in the clutch was laid; in one case the male was observed to begin within half an hour of the female's laying of the third egg. The female sat, with variations of about half an hour, from around 6.00 a.m. until 8.00 p.m., and the male consistently incubated during the night.

It is still apparently unproven whether under natural conditions sandgrouse have replacement clutches following the loss of the first, but it seems very likely. Wilkinson and Manning (1986) reported a second clutch begun by captive Pallas's sandgrouse only about 2 weeks after their first clutch was taken. Grummt (1985) found in the same species that a second clutch occurred only 9 days after the first was removed, and a third clutch was laid only 8 days after the second set was taken. Frisch (1970) reported that two captive pairs of pin-tailed sandgrouse laid three clutches each during a single breeding season, the second and third clutches being laid at average intervals of 18 days (range 9 to 30) after the previous one. Additionally, it is possible that sandgrouse may sometimes rear two broods within a single breeding season. This has been reported (without proof) for the chestnut-bellied sandgrouse in northern India (Christensen *et al.* 1964), as well as for the black-bellied sandgrouse in Israel (Urban *et al.* 1986).

Hatching and brood-rearing

The incubation periods of sandgrouse are seemingly rather variable, probably as a reflection of the influences of the extreme ambient temperatures. Thus Marchant (1961) reported a 19–20 day period for pin-tailed sandgrouse under natural conditions in Iraq. Under conditions of captive-breeding, Frisch (1970) made a similar (21·5 day) estimate, although various estimates ranging from 17 to 25 days have been reported for this species. The longest sandgrouse incubation periods are most probably those of *Syrrhaptes*. The incubation period of the alpine-nesting Tibetan sandgrouse is still unknown, but the steppe-nesting Pallas's sandgrouse has an incubation period of about 22–27 days when incubated under domestic hens (Makatsch 1976), 22–26 days when artificially incubated (Grummt 1985, Wilkinson and Manning 1986), and about 28 days under natural conditions (Cramp 1985). A surprisingly long (29–31 days) incubation period has also been estimated for the spotted sandgrouse under natural conditions (George 1969).

Hatching in sandgrouse is often nearly simultaneous, but at times may be extended over a 2- or even 3-day period. The eggshells are removed from the nest-site (as in typical shorebirds), and initial brooding of the newly hatched young occurs either in the nest-scrape itself or near it. The young are extremely cryptically coloured, and are nearly invisible in their usual habitats. The birds crouch when alarmed, but the chicks keep the neck retracted and keep the head raised well above the ground (Fjeldsa 1977).

Although the young are not fed bill-to-bill as in bustards and buttonquails, the mother begins to pick up bits of food and throw it in front of the newly hatched chicks, apparently stimulating them to eat. Unless the chicks are 'taught' to peck for food by one another or by their parents, captive-hatched chicks may starve to death (Frisch 1970).

There is no question that perhaps the most interesting aspect of parental care for the young in sandgrouse relates to the water-carrying behaviour by the male in his belly feathers, which has now been amply documented in several species. Marchant (1961) observed this first-hand in wild spotted and pin-tailed sandgrouse, thus confirming earlier accounts of this behaviour in pin-tailed, black-bellied and chestnut-bellied sandgrouse by Meade-Waldo (1896, 1897, 1906, 1922) and St Quintin (1905) that had later been questioned by many other writers. Cade and Maclean (1967) soon thereafter confirmed the occurrence of this behaviour in the Namaque sandgrouse, and reported that in all but one of the sandgrouse species (the Tibetan sandgrouse being the exception) the abdominal feathers are especially modified for holding water, presumably for this water-carrying purpose.

The development of sandgrouse chicks is very rapid. A generation of sand-coloured feathers somewhat resembling the adult remale plumage appears 4–6 days after hatching on the sides of the crop, back and wings, but with small clusters of down adhering to their tips for some time. The ten primaries emerge in descending (outward) sequence, with the mesoptile primary-coverts inserted between them nearly as long as the primaries, so that each wing may appear to have as many as 20 primaries. In *Syrrhaptes* a precocious fluttering flight may thus occur in as little as 15 days, whereas in *Pterocles* the development rate is slower, and fledging requires about 4 weeks. Not much later (at 40–45 days) a moult of the juvenal remiges begins but proceeds slowly, so that only about 4–7 primaries have been replaced when the moult is temporarily arrested in September (Fjeldså 1977).

Sandgrouse body feathers reportedly exhibit two juvenal or pre-definitive plumages following the downy stage. The first of these is not sexually differentiated, but the second is, and both plumages are carried only for a short time before being moulted and replaced by adult feathers. Although during this moulting period the juvenal secondaries initially appear in a strict sequence, later on all kinds of moult sequences develop, and asymmetric wing-moult patterns are common. Thus subadult birds typically show a uniform amount of wear on all their (juvenal) secondaries, and also perhaps retain a few outer juvenal primaries, whereas adults lack any worn juvenal primaries but typically have some bleached and worn outer secondaries evident (Kalchreuter 1979).

Sexual maturity is probably normally attained during the first year, since breeding by captive birds still in their first year of life has, for example, been reported in Pallas's sandgrouse (Grummt 1985). Breeding by birds in their first year has also been asserted without supporting documentation for the black-bellied and chestnut-bellied sandgrouse (Christensen et al. 1964). Kalchreuter (1980) considered it 'possible' that juvenile chestnut-bellied and black-faced sandgrouse in Tanzania might breed during their first year, within a period overlapping the main 'adult' breeding season of 6 or 7 months within which they had hatched.

Breeding success and productivity

No specific quantitative data appear to be available on sandgrouse nesting success and productivity. Marchant (1961) observed an apparently extremely high rate of nest loss to predators and/or destruction caused from trampling by hooved mammals; only two of twenty-eight nests (7 per cent) that he had under observation in Iraq apparently hatched successfully. Maclean (1968) believed that about 30 per cent of the eggs of the Namaqua sandgrouse were lost before hatching, and noted that parents were rarely seen accompanying more than two chicks.

Kalchreuter (1980) collected 193 specimens of the chestnut-bellied and 91 specimens of the black-faced sandgrouse over a year-long period in Tanzania. Of these birds, 70 per cent collectively were adults, suggesting an approximate recruitment (and probably annual mortality) rate of about 30 per cent. Additionally the sex ratio of all 185 adults was 1·17 males per female, whilst the collective sex ratio of adults and juveniles combined was 1·12 males per female. These figures suggest that females may be subjected to only slightly higher mortality rates than are males, and the sex ratios are similar to those that have been observed in various other groups of monogamous birds.

The total ratio of adult females (86) to fledged juveniles of both sexes (79) in Kalchreuter's collective sample indicates an average productivity rate of at least 0·9 fledged young per adult female (assuming equal vulnerability of all age-groups to being shot). Given an average clutch-size of about three eggs, this would mean that, in the absence of renesting or double-brooding, a minimum reproductive success rate of about 30 per cent would be required. However, if one assumes a maximum reproductive success rate of only 7 per cent, based on the nest success rate observed by Marchant, an average of more than four nesting efforts would be needed simply to hatch out this many young, to say nothing of any subsequent post-hatching mortality. It thus seems likely that persistent renesting and possibly also double-brooding may be important aspects of the breeding strategies of sandgrouse, and may also tend to limit their geographic distribution to areas of potentially long breeding seasons. Breeding by birds-of-the-year during the same breeding season in which they were hatched, which was suggested as 'possible' by Kalchreuter, could of course also help to increase overall sandgrouse productivity.

SUMMARY

Of all the aspects of their biology, buttonquails, bustards and sandgrouse perhaps diverge most strongly in their breeding biologies, in spite of all being ground-nesting, having precocial young and being generally adapted to breeding in similar ecological situations. Why should one of these groups have adopted a polyandrous mating system with paternal care of the eggs and chicks, another a firmly monogamous one with strongly developed biparental care, and the third a largely if not entirely polygynous system with exclusively maternal care?

It has already been suggested that polyandry in the buttonquails has allowed the birds to breed rapidly and repeatedly in tropical grassland environments as a result of their extremely short breeding cycles and the capacity of the male to care for and brood the chicks adequately by himself. The primary limitations on buttonquail productivity would seem to be the number of available males for incubating eggs and rearing young. Perhaps also there are upper physiological limits on egg production by females, but at least in captivity hens appear to be able to produce an indefinite number of eggs (each weighing about 10 per cent of the average adult female body weight) on an average alternate-day laying schedule extended over a 7-month period (Flieg 1973). Even given the possibility that only half of the laying females are able to locate an extra male to hatch their eggs and rear the young, such a trio could readily double the female's reproductive output while also maximizing the reproductive activities of each male, and thus make polyandry a worthwhile breeding system for both sexes.

The adoption of monogamous mating systems in sandgrouse is not hard to explain; if nothing else, the need for the male to participate more or less equally in incubation and to bring water to and also help protect the newly hatched young would make a biparental investment of energy an efficient mode of reproduction. Comparable kinds of mating systems involving biparental care of the young occur in other grassland or desert-adapted birds of similar size and having precocial chicks, such as quails and partridges (Perdicinae and Odontophorinae), stone-curlews (Burhinidae), seedsnipes (Thinocoridae), and nearly all species of grassland-nesting plovers and sandpipers (Charadriidae and Scolopacidae, excepting the phalaropes and a few polyandrous or promiscuous calidrine sandpipers).

The bustards present some problems in terms of their apparent trend away from monogamy and towards a polygynous mating system which is more like that of most grouse than, for example, their nearer relatives the cranes. Like the strongly monogamous cranes, they typically have such K-selected traits as a prolonged sexually immaturity period (3–6 years in the largest species), a usual clutch-size of only two eggs, and very low rates (probably about 5–10 per cent annually) of population increase. However, like the more r-selected grouse and in contrast to cranes, they exhibit distinct renesting tendencies and are usually not highly social outside the breeding season. Further, only the more social species undertake long and complex migrations that might make especially advantageous the retention of family bonds (to facilitate the learning of migratory pathways and provide additional protection of the young while *en route*). Also, at least most bustard species are large enough to perhaps permit a single adult to be able to fend off or distract most nest or chick predators, which I believe is one reason why quails, unlike grouse, have retained or evolved monogamous mating systems (Johnsgard 1973). The associated increased size of eggs and chicks of larger species might also make these more resistant to the effects of temporary cooling, and perhaps may permit the young to escape predation better by their greater speed or endurance, without so great a need for biparental brood protection.

Although it is difficult to generalize much on the evolution of bustard mating systems because of the still-uncertain systems of several species, it seems likely that monogamy was the ancestral bustard mating system, as it is in all the extant species of cranes and in the great majority of other gruiform groups. However, divergence toward polygyny has probably occurred repeatedly in bustards, not only in most or all of the larger species but also in some smaller ones where tropical environments provide a long and warm breeding season during which females can incubate and brood their young alone without danger of chilling. The breeding season in such areas may also be long enough for a female bustard to be able to re-nest effectively after losing an initial clutch. Polygyny is perhaps especially likely to develop in the larger bustard species, where the prospects for individual differences in male size, strength and associated social dominance are probably most likely to exist. At least in some of the largest species such as the Australian bustard, males seem to increase in average body mass indefinitely with increasing age (Fitzherbert 1983), which would provide a powerful age- and experience-related basis for individual differences in male social dominance during situations of sexual competition.

5 · Exploitation and conservation

One of the primary reasons for the decision to write this book is the fact that several of the included species are seriously declining in population, although this is generally rather little appreciated. For example, in the most recent edition of the International Council for Bird Preservation's (ICBP) *Red Data Book* (King 1981) only a single species out of all three families was listed as threatened or endangered, namely the great Indian bustard, which was classified as endangered and as having a late 1970s population of 'well below' a 1971 estimate of not more than 1000 birds. Today at least four more species of bustards can be added to the list of threatened species (Collar and Andrew 1988). Indeed, in 1987 a proposal was submitted to the sixth meeting of the Conference of the Parties to the Convention on International Trade in Endangered Species of Wild Fauna and Flora (CITES) that all the species of bustards not listed in their appendices be listed in Appendix II. This was based on evidence of a rapidly growing international trade in bustards, particularly between East Africa and Arabia (1987 *World Birdwatch* **12**, 11). In the area of the Arabian Gulf the use of especially the smaller bustards in training falcons has produced a startling decline in most of these species. This trend has been exacerbated by an increased interest by zoos and aviculturists in bustards, most of which have never been bred in captivity.

MORTALITY RATES AND LIMITING FACTORS

It is a sad but true fact that objective estimates of recruitment and mortality rates are virtually completely lacking for all the species in this book, in spite of that fact that, at least in the case of the bustards, several of them represent birds of considerable economic importance. Easily the best-studied species of bustard are the great bustard and little bustard, for both of which there is little or no detailed information on breeding success (Cramp and Simmons 1980). It has been estimated that in central Europe more than 50 per cent and perhaps nearly 90 per cent of the clutches of great bustards are destroyed as a result of human activities (Glutz von Blotzheim 1973); thus recruitment rates must be quite low and mortality rates correspondingly low, if the population is going to survive for any length of time. Ena *et al.* (1985) found that first-year birds in a local great bustard population in Spain made up about 6 per cent of the autumn flocks, which implies that an approximately 6 per cent annual mortality rate must occur in birds of that age and older, assuming that a stable population exists. Ena *et al.* (1987) estimated that this same population had a pre-hatching (egg) mortality rate of 50 per cent, and a pre-fledging mortality rate of 57 per cent. It was judged that 100 adult females might produce 227 eggs, from which 102 young might be expected to hatch and 44 birds fledge. It was further estimated that an adult mortality rate of about 8 per cent could be balanced by recruitment if the annual mortality rate of immatures did not exceed 18–22 per cent. Such a population model would produce a population of relatively long-lived birds once early pre-fledging mortality has been accounted for, and perhaps help to account for the apparently deferred maturity that seems to be typical of at least the largest species of bustards.

It is of course rash to assume that all bustards might have population attributes similar to those just suggested for the great bustard, but in the absence of additional information for particular species it seems worth going on these assumptions. If they are generally valid, it would seem that by far the greatest mortality factors are associated with pre-hatching and pre-fledging influences, and that conservation measures should therefore look to these portions of the life-cycle to make changes that might have the greatest near-term impact on bustard populations. Thus the identification of important nesting areas and especially ensuring their protection from human disturbance during the incubation and brooding periods would help to ameliorate the greatest single component of reproductive failures. Perhaps the search for additional ways of reducing pre-fledging mortality by whatever means may be practical may be a general secondary guideline in approaching the problem of population decline in bustards.

The reduction of human disturbance and associated vehicular disturbance by various kinds of machinery during the nesting period must be supplemented by reduced disturbance by grazing animals, as well as other kinds of intrusive agricultural practices such as irrigation and pesticide spraying. Although the planting of some crops (rape, kale, lucerne) may prove attractive to species like the great bustard, the planting of shelterbelts and hedges and erection of fences all contribute to reduced use of areas by nesting birds. On the other hand, the little bustard is sometimes attracted to spacious areas that are infrequently disturbed, such as military training fields. It is also less inclined than the great bustard to avoid areas that have associated shelterbelts, olive groves, hedgerows, or similar visual obstructions, and indeed small hillocks are often used as observation points (Cramp and Simmons 1980).

RARE, THREATENED AND ENDANGERED POPULATIONS

Sandgrouse and hemipodes

Of the three groups under consideration here, we can dispense quickly with the sandgrouse, all species of which are believed to exist in moderate to large numbers and are not in any case known to present conservation problems. And, at the species level at least, the only hemipode believed to be in real danger at present is the Australian black-breasted buttonquail. In the recently published *Atlas of Australian Birds* (Blakers et al. 1984), only in four 1-degree latilong survey blocks or (1 per cent of the total area surveyed) was the species recorded at all, and in only one of these four blocks was there a breeding record. Nevertheless, since 1968 there have been reports of its occurrence on Fraser Island, indicating a recent local expansion of range. Bennett (1985) has summarized the distribution and status of this species, noting that most reports of it are from low, closed forests, which have largely been cleared for agriculture or for pine and hoop pine (*Araucaria*) plantations. The species is now rare, with a limited and patchy distribution. By comparison, the other Australian buttonquail species appeared on the *Atlas* survey blocks as follows: red-backed, 24 blocks; chestnut-backed, 27 blocks; red-chested, 48 blocks; painted, 144 blocks; and little, 293 blocks. However, the race *olivei* of the chestnut-backed buttonquail (which was listed as a full species in the *Atlas*) was not recorded at all during the survey period, although earlier records extending to the 1970s do exist for it. Quite possibly it deserves inclusion as a rare or endangered form.

Two additional forms of buttonquail deserve mention here, inasmuch as they are sometimes regarded as full species and are known from only a few specimens. One of these is *T. p. everetti*, which apparently is still known only from the type and two additional juveniles from Sumba Island (eastern Indonesia) which were taken during the 1890s and in 1949 respectively (Sutter 1955a). The other questionably distinct species is *T. p. worcesteri*, which had been described in 1904 from a few specimens taken on Luzon Island, all of which were lost during World War II. However, the species has since been rediscovered and a small series of specimens has been collected from the area around Manila, mostly at about 100 elevation (Amadon and du Pont 1970). Additionally, the Australian black-breasted buttonquail is now found only in some remnant forests of southeastern Queensland, and the *olivei* race (sometimes considered a distinct species) of the chestnut-backed buttonquail has only rarely been seen since it was originally discovered in 1894.

Doubtless several additional insular or otherwise apparently highly restricted races of buttonquail might be listed as potentially vulnerable, based simply on the fact that their ranges are intrinsically small and probably highly susceptible to ecological changes. Thus *T. maculosa furva*, of Huon Peninsula, New Guinea, *T. sylvatica suluensis* of Sulu Island, and *T. s. masaaki* of Mindanao are all known only from their type specimens (Sutter 1955a). Additionally, *T. s. celestinoi* was originally known only from a male collected on Bohol Island, in the Philippines, but subsequently additional specimens were obtained on Mindanao (Sutter 1955a). However, its status there is still uncertain (C. M. N. White and Bruce 1986).

Bustards

As noted earlier, King (1981) included only the great Indian bustard in the ICBP's updated world list of vulnerable, threatened and endangered bird species, based on a comprehensive review in the 1970s. However, in 1972 the ICBP established a 'Bustard Group' of specialists for planning, promoting and coordinating research and conservation activities on bustards worldwide, and that group subsequently identified six bustard species as the world's most threatened forms. These six species include all three of the bustards endemic to the Indian subcontinent—the great Indian bustard, Bengal florican and lesser florican—and all three of the more generally widespread species of the Palearctic region, namely the great bustard, little bustard, and houbara (Osborne et al. 1984).

The Australian bustard is still relatively widespread, albeit certainly not generally common in that continent, having been reported on 518 (or 64 per cent) of the 1-degree latilong survey blocks of the recently published Australian *Atlas*. It may actually have undergone some recent increases in range and abundance locally, especially in eastern Queensland

(Blakers *et al.* 1984). Its current status in New Guinea is much less well known.

Similarly, none of the endemic African bustard species was identified by Osborne *et al.* (1984) as being of special concern (although the Nubian bustard was listed as 'much threatened by hunting' and 'many' African species were considered as perhaps 'perilously uncommon'). Likewise, no bustard species was described as endangered by Urban *et al.* (1986) in their general review of the African avifauna. Collar and Stuart (1985) have provided a more general continent-wide assessment of threatened African birds species that included the endangered Canary Islands race (*fuertaventurae*) of the houbara bustard, but no full species of bustards. Brooke (1984) listed the Ludwig's, Denham's and kori bustards as all having vulnerable populations in South Africa. It is worth remembering that most of the African bustard species have scarcely been studied at all, and their population status or trends can at present only be guessed at. For example, the Nubian, little brown and Rüppell's bustards all have limited ranges in desert, or near-desert, habitats, and possibly are far rarer than is currently realized.

With these generalities and considerable limitations of available data in mind, we may now survey the status of the world's six apparently rarest species of bustards.

Great Bustard. The world breeding status of the great bustard, as of the mid-1980s, has very effectively been reviewed by Collar (1985), making this quick overview very simple. Unless otherwise indicated, the discussion that follows is directly based on his paper. Collar had previously (1979) tentatively estimated the world population in the 1970s as between 25 825 and 27 185 birds, although no quantitative data were then available for China, Mongolia, Iraq or Syria. In his more recent review, he estimated a total world population of between 14 154 and 21 169 birds, again in the absence of specific data from these four areas, although the species was judged by then to probably have been extirpated from Syria and few if any great bustards were believed to still be present in Iraq. North Korea was also excluded from consideration, since it appears likely that the species has perhaps never bred there.

Collar organized his discussion on a west-to-east basis across the species' breeding range, and judged that Africa's sole population of breeding great bustards was in northern Morocco, where it existed as recently as the late 1970s and early 1980s as some local scattered groups in the Tangier region, the Rharb, the Habt, and 'le Moyen Sebou' (Sefrou?). Collar estimated that perhaps 100 birds still existed in this part of Africa.

Collar's surveys of Portugal in the period 1969–1979 convinced him that at least 650 and probably upwards of 1000 birds existed there, and in 1981 a total of 1015 birds were counted. This healthy population is largely a reflection of the antiquated agricultural methods of that area, and modernization in this area would certainly impact the bustard population adversely. In adjacent Spain the population is easily the best in Europe, Collar estimating 5000–8000 birds present. This number is in part based on extensive surveys by Otero Muerza (1985a,b), who found the largest numbers and/or densest concentrations in Tierra de Campos (4421 birds, 0·35 birds per km^2), Castilla-La Mancha (1856 birds, 0·39 birds per km^2), and Baja Extramadura (813 birds, 1·05 birds per km^2).

Surveys in Hungary suggest that its great bustard population is perhaps now second only to that of Spain in Europe, with annual surveys made between 1971 and 1980 indicating a minimum Hungarian population of 2982 (in 1972) and a maximum of 3442 (in 1980). This is a substantial increase from an estimated low of 2300 in 1969, but well below an estimated 1941 population of 8500. The only remaining bustard population in adjacent Austria occurs along its common boundary with Hungary. The Austrian component of this combined population consisted of 151 birds in 1979. This figure is somewhat higher than those of 1977 and 1978, but over a longer period in the past there has in general been an inexorable decline in bustard populations in Austria, with populations of 700–800 birds apparently typical in 1939–40 (Cramp and Simmons 1980). Great bustards are completely protected in both Hungary (since 1970) and Austria (since 1969).

Although the great bustard became extirpated from what is now West Germany in 1910, it persists in small numbers in East Germany. These are the remnants of a former population that in the 1930s may have amounted to about 3000 (Glutz *et al.* 1973) to 4000 (Dornbusch 1983). This German decline has been extensively documented by Gewalt (1959) and many other authors (see references given by Collar), and include country-wide population estimates of about 1200 birds in 1965, 900 in 1970, and 800 in 1975. Survey counts for 1979 and 1980 suggest that only 500–560 bustards then occurred in East Germany, in spite of protection from hunting since 1949 and listing as an endangered and fully protected species since 1955 (Dornbusch 1983, Weinitschke 1983). The bird occurs east of the River Elbe, and is most abundant in the area to the north of Brandenburg and in Welsebruck (Dobai 1983).

The history of the Czechoslovakian population of great bustards has been documented by Dobai

(1983), who judged that at the turn of the century the population may have consisted of about 2400 birds; it was nearly extirpated in the early 1920s, but had reached 2000 birds by 1936. Since then it has gradually declined, with a 1979 population estimate of 315 birds, mainly in the southern part of western Slovakia in an area between the Danube and Little Danube rivers comprising some 167 000 hectares, or an average of 2·35 birds per 1000 ha. A few birds also occur in southern Moravia. Legal hunting was terminated in 1947, and the species has been completely protected since 1970.

In Poland, the great bustard's population has dropped steadily from about 600–700 in the late 1930s to only 16 as of 1980 (Collar 1985). It is now confined to western Poland, and has most recently been found persisting in the areas around Tódz, Warsaw and Bialowskok (Dobai 1983). It is now fully protected in that country.

In Yugoslavia, the species is now confined to four localities along the Hungarian and Romanian borders, located within a triangular area of Vojvodina Province consisting of some 1000 km² and in which 30–40 birds were believed present since 1965 (Collar 1985, Dobai 1983).

Bulgaria has a similarly small if not already extirpated population: only about 40 birds in the late 1960s (Dobai 1983), which Collar (1985) judged perhaps to have vanished by the late 1970s. These birds were in the Dobruja area shared with Romania. The species is fully protected in Bulgaria.

Romania's population of great bustards may have numbered about 2000 in the mid-1960s, but by 1970 there were about 500–600 present, and by 1978 only some 300–350. The vast majority were in national reserves or shooting reserves in the provinces of Olt, Teleorman and Timiş, but a few also were reported from Arad, Argeş and Dolj (Collar 1985).

Turkey's population of great bustards has generally been considered fairly secure; Goriup and Parr (1985) estimated that perhaps as many as 3000–4000 birds were present there in the early 1980s. However, they actually counted only 145 birds at twelve widely dispersed sites during a brief 1981 spring survey. Furthermore, Kasparek (1989) could find evidence of only a few breeding sites in recent years, and an absence of recent winter records outside the probable breeding range.

Syria and Iraq were both considered by Collar (1985) to have perhaps extirpated their populations of great bustards, with the possibility however that small breeding populations still remain in Iraqi Kurdistan. Birds wintering in the Tigris basin were presumed to come from breeding areas in south-eastern Turkey and perhaps in Iraqi and Iranian Kurdistan. Recent political unrest in the area has probably reduced the likelihood that birds still survive there. However, Collar concluded that Iran might have supported a population of 100–200 great bustards as of 1978. These occurred in relatively well-watered areas of western Iran, which also have probably been severely impacted by war.

Based on surveys conducted in the USSR during 1978–1980, the estimated Soviet great bustard population was only about 3000 adults (Isakov 1982, Ilicek and Flint 1989), or less than half the numbers found only a decade previously. More than two-thirds of these were found in the middle and lower Volga basin, with substantially fewer numbers along the Black Sea and in the Crimea, the Donets and Don basins, the lower Ural basin and north Caspian Sea region, Kazakhstan, the steppes of Minusinsk and Tuva, and the Transbaikal and Amur regions. The species is included in the USSR's 1983 list of threatened species.

Collar found it impossible to make numerical estimates of great bustard populations (primarily of *dybowskii*) for Mongolia and China, noting that in Mongolia the populations had declined as a result of overgrazing and pesticide use, although the species might still be of widespread if localized distribution. As to China, the nominate race probably breeds near the USSR border in Xinjiang Province, while elsewhere breeding by *dybowskii* may still be occurring in both north-eastern Nei Monggol (Inner Mongolia), and possibly also in eastern Heilongjiang (previously Manchuria) in the vicinity of Lake Khanka (Xingkai Hu), near the Korean and USSR borders. China also supports wintering *dybowskii* from the USSR.

In summary, Collar judged that in order of their probable numerical populations, the most important countries for survival of the great bustard are Spain, Hungary, USSR, Turkey, Portugal and probably Mongolia and China. Of the European countries, probably only Hungary, Portugal and Turkey have avoided population declines since the mid-1970s. Serious problems exist in Iran, Austria, Czechoslovakia and Romania. A critical situation may exist for the great bustard in Poland, Yugoslavia and Morocco, and possibly also Iraq, whilst in Bulgaria and Syria the species may have already been lost.

Little bustard. An excellent review of the world status of the little bustard as of the early to mid-1980s has been provided by Schulz (1985a), and is the general basis for the following discussion. Of the European and North African countries that once represented breeding areas of the little bustard, it is now extirpated from at least ten, including Algeria, Tunisia, Germany, Austria, Czechoslovakia,

Yugoslavia, Romania, Bulgaria, Poland, and Greece. Additionally it is nearly or completely gone from Morocco (the three most recent nesting-area records are from along the Atlantic coast and along the Mediterranean near the Algerian border), Hungary (last nesting record 1973), and Turkey (no recent nesting records).

Starting with the south-western end of the European range, Portugal still supports good populations of bustards on grazed steppe-like lands where pesticide use is still not prevalent. Areas of good habitat may support as many as 9–13 males per 100 hectares, the highest densities measured for the species anywhere. There is no evidence yet of significant population declines, according to Schulz, who however did not estimate a total population size. At most only about a third of eastern Portugal is occupied by little bustards, however, and if average population densities are at all comparable with those of Spain it seems unlikely that more than 10 000 birds are present.

Spain's population of little bustards is similar to that of Portugal, the birds still occurring widely over the central steppe habitats of that country. A total Spanish population of 50 000–70 000 birds has been estimated (J. Garzon, in Cramp and Simmons 1980). Schulz suggested that should Spain and Portugal both join the European Community, a great intensification of agriculture would certainly follow, with associated declines in the Iberian bustard populations.

The population of little bustards in France has been reviewed by André (1985), who estimated that in 1978–9 it consisted of a minimum of 7200 males. That was a considerably more generous projection than one made by M. Metais (in Schulz 1985b), who estimated a total population of more than 7000 birds of both sexes. The species' primary French breeding areas consist of the Champagne region, the Poitou plains in the Avanton region, the La Beauce area (between the rivers Loire and Seine) and the neighbouring plains west and south-east of Paris, and the Crau in the south. André's estimates for the Champagne region were of 1250–2000 males. Another 2000 males were perhaps present over a vast area of about 420 000 hectares in central France (Issoudun and Levroux regions). He judged that about 2600 males occurred in the Poitou plain, Charente (Ruffec area) and Vendée region. His estimates for the La Beauce area were of about 600 males, and about 500 males in the Crau. Several small populations were recorded elsewhere (Auvergne, Causses, Dordogne, Languedoc, Rhône-Alpes, and Alsace). However, the estimated breeding densities of all these areas were considerably lower than those in Iberia, and typically ranged from about 1 to 6·3 males per 100 ha in the Crau and from 0·1 to 1·5 per 100 ha in other areas.

Although the situation in Italy and Sardinia is not comparable with earlier times, when the little bustard was widely distributed over both areas, relict populations exist on Sardinia and mainland Italy, and the status of species nationwide is considered as 'vulnerable'. Sardinia represents the most important Italian stronghold for the species, with protection there since 1953, and for Italy as a whole since 1978. Although there are no national reserves in Sardinia, nor is any special research attention given to the species, the Sardinian population may have ranged from about 1400 to 2100 birds during the period 1971–82, with densities of up to about 3 birds per 100 ha fairly frequent, and local densities as great as 20 birds per 100 ha (Schenk and Aresu 1985). On the Italian mainland the birds are now limited to Apulia, in southern Italy, where the birds are locally clustered in groups of up to about 1·6 males per 100 ha (Petretti 1985). Some ten to fifteen birds are killed each year, which may represent 10–30 per cent of the total population (Petretti 1988).

In Germany the little bustard is now only a very rare winter visitor, although it bred in what was East Germany until about the start of the twentieth century, Similarly, the last known nesting in Czechoslovakia and Poland occurred around the turn of the present century, in Austria during the early 1920s, in Yugoslavia during the 1940s, and in Hungary during the 1970s, although some wintering may still occur in both countries, probably from a USSR breeding population. There are nineteenth-century breeding records for Romania and Bulgaria, but the species is now extirpated from both these countries (Schulz 1985a).

As noted above, the Greek population of little bustards is probably extirpated, with no breeding records since the mid-1940s. However, in Turkey the species probably still breeds locally in the interior, such as in the southern Anatolian highlands and in northwestern Karacabey, although definite information is very limited. These populations are probably very small and have few if any chances for long-term survival (Schulz 1985a).

Although the breeding population in the USSR at one time was doubtless extremely large, and the birds evidently bred in high densities over the vast steppes as far east as western Siberia, there has been a general severe decline in numbers and extirpation over wide areas of the entire Soviet Union. In the Crimea there were only about twenty pairs by the 1970s, and in Kazakhstan the species had almost disappeared from the developed areas but still occurred locally in semi-desert and steppe areas

(Borisenko 1977, Sludski 1977). In the Ural valley, the species' best remaining area, maximum local densities of up to about 2 pairs per 100 ha were found in a few locations (Samarin 1977). Intensified agriculture, overgrazing, and hunting pressures have evidently all contributed to this massive decline in numbers (Kostin 1978, Schulz 1985a).

Houbara bustard. The worldwide status of the houbara was reviewed in 1978 by Collar (1980), and thus some preliminary efforts have been made toward estimating possible world populations. Including the Canary Islands, but excluding the western Sahara countries, Collar documented declines in houbara populations in fifteen out of eighteen countries and, of the three exceptions, two (Afghanistan and Egypt) were too little known for conclusions to be drawn, and the third (Oman) has never been known to be a significant breeding area. In general, there has been a general marked reduction in all areas of the species' (Cramp and Simmons 1980), with one of the races (*fuertaventurae*) known to be highly endangered and the others of uncertain but perhaps highly vulnerable status.

On the Canary Islands, the endemic race of the houbara was extensively studied during 1979. At that time a minimum of forty-nine bustards were encountered, made up of forty-two on Fuerteventura (including two young) and seven on Lanzarote (including two young) (Collar *et al.* 1983).

In extreme north-western Africa (Mauritania and Morocco, including Western Sahara, formerly Spanish Sahara) the status of the houbara is rather poorly documented, but at least in Morocco it is confined to the southern arid plains adjacent to the Sahara, especially desert areas having a low annual precipitation and halophytic vegetation such as *Artemisia*. In spite of official protection in Morocco, there has been a decline in houbara populations as a result of habitat deterioration, human disturbance, overgrazing by goats and camels, drought, and hunting (by nomads, soldiers, and groups of Arab falconers). In 1983 the Morocco population of houbaras was estimated at only about 2000–3000 (Haddane 1985). Hunting and habitat changes have probably had generally drastic effects on the species throughout northern Africa (Mayaud 1982).

In Algeria there are no recent authoritative accounts of the houbara's status, but there have been apparently serious declines in the species, which at least once was fairly common in areas of stony plateaux and grassy plains. In part this decline has resulted from the search for oil in the Algerian desert. Little recent information exists for Tunisia and Libya, but there is some indication that at least in the late 1970s Tunisia may have still supported a very healthy population of houbaras (Collar 1980).

In Egypt, where the two races *undulata* and *macqueenii* are separated by the Nile valley, the situation of a very limited amount of available and useful information is similar to that for further west in Africa. However, Urban *et al.* (1986) list *macqueenii* as of uncertain status (resident or winter visitor) in eastern Egypt, and Arab falconing parties have been reported making substantial kills of houbaras in Egypt, as they also have in Morocco (Collar 1980).

The status of the houbara on the Arabian peninsula is probably quite precarious, because of both extensive hunting and perhaps relatively little remaining suitable breeding habitat. In Saudi Arabia the species is primarily a winter visitor, with only a few breeding records from the early 1900s. The species has been protected from general hunting by royal decree since 1969, although the King still takes considerable numbers by falconry during winter. There is probably a small breeding population in Oman, south of Ibri, while in the United Arab Emirates small numbers of houbara pass through in winter (Collar 1980).

In Israel, the houbara occurs in an area of about 3000 km^2 in the southern arid Negev, probably in numbers of 150–200, or perhaps more. Most of the population nests in the northern part of this region, where rainfall is higher, but in recent years the population has increased in the more southerly Arava valley (Mendelssohn 1983). Heavy grazing and intensive cultivation are posing a serious threat to the population of the northern Negev, especially in terms of reducing available nesting habitats (Lavee 1985).

In Syria the houbara was probably still a fairly widely distributed if declining nester in the 1960s, but only a single bird was observed during searches made in the mid-1970s. In Iraq the situation is also quite unclear, the birds apparently becoming quite rare by the 1960s, although perhaps persisting in the southern parts of that country (Collar 1980).

In Iran there is a wintering population from the USSR that uses eastern Iran, and a residential or partly migratory population exists in southern Iran. The total population is almost impossible to judge but has apparently suffered a severe decline throughout the country. Hunting is banned in the country, but the war activities have probably had devastating effects in at least some areas (Mansoori 1985).

In Afghanistan the houbara is at present probably only a very rare or accidental winter visitor, but the breeding populations of the USSR have traditionally wintered in large numbers in Pakistan.

Evidently hunting by Arabs began there around 1966, and by the late 1960s annual kills of up to several thousand birds by single hunting parties were not uncommon. In more recent years these numbers have been decreasing, as apparently also have the numbers of Arab hunting parties (Collar 1980).

In Pakistan, a critical wintering region, houbaras mainly occur in the semi-desert plains of Baluchistan, Sind and Punjab provinces. In Baluchistan, which is a major wintering area, significant numbers of the birds are killed both by local hunters (perhaps 1000–1500 per year) and by visiting falconers (perhaps 4000–5000 per year) (Mian and Dasti 1985). The bustards also breed locally in Baluchistan, at least in the vicinity of Chagai, Yakmuch and Kharan, and some sanctuaries have been established in these regions (Shams 1985). A major part of the wintering houbara population is centred in Sind, and this area has come under heavy hunting pressure from falconry parties, who for example during the 1981–2 winter season may have taken about 1500 birds (Surahio 1985). In the Cholistan desert area of the Punjab region, the birds winter in rather small numbers, and at an average density of about 1 bird per km^2 (Mirza 1985). The government authorities of Punjab have recently launched a major conservation programme of education, protection, and education towards saving the species (Ahmed 1985). Finally, in the North West Frontier Province of Pakistan, migrating houbara flocks pass through every year, and the area retains a few migrants all winter. There all bustard hunting, including falconry, has been completely banned since 1975, although grazing, poaching, increasing habitat loss to agriculture and similar pressures have all had their effects (Malik 1985).

In India, wintering by the houbara occurs on the Thar desert of the north-west, and these birds apparently escaped the effects of Arabian hunters until the early 1970s, when extensive kills began. However, in more recent years such kills have declined, either because of governmental restrictions or declining bustard populations (Collar 1980).

Although an enormous area of about 300 million hectares of probable or potential houbara breeding habitat once occurred in the USSR (mostly in Uzbekistan, Kazakhstan and Turkmenistan), vast areas of these semi-desert steppes have been converted to agricultural use, overgrazing has been rampant, and road construction, industrialization, and similar processes have all had devastating effects on the houbara population. Thus by the late 1970s the houbara was listed as 'very rare, with a sharp decline in numbers' in the Soviet Red Data Book, and in 1978 its population was estimated at only a few thousand birds (Collar 1980). According to Ponomareva (1985), the most favourable remaining breeding area for the species is in Uzbekistan, where the flat expanses of the Kyzylkum steppe once covered about 250 000 km^2. However, during the period between 1956 and 1979 the houbara population decreased about 75 per cent in north-western Kyzylkum, from such factors as the loss of steppe grasslands to agriculture, increased disturbance, illegal shooting, and large-scale hunting on the houbara's wintering areas. By 1980 only a few areas supported more than 100 breeding pairs (Ilicek and Flint 1989).

Great Indian bustard. In contrast to the three preceding species, the great Indian bustard's future depends entirely upon its conservation and management within a single country, India. It once occurred in what is now eastern Pakistan but is now virtually extirpated, and it has been reported only very infrequently in the vicinity of Dera Ghazi Khan (Baluchistan) and Kacchi (Sind) (Shams 1985). In India it once occurred widely from Uttar Pradesh in the north to Tamil Nadu in the south, and from Sind (now Pakistan) in the west to Orissa in the east, but by the 1960s its total numbers were estimated at only around 12 000. It has been protected since 1951, but enforcement of this law was inadequate, and at present it survives only in a few areas within six Indian states: Gujarat, Rajasthan, Madhya Pradesh, Maharashtra, Karnataka, and Andhra Pradesh. During the 1980s it was estimated that about 1500–2000 bustards still survived in India, with Rajasthan supporting more than half of the total population (Rahmani and Manakadan 1988). During the 20-year period between the mid-1960s and the mid-1980s the birds were probably extirpated from the Indian Punjab, Haryana, Uttar Pradesh and Orissa and in the late 1980s the estimated populations of birds in the remaining states were: Rajasthan 500–1500, Maharashtra 70–100, Karnataka 30–40, Andhra Pradesh 100–150 and Gujarat 20–30 (Rahmani 1989).

In Rajasthan the primary stronghold of the great Indian bustard is in the western Thar Desert area, including Barmer, Bikaner, Jaisalmer, Jalor and Jodhpur districts, where agricultural activities are minimal and human populations are low. Some birds also are found in Ajmer, Bhilwara and Kota districts of eastern Rajasthan (Rahmani 1986). In western Rajasthan the Desert National Park in Jaisalmer and Barmer districts was initially named as a protected area in the late 1970s, and became a national park in 1981. This area of 3161 km^2 supports about 200–400 bustards, the largest group known anywhere. Other protected areas in Rajasthan include Sorson (near Kundanpur, Kota district), of about 10 km² and holding some ten to fifteen bustards, and Shokhaliya (Ajmer district), of 17 km² and supporting about eighty birds.

There are no bustard sanctuaries in Gujarat, although three have been proposed, but in Madhya Pradesh there are two existing sanctuaries and an additional proposed one. The biggest existing sanctuary is Karera (Shivpuri district), of 202 km² and with at least thirty bustards. The other existing sanctuary is Ghatigaon (Gwalior district), which supports fifteen to eighteen birds. There is a single sanctuary (Nanaj) in Maharashtra, in Ahmadnagar and Sholapur districts, and it supports fifty to sixty birds in an area of 7818 km². A still undeveloped sanctuary of comparable importance in Andhra Pradesh is Rollapadu (Kurnool district), which currently consists of 5·0 km² of protected 'core areas' within a much larger buffer zone whose size is yet to be determined. This area also supports about sixty bustards. Finally, in Karnataka (formerly Mysore), there is the Rannebennur blackbuck sanctuary (Dharwar district), of 123 km² which supports ten to fifteen bustards. In most of these areas the bustard population is increasing, and such reserves probably hold the best hope for saving the species (Rahmani and Manakan 1988).

Lesser florican. Like the great Indian bustard and the Bengal florican, this species is considered by Indian authorities to be endangered and is protected by the Wildlife (Protection) Act of 1972. Its total range is even more restricted than is that of the great Indian bustard and is now virtually confined to India, although it also probably still occurs very locally in the grassland terai of southern Nepal (Inskipp and Inskipp 1983).

Probably its original breeding range was centred in the grasslands of Gujarat (especially the Kathiawar peninsula), eastern Rajasthan, eastern Haryana, western Madhya Pradesh, central Maharashtra (Ahmednagar, Nasik and Sholapur districts) and the Deccan plateau region of western Andhra Pradesh and northern Karnataka. However, recent surveys indicate that it is now extremely rare in Andgra Pradesh, where there has been only a single recent sighting of the species (Lachungpa and Rahmani 1985). Furthermore, a 1981 visit to Karnataka indicated that no suitable habitat for it remains in that state's most likely site, the Tungabhadra Wildlife Sanctuary (Goriup and Karpowicz 1985). A survey undertaken in both of these states during March of 1985 revealed only seven possible remaining habitats for the species (Lachungpa and Lachungpa 1985).

In Maharashtra the lesser florican was once a common monsoon breeder, but a preliminary survey in 1982 revealed no definite records, although there were reports of its occurrence in the Vidharbha region, and the species may also exist within the boundaries of the proposed great bustard sanctuary in Sholapur and Ahmadnagar districts (Rahmani and Yahya 1985). There are evidently few records from Rajasthan for recent decades. Goriup and Karpowicz (1985) were unable to establish any sightings since 1981 at a traditional florican site near Ajmer, but Saxena and Meena (1985) observed a displaying male and two females at Athun Geneshpur (between Nasirabad and Kekri). A 1984 Rajasthan survey by Haribal *et al.* (1985) located only seven birds, all in private grasslands at Kalsanse, Bhilwara district.

During the same survey of floricans in the grasslands of western Madhya Pradesh, forty additional birds were located. These were in Ratlam, Jhabua, and Dhar districts, and perhaps a total of about 150 birds remained in that region during the mid-1980s. One of the best remaining breeding populations is at Sailana Kharmor Sanctuary, in Ratlam district, where at least fifteen males and three females occurred on an area of 354 ha of grassland.

The species once occurred very commonly in Gujarat, especially in the grasslands of the Kathiawar Peninsula, and this area probably remains the species' major breeding stronghold. A recent survey by Rahmani *et al.* (1985) found a considerable number of known or potential florican habitats in Amreli, Bhavnagar, Jamnagar, Junagadh, Kutch and Rajkot districts of Gujarat, but saw or obtained evidence of very few birds. In a survey of the Kathiawar peninsula, Magrath *et al.* (1985) observed seventy-seven floricans on twenty-one grassland 'vidis', and estimated a total population of 362 birds in the three western districts of Jamnagar, Junagadh and Rajkot. Besides these critical areas the birds evidently still occur locally in the grassland terai valleys of southern Nepal (Inskipp and Inskipp 1983). They presumably also survive in adjoining northern Uttar Pradesh, so perhaps this region should also be considered as part of the species' remaining breeding range (Goriup and Karpowicz 1985).

In their review of the status of this species, Goriup and Karpowicz (1985) urged that it be listed on Schedule 1 of India's 1972 Wildlife (Protection) Act, and that an extensive area of natural grassland and sanctuaries be established through its historic breeding range in order to support a minimal breeding population of this elusive species, whose highly mobile tendencies make conservation management more difficult than for sedentary bustards.

Bengal florican. Of all the world's bustards, this is

probably the rarest and most endangered species. Inskipp and Inskipp (1983, 1985a) surveyed the birds in Nepal and Dudwa National Park, Uttar Pradesh, while Rahmani *et al.* (1988a) surveyed Uttar Pradesh, West Bengal, Bihar and Assam, which collectively include nearly all the reported historic range of the species except Bhutan, which probably lacks suitable lowland grass habitat. The somewhat mysterious Kampuchean subspecies *blandini* has apparently not been reported again since its original description, and in Bangladesh the species has probably been extirpated since about the 1950s (Karim 1985). It was apparently only a rare straggler to Bihar, and probably once occurred only marginally in Arunachal Pradesh, and perhaps still does (Inskipp and Inskipp, 1985a).

The only area in India where the florican can now be regularly seen is in Dudwa National Park, Uttar Pradesh (Sankaran and Rahmani 1988). In a 1988 survey, Rahmani *et al.* (1988a) also saw only sixty to eighty floricans throughout Uttar Pradesh, Bihar, West Bengal and Assam. Nearly all were in various sanctuaries, national parks or wildlife reserves, and were mainly seen in Dudwa National Park, Manas Wildlife Sanctuary (Assam) and Orang Wildlife Sanctuary (Assam), which together accounted for most of the birds found. It is likely that Assam and Uttar Pradesh hold virtually all of India's remaining population of Bengal floricans, which has been variously estimated as between 195 and 300 birds (Inskipp and Inskipp 1985a).

In Nepal the species is ecologically restricted to the narrow belt between the Himalayan foothills and the Indian frontier, and there it is fully protected. It has been observed most often in Chitwan National Park, but has also been seen in western Nepal at the Royal Sukla Phanta Wildlife Reserve and the Royal Bardia Wildlife Reserve. Loss of the eastern Nepalese grasslands to agriculture has perhaps already eliminated the species from that part of Nepal, although it was reported as recently as 1981 from the Kosi area. Based on their surveys, Inskipp and Inskipp (1985a) estimated a probable total 1982 Nepalese population of thirty-five to fifty birds, and a maximum of 100 birds, with the majority in the already protected Chitwan, Sukla Phanta and Bardia Royal Reserves. Adding to this total the approximately 100–300 birds in India, the world population is likely to be no more than 500. Thus the species should be immediately added to the ICBP–IUCN world list of endangered species, and urgent measures taken immediately to help protect it (Inskipp and Inskipp 1985a).

Future outlook. This rather brief review of the six rarest bustard species is generally a depressing one, with the future of several species largely dependent upon the vagaries of the economies of various third-world countries with serious problems of inadquate food production for growing human populations, such as India and Pakistan, or ones that are struggling to modernize and maximize their agricultural production, such as Portugal and Spain. In addition there are the problems of inadequate protection of bustards from hunting in several critical wintering areas, and the general impression in terms of bustard conservation must be one of pessimism. There are a few reasons for limited optimism, such as Pakistan's recent establishment of a National Council for Conservation of Wildlife, and its consideration of a 5-year ban on the hunting of houbara bustards, as was strongly recommended by the 1983 International Symposium on Bustards. India has also taken special efforts to protect the great Indian bustard, through total protection and the establishment of the Desert National Park and special bustard sanctuaries.

II · SPECIES ACCOUNTS

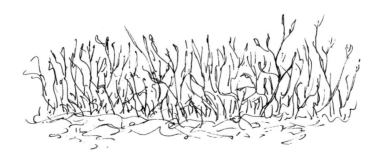

A · Hemipodes (Family Turnicidae)

KEY TO THE GENERA AND SPECIES OF HEMIPODES

A. Most upper wing-coverts and tips of remiges white, retrices of normal texture; iris brown: *Ortyxelos*, Larkquail (only sp.)

AA. Upper wing-coverts not predominantly white; recrtices unusually soft; iris usually whitish to straw (rarely reddish) in adults: *Turnix* (12 spp.)

  B. Tarsus longer than middle toe and claw; mostly (all but one) non-Australian forms

    C. Entire breast barred with black; throat also black in females; legs bluish slate to greenish: Barred buttonquail

    CC. Breast not transversely barred with black; legs not bluish

      D. Upper wing-coverts marked with distinct black ocelli

        E. Legs greenish; sides of neck flecked with black and white; tail over 40 mm: Spotted buttonquail

        EE. Legs bright yellow; sides of neck not flecked with black and white; tail under 40 mm and hidden by upper tail-coverts: Yellow-legged buttonquail

      DD. Ocelli lacking on upper wing-coverts

        E. Legs bright yellow to greenish yellow; greater secondary coverts edged with buff; Australian: Red-backed buttonquail.

        EE. Legs flesh-coloured to pale yellowish; greater secondary coverts usually white-edged; not Australian

          F. Central crown stripe poorly developed and rump blackish; sides of breast transversely barred or spotted with black, African: black-rumped buttonquail

          FF. Central crown stripe well developed; rump not blackish; breast markings more crescent-shaped or heart-shaped; Africa, Eurasia and Pacific islands

            G. Wing 80–101 mm; bill slender (less than half as deep as long); widespread: Striped buttonquail

            GG. Wing 62–73 mm; bill stout (about half as deep as slong, at least 5 mm deep at angle of gonys); southwestern Pacific islands

              H. Lighter above, the auriculars black-flecked; Luzon I.: Worcester's red-chested buttonquail (sometimes considered a full species, *T. worcesteri*)

              HH. Darker above, the auriculars black-streaked; Sumba I.: Everett's red-chested buttonquail (sometimes considered a full species, *T. everetti*)

  BB. Tarsus not longer than middle toe and claw; Australian and Madagascan species

    C. Breast black or extensively spotted with black in females; males darkly barred on breast; both sexes with white spotting on upper wing-coverts

D. Wing over 100 mm; Australian: Black-breasted buttonquail
DD. Wing under 90 mm; Madagascan: Madagascan buttonquail
CC. Not with a black or darkly barred breast; upper wing-coverts sometimes spotted with white; Australian
D. Breast grey to olive-grey, often faintly dotted with black-edged white spots; upper wing-coverts also spotted with white; bill relatively weak (about three times as long as it is deep)
E. Back purplish grey with some chestnut barring; iris reddish; breast grey, spotted with whitish: Painted buttonquail
EE. Back and rump chestnut-toned; iris whitish; breast olive, spotted with white (*castanota*) or unspotted pale olive (*olivei*): Chestnut-backed buttonquail
DD. Breast white to reddish, never spotted with white; no white spots on upper wing-coverts; bill relatively heavy (almost half as deep as long)
E. Crown not distinctly striped in centre; more chestnut above; malar area and flanks not spotted with black; breast pale reddish buff (females) to nearly white (males): Little buttonquail
EE. Crown with whitish or tan stripe; malar area and anterior flanks both spotted with black; breast tinted with reddish ochre, especially in females: Red-chested buttonquail

LARK-QUAIL (Plate 1)

Ortyxelos meiffrenii (Vieillot) 1819

Other vernacular names: bush lark-quail, quail-plover; caille-pluvier, turnix de Meiffren (French); Lerchen-Kampfwachtel, Lerchenlaufhühnchen (German)

Distribution of species (Map 1)

Africa, from Senegal east to central Sudan and northern Kenya, with some coastal records of wintering or perhaps locally breeding birds. No subspecies recognized. Probably nomadic.

Measurements (mm)

Wing, males 72–76 (av. of 4, 73), females 76–80 (av. of 4, 77·8); tail, males 29–32 (av. of 4, 30·3), females 33–36 (av. of 4, 34·8); tarsus, males 17–19 (av. of 4, 18), females 18–20 (av. of 4, 19); exposed culmen, male 8, females 7·5–8 (av. of 4, 7·9) (Urban *et al.* 1986). Egg, c.17·5 × 14·5 (Mackworth-Praed and Grant 1952).

Weights (g)

Two males 15·7 and 19·5 (Urban *et al.* 1986). Egg 2·3 (estimated).

Map. 1. Distribution of the lark-quail. The presence of some extralimital records (indicated by dots) suggests that seasonal movements and possible local coastal breeding may occur.

Description

Adult female (after Urban et al. 1986). Like the male, but the breast deeper rufous-brown and outer three rectrices narrowing toward tips, with dusky brown submarginal marks and broad *white fringes*.

Adult male. Forehead, crown, nape and hindneck rufous-brown, the feather edges cream with black inner border. Lores, face and supercilium cream, with golden buff tinge. Streak from eye back to side of neck rufous-brown; ear-coverts washed pale rufous-brown. Chin and throat white, the throat feathers tipped pale golden buff. Mantle, scapulars, and back rufous-brown, paler on upper mantle, the feathers broadly fringed with black-bordered cream. Upperpart pattern variable, some mantle and back feathers with black and cream blotches, spots, or bars. Rump and upper tail-coverts paler than back, uniform pale rufous-brown, and narrowly tipped with buff. Tail pale rufous-brown, the two outermost rectrices with buff to white fringes and the outermost rectrix with a dusky brown submarginal area on outer web. Other rectrices tipped cream, with two or three more transverse cream bars on distal half, variably bordered with black. Breast golden buff, the feathers with rufous-brown and white tips, producing a spotted effect. Sides of breast darker and more rufous-brown, with black-bordered cream spots; the lower breast and belly cream, becoming white on flanks and under tail-coverts. Primaries blackish, the outermost (tenth) with outer web, central half of inner web and most of shaft white; next four primaries with central patch, tip and narrow edge of outer web buff, becoming more rufous at tip and with dark subterminal markings; the five innermost primaries with broad white tip and rufous-buff spot at base of inner web. Secondaries blackish with broad white tips and narrow white edges. Tertials rufous-brown with cream tips and transverse bars, bordered with black. Upper wing-coverts except for the greaters rufous-brown and cream, the greaters white. Under wing-coverts cream, axillaries white. Iris pale to rich brown; bill yellowish horn to pale green, with bluish brown culmen; tarsi and toes whitish flesh to flesh or creamy yellow.

Immature. Like the adult, but more spotted than streaked above (Praed and Grant 1952). The upperparts also considerably paler, more sandy and less rufous, the feathers more vermiculated and broadly fringed with white, the rufous at sides of breast paler, and the wing markings less regular (Urban et al. 1986).

Identification

In the hand. Distinguished from the typical *Turnix* buttonquails by the relatively long and white-tipped tail, and the mostly white upper wing-coverts.

In the field. Easily told in the field from the African buttonquails by the white upper wing-coverts, contrasting with the otherwise blackish wing.

General biology and ecology

To a greater degree than perhaps any of the other hemipodes, this species is distinctly adapted for life in arid habitats. It is largely found in arid to semi-arid grasslands, bushed grasslands, thin scrub, and acacia savanna, but extends locally into quite dense bushland, sometimes with a few trees present, and into wetter coastal grasslands. In Sudan and Chad it is associated with sandy grasslands dominated by sandburs (*Cenchrus*) and needlegrass (*Aristida*) respectively, often in the absence of any source of surface water. Its altitudinal range extends from sea level to 2000 m, and in much of its range is apparently resident. However, in some areas such as coastal Gambia and Ghana it is present only in the cool, dry season, and is clearly migratory. Similarly in northern Nigeria it is present mostly in the dry season, but in other parts of its range it is apparently present only during the wet season (Snow 1978, Urban et al. 1986).

Its foods are still essentially unstudied but probably include grass seeds as well as termites and other insects. Apparently it can obtain sufficient moisture from its foods, such as termites, so as to be able to survive in the absence of water (Someren 1926, Bannerman 1931). Outside the breeding season it occurs only singly or in pairs (Urban et al. 1986).

Social behaviour

Almost nothing is known about the social behaviour of this elusive species, which is extremely difficult to observe because of its cryptic colouration and inconspicuous activities. In contrast to the typical buttonquails, there is no direct evidence (yet) of a polyandrous mating system, although reported incubation by the male would suggest the possibility of its occurrence. The species' (sex?) low whistling call apparently lacks the well-resonated, booming quality of typical buttonquails, and additionally the presently available although still limited evidence (Lowe 1923) suggests that no special resonating adaptations exist in the species' trachea of esophagus.

Reproductive biology

The limited evidence now available indicates that this species breeds during the cool, dry season, both inland as well as in coastal locations. Laying dates include January, March, and September to December in Senegambia, and January in Sudan, with birds in breeding condition from November to February, and birds in breeding condition during December and January in Kenya. Breeding probably occurs during March in Ethiopia, and during winter in Ghana (Urban *et al.* 1986).

Almost nothing is known of nesting, but Lynes (1925) found two nests. Both were in shallow depressions in firm sand, near the food of an herb, and in light, open vegetation. Both nests had two slightly incubated eggs, and at one of them the incubating bird was collected and proved to be a male. The incubation period is unknown, and the downy young are still undescribed.

Evolutionary relationships

The only real evidence currently available on this point comes from the anatomical studies of Lowe (1923), who rather conclusively showed that the affinities of *Ortyxelos* are with *Turnix* rather than the shorebirds or galliform birds. It is tempting to suggest that *Ortyxelos* is less specialized both morphologically and behaviourally than *Turnix*, but current information does not provide much support for this position. Perhaps when DNA hybridization data become available more can be said on this subject.

Status and conservation outlook

There seems to be little if any reason for concern about this species; indeed the southward spreading of desert-like conditions in the sub-Saharan zone probably favours rather than harms this species. A local extension of the species' range has also been reported in the area of Tsavo East National Park, Kenya (Lack 1975).

MADAGASCAN BUTTONQUAIL (Plate 2)

Turnix nigricollis (Gmelin) 1789

Other vernacular names: black-necked buttonquail, Madagascar bustard-quail, caille de Madagascar, hemipode à cou noir, hemipode de Madagascar (French); Madagaskar-Laufhühnchen, Schwarzkehl-Laufhühnchen (German); kibo, rakibo (Madagascar).

Distribution of species (Map 2)

Endemic to Madagascar, where widespread in fairly open habitats. Also reportedly introduced (possibly self-introduced) on Mauritius (since disappeared), Réunion (still present) and Iles Glorieuses (still present) (Long 1981). No subspecies recognized.

Measurements (mm)

Wing, males 72–82 (av. of 10, 77·7), females 81–88 (av. of 10, 83·8) (Benson *et al.* 1976); tail, both sexes 34–36; tarsus, both sexes 20 (Ogilvie-Grant 1893). Egg, av. 26·8 × 19·8 (Rand 1936).

Weights (g)

Females 67–84 (av. of 4, 70) (Benson *et al.* 1976). Egg 5·8 (estimated).

Description

Adult female. Feathers of the lores and forehead black barred with white, those on the sides of the face and neck white tipped with black; a white stripe from the base of the lower mandible runs down either side of the throat, which, together with the chin and middle of chest, is deep black; crown brown, dotted with white; nape dark grey; upper back and scapulars black, barred with rufous and margined on either side with whitish or buff; lower back, rump and upper tail-coverts brownish grey, somewhat mixed with rufous, and with wavy black bars and vermiculations, some of the feathers with one or a pair of marginal buff spots near the tip; remiges and primary-coverts blackish brown, the remiges margined on the outer web with buff; tertials and rest of the wing-coverts rufous, vermiculated with black and ornamented with irregularly-shaped black and white spots situated mostly on the outer web; chest, breast and underparts dove-grey, paler on the abdomen; sides of chest and 'shoulders' rust-red. Iris bright to pale yellow; bill bluish to grey; tarsi flesh-colour to pale grey.

Adult male. Differs from the female in having the feathers of the forehead black widely edged with buff; the nape like the upper back; the chin and middle of throat white; the sides of the chest washed with pale rufous; the middle of the chest, breast and flanks buff, these areas barred with black; abdomen paler.

Juvenile. Outer webs of the secondaries rufous, vermiculated with black and ornamented marginally with black-edged white spots; otherwise similar to older males.

Identification

In the hand. Separated from the other buttonquails by the combination of a tarsus that is no longer than the middle toe and claw, a forehead that is black (edged with buff in males), a chest that is barred with black (males), or is black from the chest to the throat (females), and a nape that is barred with blackish (males) or is greyish black (females).

In the field. This is the only buttonquail found on Madagascar, and is unlikely to be confused with anything else except the two species of *Coturnix* (*coturnix* and *delegorgei*) that have been reported from Madagascar. The distinctive blackish colouration on the chest of *Turnix* and absent in *Coturnix* should help to separate these genera.

General biology and ecology

Little can be said of this species' ecology, which appears to be quite adaptable. Rand (1936) encountered the species from sea level to 1900 m, in subdesert areas, in grasslands of the treeless central plateau, around manioc fields, in grassy savannas, and in more open woodlands. Benson *et al.* (1976) collected a number of specimens in dense, dry forest habitats. Judging from its rather dark colouration, one might imagine that it is adapted to a rather shaded environment. Interestingly, Butler (1905a) reported that birds in a German aviary preferred to remain in areas of dark recess.

Social behaviour

According to Rand (1936), these birds occur singly in groups of up to four. Translating from German accounts, Butler (1950a) noted that within 3 weeks of a pair's establishment in an aviary, a nest with two eggs was found. From this time on, during the female's egg-laying period, she 'strutted with long strides over the whole territory from dawn to eve; and frequently and ever louder uttered its peculiar pairing note'. As no other pairs of buttonquails were present in the aviary, no territorial fighting or mate-changing was possible. Courtship is apparently very similar to parental feeding, with the female offering a seed to the male but uttering its pairing note rather than the soft parental call.

Reproductive biology

Judging from female specimens in breeding condition collected during December and January by Ben-

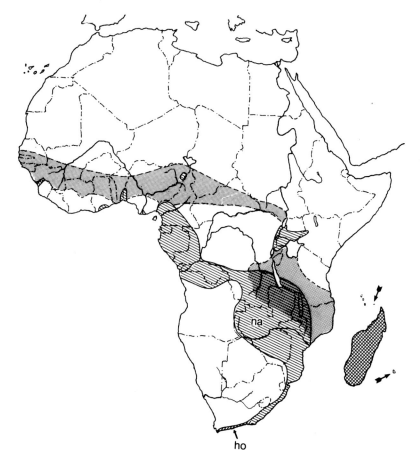

Map. 2. Distribution of the Madagascan buttonquail (cross-hatched), with the locations of apparently introduced populations indicated by arrows, and of the black-rumped buttonquail, including its breeding (hatched) and nonbreeding (stippled) ranges, and the locations of its races *hottentotta* (ho) and *nana* (na).

son *et al.* (1976), and the sighting of a brood in December, this general period would appear to be the breeding season. Additionally, Rand (1936, p. 369) located nests with eggs or collected reproductively active females between September and November, while downy young were seen between November and March. A captive female in Germany reportedly laid eight clutches in a single year, beginning in April, although young were hatched from only three of these clutches (Butler 1905a).

The nest-building behaviour of this species, based on these captive observations, was described in Chapter 4, and need not be repeated here. Several completely arched-over nests were built by one pair, mainly by the female. One of the nests was constructed entirely of moss, with a tunnel about 20 cm long and having a round opening 5 cm wide, which led to the nest scrape about 2 cm wide. In another pair a similar arched-over nest was made with a tunnel of ribbon-grass and moss some 30 cm long (Butler 1905a).

According to these observations the incubation periods is 13–16 days, with the male sitting (incubation in one case apparently starting after the laying of the fourth egg) and being guarded by the female. In exceptional cases the female relieved the incubating male, and 'in individual cases and under special circumstances' only the female incubated, and also fed and led the young about. Normally, however, the female appeared 'spiteful' to the young during the first few days after hatching, and the male exhibited protective behaviour towards them. Nevertheless, within about a week the female was leading and feeding the young with 'equal solicitude to that of the male'. By the second week the young were able to pick up a little food for themselves, and by the third week they ate almost unaided and were brooded little, but were still being led about.

By the time the young were 3 weeks old they were fully feathered, and frequently tested their flying ability. By the fourth week the young appeared to separate from the adults, and the male was observed to chase them aggressively. However, in another case a male took five-week-old young under his wings (Butler 1905a). Only males were observed incubating or leading young by Rand (1936), although Benson *et al.* (1976) mentioned seeing a female with three young.

Evolutionary relationships

Benson *et al.* (1976) believed that although there were plumage similarities between this species and *suscitator*, an African origin was very probable from zoogeographic considerations, the species most probably originating from ancestral *hottentotta* stock. This seems quite possible, although I have tentatively phyletically associated *nigricollis* with *suscitator*, also on the basis of plumage similarities in these two species. Ogilvie-Grant (1889) included *nigricollis* in a group of six otherwise strictly Australian species, all of which have a tarsal length that is no longer than the middle toe and claw.

Status and conservation outlook

Apparently this species survives well in diverse habitats such as dry forests, second-growth woodlands and along the edges of cultivated lands, so it seems likely that it is in no danger from a conservation standpoint. Rand (1936) considered it to be one of the most widely distributed of all Madagascan birds, and Dee (1986) reported it as widely distributed through a variety of habitats and generally considered it common, but noted that considerable gaps in distributional information still existed. There are fairly few specific records from the eastern, western or northern parts of the island, but it probably has followed clearings into eastern forests, and has also occupied areas of severely degraded secondary woodlands or afforestation (Dee 1986).

SPOTTED BUTTONQUAIL (Plate 3)

Turnix ocellata (Scopoli) 1786

Other vernacular names: chestnut-breasted bustard-quail, ocellated buttonquail, Philippines buttonquail; turnix de Luzon (French); Riesenlaufhühnchen (German).

Distribution of species (Map 3)

Central and northern Luzon Island, Philippines. Probably sedentary.

Distribution of subspecies

T. o. ocellata (Scopoli): central Luzon, around Manila.

T. o. benguetensis Parkes 1968: highlands of northern Luzon (vicinity of Monte Data, Benguet Subprovince, Mountain Province). Probably more widespread; a specimen intermediate in measurements between the two known subspecies has been collected in northeastern Luzon (San Mariano, Isabela Province).

Measurements (both races, mm)

Wing, males 88–102, females 97–111; tail, male 41, females 43–46; tarsus, male 28, females 28–30; culmen (from base), males 16–19.5, females 16–22 (Ogilvie-Grant 1893, Parkes 1968, du Pont 1971). Egg, av. 23.5 × 18.4 (Schönwetter 1967).

A. Hemipodes (Family Turnicidae)

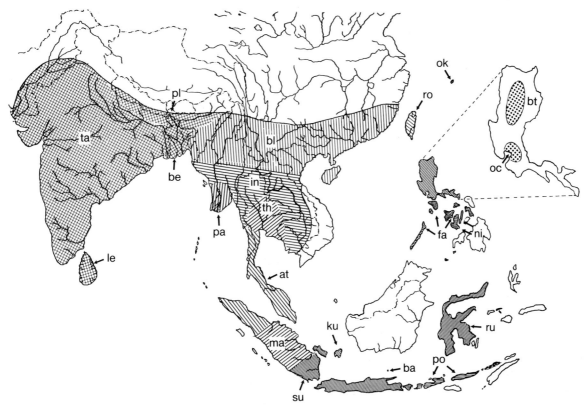

Map. 3. Distribution of the spotted buttonquail, including its races *ocellata* (oc) and *benguetensis* (bt), and the barred buttonquail, including its races *atrogularis* (at), *baweanus* (ba), *bengalensis* (be), *blakistoni* (bl), *fasciata* (fa), *interrumpens* (in), *kuiperi* (ku), *leggei* (le), *machetes* (ma), *nigrescens* (ni), *okinavensis* (ok), *pallescens* (pa), *plumbipes* (pl), *powelli* (po), *rostrata* (ro), *rufilata* (ru), *suscitator* (su), *taigoor* (ta) and *thai* (th).

Weight (g)

No information. Egg 4·3 (estimated).

Description

Adult female. General colour above ashy brown; crown black, each feather margined with rufous brown; a white band down the middle of the crown and over each eye; sides of the head and throat black, irregularly spotted with white and becoming almost uniform black in older individuals; a fairly well-developed rufous nuchal collar; rest of the upperparts with irregular, narrow, wavy black bars and vermiculations; some of the feathers of the upper back and scapulars with subterminal black blotches, edged or mixed with rufous. Remiges and primary coverts blackish brown, the outer web edged with pale buff; secondary coverts brownish buff; rest of the wing-coverts the same, but with a subterminal black spot on the outer web of most feathers, and edged externally with whitish buff. Neck, chest and breast uniform rufous chestnut; rest of the underparts dirty buff, paler on the abodomen. (Subadults differ from the adult in having the black subterminal blotches on the feathers of the upper back and scapulars margined laterally with pale buff, the greater part of the chin and throat white, and the upperpart markings more distinct.) Iris pale yellow or white; bill and tarsis light greenish yellow; claws grey.

Adult male. Differs from the adult female in having the chin and middle of the throat white or lightly spotted with black, many of the feathers of the chest with a subterminal black dot (sometimes absent), no rufous nuchal collar, and the black ocelli on the wing-coverts larger and more numerous. Iris yellowish white; bill greenish; tarsi greenish yellow; claws grey.

Immature. In young males the throat is more or less spotted with black, and many of the breast feathers are subterminally spotted with black. In young females the throat is variably spotted with white (Hachisuka 1932). Juveniles probably also have darker eyes than adults.

Identification

In the hand. Separated from the other buttonquails by the combination of having a bright, unbarred rufous breast and foreneck, the upper tail-coverts hiding the tail, and with distinct black ocelli on the upper wing-coverts. The head is greyish, with conspicuous white superciliary and median crown stripes.

In the field. Limited to Luzon Island, where the other buttonquails are *sylvatica* (the distinctive blackish-backed race *whiteheadi*), the unusually heavy-billed *pyrrhothorax worcesteri*, and *suscitator fasciata*, from all of which *ocellata* can be separated by its unspotted rufous breast, the large black spots on its upper wing-coverts, and its conspicuous white superciliary and median crown stripes.

General biology and ecology

Little specific information. The northern subspecies was collected at an elevation of about 2000 m on Monte Data, a mountain averaging about 2200 m high and covered by oak and pine forests. The annual rainfall in this area is about 350–400 mm. The nominate subspecies occurring around Manila occurs near sea level in an area receiving about 200–250 mm of rainfall annually. It apparently occurs in or around rice fields, where it is often captured during the harvest period.

Social behaviour

No information.

Reproductive biology

No information.

Evolutionary relationships

This is a rather distinctive species, but the ocelli on the wing-coverts also occur in *tanki*, which occurs on mainland China and from a zoogeographic standpoint would seem to be a likely candidate for its nearest relative.

Status and conservation outlook

No specific information.

YELLOW-LEGGED BUTTONQUAIL (Plate 4)

Turnix tanki Blyth 1843

Other vernacular names: Blanford's bustard-quail (*blanfordii*), Burmese (or Chinese) yellow-legged buttonquail (*blanfordii*), Indian yellow-legged buttonquail (*tanki*), Nicobar bustard-quail (*tanki*); hemipode moucheté; turnix de Tank, turnix indien (french); Rotnacken-Kampfwachtel, Gelbfusslaufhühnchen (German).

Distribution of species (Map 4)

Northern and eastern Pakistan and India east through Burma to southern Thailand, and north through eastern China to Heilongjiang Province and adjacent USSR, plus the Nicobar and Andaman Islands. Variously migratory, nomadic and sedentary.

Distribution of subspecies

T. t. tanki Blyth: Indian subcontinent, north to Nepal and west to Pakistan (Sind to North-West Frontier Province); also the Nicobar and Andaman islands and probably western Burma.

T. t. blanfordii Blyth 1863: From north-eastern China (Heilongjiang Province) and adjacent southern USSR (central Amurland and Maritime Territory) south through eastern China to Thailand and Burma, reaching southern Thailand and intergrading with *tanki* in Assam or western Burma.

Measurements (mm)

Of *tanki*: wing, males 71–84, females 79–93; tail, males c.26, females 30–40; tarsus, males c.23, females 25–28 (Ali and Ripley 1983). Of *blanfordii*: wing, males 85·5–91, females 96–100; tail, males 30–33, females 33–34; tarsus, males 23–25, females 26–27. Egg of *tanki*, av. 23 × 18 (Ali and Ripley 1983); of *blanfordii*, av. (China) 25 × 21 (Cheng Tsohsin 1963), av. (USSR) 26·7 × 21·5 (Ilicek and Flint 1989).

Weights (g)

Both sexes (of *tanki*) c.36–43 (Ali and Ripley 1983), males (of *blanfordii*) 35–77 (av. of 4, 51), female 93 (Cheng Tso-hsin 1963); males (of *blanfordii*) 65–78, females 105–113 (Ilicek and Flint 1989). Egg, *tanki* 3·9, *blanfordii* (China) 4·2, *blanfordii* (USSR) av. of 4, 6·7 (Ilicek and Flint 1979) (all estimated.)

Description (of *blanfordii*)

Adult female. Breeding plumage: General colour above greyish brown, with a well-defined rufous

A. Hemipodes (Family Turnicidae)

Map. 4. Distribution of the yellow-legged buttonquail, including the residential range of its race *tanki* (ta), and the breeding (hatched) and residential (cross-hatched) ranges of its race *blanfordii* (bl).

nuchal collar; lores, forehead, superciliary stripes and sides of the head buff, tipped with black; crown feathers black, edged with buff; sometimes with a line of buff feathers down the middle of the head; back, rump, upper tail-coverts and scapulars irregularly barred and vermiculated with darker brown or black, mixed here and there with traces of rufous; some feathers may retain traces of the whitish-buff lateral margins characteristic of immature birds; remiges and primary coverts blackish brown, narrowly margined on the outer web with buff; rest of the wing-coverts buff, some of the outer secondary and most of the lesser and median coverts have a small subterminal black ocellus; chin and middle of the throat whitish buff; sides of the throat, chest and breast rufous buff, becoming paler on the flanks and white on the belly; most of the feathers of the sides of the chest and breast ornamented by a small subterminal round black spot. Iris straw-yellow to creamy white; lower mandible, gape, and base of upper mandible chrome-yellow; rest of upper mandible reddish brown; tarsis, toes and claws chrome-yellow, the joints of the toes; more brownish. A duller, more male-like non-breeding plumage is also present, during which the chestnut collar is completely lost (Seth-Smith 1903, Sutter 1955a).

Adult male. Differs from the female in having no rufous nuchal collar; black spotting on the upperparts may also be more prevalent. Softparts similar to the female, but the iris is whitish, the bill pale horny brown, with only a tinge of yellowish, the tarsi, toes and claws are variably yellowish.

Immature. Both sexes differ from adults in having all the feathers of the upper surface with a patch of rufous at the extremity, and narrowly margined on one or both webs with buff; the round black spots on the wing-coverts and sides of the chest and breast are also more numerous.

Identification

In the hand. Separated from other buttonquails by the combination of the rectrices extending beyond the upper tail-coverts, the middle rectrices not lengthened or edged with white, the back feather pattern not especially scale-like nor the scapulars edged with golden buff, but usually having a rufous collar and bright yellow toes. Round black ocellus-like spots are common on the breast and upper wing-coverts. Distinguished from the *Coturnix* quails by the lack of a hind toe.

In the field. The bright yellow leg and bill colour are distinctive, and additionally the birds are distinctly greyish rather than brownish dorsally, as are *sylvatica* and *suscitator*. This latter species is barred with black on the breast (becoming all black on the throat of females), and the flank and back markings of *sylvatica* are more streaked and barred rather than rounded and ocellus-like. In flight the chestnut collar and whitish underparts are apparent. Similar to several *Coturnix* species, but having generally less patterned heads, and lacking the white dorsal shaft-streaking of most of these.

General biology and ecology

In India and Burma this species occurs widely, from elevations near sea level through intermediate elevations and reaching in areas such as the Nepal and Travancore hills up to about 1200 m. In Sikkim it has been recorded up to as high as about 2200 m. In general it especially seems to prefer areas that are cultivated, in the close vicinity of cultivation, or of dense secondary growth on deserted cultivated lands. It reportedly prefers grassland habitats to bush jungle, but is often found in the latter, or in bamboo jungle with an undergrowth of grass. It also occurs in all kinds of croplands, from the shortest to the tallest, such as young jute and dried sugarcane fields (Baker 1930). In the USSR as many as three or four females may be heard calling in isolated areas of habitats only 0·5–0·7 ha in area (Ilicek and Flint 1989).

According to Seth-Smith (1903), the birds tend to avoid long grass, and instead prefer open, sandy ground, with patches of short, rough grass present, and seek out areas of sand for dusting.

The birds rarely occur in groups, other than single adults leading young, and forage on grain, grass seeds, crop greenery, and insects, especially ants. Minute insects are regularly fed to the young by males. However, in the stomachs of four chicks only 2–3 days old, only plant seeds were found. Small beetles and grasshoppers made up about a quarter of the stomach contents of four adult birds collected during summer; the rest consisted of seeds (Ilicek and Flint 1889).

Social behaviour

Apparently sexual activity begins in about March; Seth-Smith (1903) noted that during that time captive females regained their rufous collars, and Cheng Tso-hsin (1963) observed that in China the calls change to a booming, low-pitched whistle during that month. Seth-Smith described the call (apparently referring to buttonquails in general) as similar to that of a bronze-winged pigeon (*Phaps chalcoptera*). In another description the call is described as a repeated *guuk-guuk, guuk*, and audible for a distance of 100 m (Ilicek and Flint 1989). Taka-Tsukasa (1967) stated that the female's call consists of a 'hrek' and 'boom', which at first is low-pitched and resembles the distant hoot of a steamer's whistle. Then there is a short pause, followed by a slightly louder but similarly ventriloquistic hoot. Other similar notes follow after short pauses, and after about the fifth call-note they change to a weird moan. These calls are also repeated at short intervals, the last three or four notes being very human-like and heart-rending moans that increase in intensity. As the female makes each of these calls, which total about eight, she bows her head and slightly depresses her wings.

Seth-Smith (1903) described the female's courtship behaviour as squatting in a corner of the aviary, with her breast on the ground and tail pointed upward, and making a peculiar soft clucking noise. The male would then replace the female at that location, the female meantime often standing nearby and rocking her body backwards and forwards in a peculiar manner. Seth-Smith observed no other obvious direct form of courtship, but the female would often hold an insect in her bill and while standing motionless and stretching out her body, would wait for the male to run up and take it. This was done in precisely the same way that the male presented food to his chicks.

Reproductive biology

Nesting of the nominate race *tanki* in the Indian region is fairly extended, although there are no records for the colder period of November to March. Apparently breeding normally begins with the onset of rains in the middle of June, and continues until early October. There are fewer records for the race *blanfordii*, but in the Cachar and Khasia hills it apparently begins breeding at the end of April and continues until the end of August, with a few birds continuing on to late September (Baker 1930). It is believed that the female may often nest two or three times in a single breeding season (Ilicek and Flint 1989).

A captive pair described by Seth-Smith (1903) began nesting in late April, the eggs being laid on a slight nest of hay. A second clutch was begun in late June. In this case the nest was composed entirely of hay, which the birds carried from some distance away. Nests of wild birds in India and Burma are likewise sometimes rather meagre accumulations of grass in roots in some natural hollow, or may be roughly domed and consist of a well-made pad. Perhaps on average they are not so well finished or so often domed as is the case with the barred buttonquail (Baker 1930). Taka-Tsukasa (1967) stated that his captive females began to boom in April and stopped in June, each pair typically having two broods in a season.

In the case of Seth-Smith's captive pair, three eggs of an initial clutch were laid during a period of 4 days, when the male began incubation, and hatching occurred only 12 days later. The female took no notice of the eggs or of the young. Only one of the chicks, a female, survived to adulthood, but by the age of 7 weeks it was in virtually fully adult plumage, and the iris had turned from blackish to light-coloured. A second clutch of three eggs was laid in a period of 3 days, and it too hatched after about 12 days of incubation. Reportedly the young are able to fly at the age of 10 days (Ilicek and Flint 1989).

Evolutionary relationships

Little has been written on the possible relationships of this species, although Sutter (1955, p. 106) suggested that it may be an allospecies with *maculosa*, whose relationships have more generally been assumed to be with *sylvatica* if not conspecific with that form. This position is in agreement with Ogilvie-Grant (1889), who placed *tanki* and *maculosa* in the same subgroup of the genus *Turnix*, a subgroup that he characterized as having the tarsus longer than the middle toe and claw, distinct sexual dimorphism present, the central rectrices not elongated or pointed and edged with buff, and the upperpart feathers not patterned in a scaly manner.

Status and conservation outlook

The broad geographic range of the species, and its apparent preference for cutover or cultivated areas, should allow it to survive indefinitely in the presence of humans.

BLACK-RUMPED BUTTONQUAIL (Plate 5)

Turnix hottentotta Temminck 1815

Other vernacular names: African bustard-quail, black-rumped buttonquail, Hottentot buttonquail, Natal buttonquail (*nana*), South African bustard-quail (*hottentotta*); Kaapse kwarteljie (Afrikaans); turnix hottentot (French); Hottentoten-Laufhühnchen (German).

Distribution of species (Map 2)

African grasslands and savannas from Ghana (possibly west locally to Senegambia) and Cameroon east to Uganda and Kenya, and south to north-eastern Angola and western South Africa (Cape Province). Migratory and nomadic, wintering north of known breeding range to the Sahel zone immediately below the Sahara.

Distribution of subspecies

T. h. nana (Sundevall): Ghana to Uganda and Kenya, south to Angola and south-east Cape Province. Includes *luciana* Stoneham and *insolata* Ripley and Heinrich (Urban *et al.* 1986).

T. h. hottentotta Temminck 1815: southern Cape Province, east to Port Elizabeth, South Africa.

Measurements (mm)

Wing (of *nana*), males 71–77 (av. of 7, 73·9), females 77–86 (av. of 8, 81·1); tail, males 21–27 (av. of 6, 24·7), females 28–32 (av. of 7, 29·7), tarsus, males 19–21 (av. of 7, 19·8), females 19–21 (av. of 8, 20·4), exposed culmen, males 9–11 (av. of 7, 10), females 10–12 (av. of 6, 11·2) (Urban *et al.* 1986). Egg, 21·5–24·5 × 17·3–20 (Roberts 1957).

Weights (g)

Two males 40 and 40·2, two females 57·5 and 62·4 (Urban *et al.* 1986). Egg 4·1 (estimated).

Description (of *nana*)

Adult female. General colour above black, barred and mottled with rufous, most of the feathers of the crown and back in completely edged with white; rump and upper tail-coverts nearly uniform black; scapulars edged with golden buff; rest of the head, sides of the throat, chest and upper breast bright buff, the ends of the feathers along the sides of the neck and breast variably barred with black and white, the centre usually unbarred; remiges and primary-coverts brownish black, the outer primaries margined on the outer web with buff; rest of the wing-coverts irregularly mixed with light red, buff and white, with a more or less regular subterminal black bar and white tip; chin and centre of throat whitish; breast and belly buff anteriorly to white posteriorly (heavily marked anteriorly with round blackish brown spots in *hottentotta*, unmarked in *nana*); flanks and under tail-coverts buff; middle pair of rectrices elongated, pointed and, like the scapulars and upper tail-coverts, edged with golden buff. Iris pale brown (yellowish to white in *hottentotta*); bill greyish; tarsi and toes whitish flesh-colour (Clancey 1967).

Adult male. Resembles the female, but the forehead is like the crown, the superciliary stripes and sides of the head and throat are much paler than the chest, and the black and white barring on the sides of the chest and breast extends farther onto the chest so that two or three rows of feathers instead of one have the tips black and white. (Males of the desert form '*insolata*' are reportedly more greyish and paler than other forms, and lack any solid black spots.) Iris grey; upper mandible brown, lower mandible and tarsi flesh-colour.

Juvenile. Entire underparts white, except for the breast, where scattered buffy adult feathers occur. Entire chest and breast barred with black. Iris pale brown, legs and toes dark flesh-coloured.

Identification

In the hand. Separated from other buttonquails by the combination of having the rectrices extending beyond the upper tail-coverts, the middle rectrices lengthened, pointed and edged with buff, the sides of the breast barred or spotted with black and white, and the rump area and tail mostly blackish, with lighter barring. Distinguished from *Coturnix* species by the lack of a hind toe.

In the field. Distinguished from *sylvatica* by its darker rump and tail colouration (visible only in flight), and by its generally richer brown upperparts, including the upper wing-coverts, and more brownish on the sides of the head. The female's advertisement call is a series of low-pitched and resonating *hoo* notes similar to but faster (c.1·6 calls per second) and shorter in length than those of *sylvatica* (Urban *et al.* 1986). It is also said to be lower in pitch. Quite similar to several of the true *Coturnix* quails (especially females), but generally having less patterned heads and lacking the buffy shaft-stripes typical of these.

General biology and ecology

This is a generally less common and more arid-adapted species than is the striped buttonquail in Africa, and tends to occupy open, short grasslands rather than the ranker grasslands frequented by the latter species (Snow 1978). It generally occurs from about sea level to around 1800 m elevation, and seems to favour grasslands that have been well grazed or disturbed by cattle. Whereas in southern Africa *sylvatica* sometimes occurs in thornveld and light woodlands as well as its favoured grassveld, *hottentotta* is generally found in grasslands, including the moist fringes of marshes where the grass is not too long and dense, in fallow fields or native gardens, and on scrublands having thin grass and low bushes. In such areas it feeds on grass and weed seeds, insects and their larvae, and some other invertebrate foods (Clancey 1967).

Areas favoured for nesting in Zimbabwe are those with dark clay soil, on which the grass if 25–50 cm high, and growing in such a way that the ground itself is clear of obstructions preventing easy movement by the birds below the vegetation. Both wet and marshy areas with standing water and dry bushy areas supporting taller grasses are generally avoided, although nesting has been observed in a cut sugarcane field, where the conditions were rather marshy (Masterson 1973).

Social behaviour

This species has only rarely been bred in captivity (in 1982 at the Frankfurt Zoo, according to the *International Zoo Yearbook*) and as a result a very limited amount of information on its behaviour is available. The birds usually occur as solitary individuals or pairs, but sometimes occur as larger loose aggregations. There is some information suggesting nomadic or migratory movements in some parts of its range (Malawi, Nigeria), the birds apparently arriving to breed during or at the end of the rains, whereas in others areas such as Kenya, Uganda, and extreme southern Africa it is residential (Urban *et al.* 1986).

The female's advertisement calls are very much like those of the striped buttonquail, consisting of a series of low-pitched, resonant *hoo* notes that are uttered at the rate of about 1·6 per second (Urban *et al.* 1986).

Reproductive biology

Nesting is evidently timed to coincide with the rainy period or at its termination, which means that the overall breeding period of this species is substantially shorter than that of the striped buttonquail, and that the bird is more likely to show migratory or nomadic movements coinciding with the rainfall pattern. Egg dates for South Africa are from October to January, for Zimbabwe from September to February, for Malawi in April, and for Nigeria in December and January. Kenya records are from May to July, plus October (Urban *et al.* 1986).

Typically the nest consists of a scanty pad some 5

cm in diameter, made of dried grass underneath the spreading blades of a grass tuft. Occasionally blades of growing grass are pulled down to form a loose canopy above the eggs, although at other times nests are placed under a sheaf of dried grass without any specially constructed canopy (Masterson 1973).

Clutch-sizes range from two to six eggs, but are usually three, as was the case in all eight nests found in Zimbabwe by Masterson (1973), who noted further that groups of three fledglings have sometimes been flushed. Incubation is by the male only, and is reportedly of 12–14 days' duration (Urban *et al.* 1986).

Evolutionary relationships

Snow (1978) noted that this species is strikingly similar, at least superficially, to *T. pyrrhothorax*, although it is also 'very closely related' to *T. sylvatica*. Ogilvie-Grant (1889) placed it within the same generic subgroup as *sylvatica*, noting that in this group there is slight sexual dimorphism, the central tail feathers are elongated and pointed and edged with white or buff, and the upper back feathers have a scaly appearance as a result of buff or white edging. This group was believed by him to be most nearly related to the subgroup that included *tanki* and *maculosa*. I have tentatively shown *sylvatica* as the probable nearest relative of *hottentotta* (Figure 1).

Status and conservation outlook

The association of this species with dry and/or heavily grazed grasslands and similar semi-arid habitats should mean that no conservation problems are likely to arise for it in the foreseeable future.

STRIPED (SMALL) BUTTONQUAIL (Plate 6)

Turnix sylvatica (Desfontaines) 1787

Other vernacular names: Andalusian hemipode, common buttonquail, David's buttonquail (*davidi*), Kurrichaine buttonquail (Africa), little buttonquail, little bustard-quail (*dussumieri*), Smith's bustard-quail (*lepurana*), Sulu buttonquail (*suluensis*), Whitehead's buttonquail (*whiteheadi*); bosveldkwelteltjie (Afrikaans); turnix d'Andalousie, turnix sauvage (French); Spitzschwanz-Laufhühnchen, Rostkehl-Kampfwachtel (German).

Distribution of species (Map 5)

From the southern Iberian Peninsula and coastal north-western Africa south through most of the African continent except the extreme desert regions and Congo basin to South Africa, and from Pakistan east through India and Indochina to south-eastern China, as well as on Hainan and Taiwan. Also locally and discontinuously distributed in the south-western Pacific, including Java, Bali, the Sulu archipelago and the central Philippines (Luzon, Mindanao and Negros islands). Apparently sedentary (Vaurie 1965).

Distribution of subspecies

T. s. sylvatica (Desfontaines): Southern Iberia (southern Portugal and southern coastal Spain from Malaga to Cadiz) and north-western Africa (coastal areas of Morocco and Algeria).

T. s. lepurana (A. Smith) 1836: Sub-Saharan Africa from Senegal to Sudan, south (except the Congo basin and horn of Africa) to northern Angola and eastern South Africa. Includes *alleni* Clancey 1978 (Urban *et al.* 1986). Also includes *arenaria* Stresemann 1938.

T. s. dussumier (Temminck): Pakistan (east of the Punjab) and India (south to Kerala, possibly to Sri Lanka), east to Nepal, Sikkim and central and southern Burma. Also reportedly occurs in eastern Iran, but no recent information.

T. s. davidi (Delacour and Jabouille 1930: Indochina (north of the Malay peninsula) and south-eastern China, from Yunnan east to Guangdong (Kwantung), plus Hainan and Taiwan. Includes *mikado* Hachisuka 1931 (Vaurie 1965).

T. s. bartelsorum Neumann 1929: Java and Bali.

T. s. suluensis Mearns 1905: Jolo Island (Sulu archipelago).

T. s. whiteheadi Ogilvie-Grant 1897: central Luzon Island (Philippines).

T. s. masaaki Hachisuka 1931: Mindanao (Philippines).

T. s. nigrorum du Pont 1976: Negros Island (Philippines).

Measurements (of *sylvatica*, **mm)**

Wing, males 83–92 (av. of 12, 88), females 91–101 (av. of 17, 97·3); tail, males 36–43 (av. of 10, 39·4), females 40–46 (av. of 14, 42·6); tarsus, males 22–25 (av. of 11, 23·4), females 22–26 (av. of 21, 24); culmen, males 10–12 (av. of 12, 11·2), females 11–14 (av. of 20, 12·7). Egg, av. 26 × 21 (Cramp and Simmons 1980). Measurements of other races: wing, males 62–72, females 68–79·5; tarsus, males 17·5–20·5, females 19–22·3; bill, males 10·1–12, females 10·7–13·5 (Sutter, 1955).

Weights (g)

Males (*lepurana*) 32·2–43·5 (av. of 10, 36), females 39–53·7 (av. of 12, 51) (Urban *et al.* 1986). Males of

80 *Bustards, Hemipodes, and Sandgrouse*

Map. 5. Distribution of the striped buttonquail, including the ranges of its races *bartelsorum* (ba), *davidi* (da), *dussumier* (du), *lepurana* (le), *masaaki* (ma), *nigrorum* (ni), *sylvatica* (sy), *suluensis* (su), *whiteheadi* (wh).

the larger *sylvatica* may average c.60, females c.70 (Cramp and Simmons 1980), and both sexes of the smaller *dussumier* range from c.36–43 (Ali and Ripley 1983). Egg 6 (estimated).

Description (Of *sylvatica*)

Adult breeding female. General colour above dull light red, the feathers of the top of the head, back, rump, upper tail-coverts, rectrices and scapulars barred and margined internally with black and externally with white or grey, giving the back a scaly appearance, which is often increased by the pattern of black and dull light red following the shape of the feathers; a white band down the middle; lores, superciliary stripes, and sides of the head and neck white tipped with black; remiges and primary-coverts brownish black, the outer web of all edged and in the outer secondaries also vermiculated with buff and spotted with white; rest of the wing-coverts dull light red, obliquely barred on the outer or both webs with black and white; chin and middle of throat white; sides of the chest and breast whitish buff, each feather with a heart-shaped subterminal black spot; centre of the chest bright rust-colour, paler on the flanks and under tail-coverts; rest of underparts whitish buff; centre pair of rectrices elongate and pointed. Iris white, with yellow tinge, and bluish orbital skin, bill blue-grey; tarsi and toes flesh-coloured with yellow or blue tinge (Cramp and Simmons 1980).

Adult non-breeding female. Like the breeding plumage, but with less coarse black vermiculations and narrower off-white streaking in the upperparts. Also similar to the non-breeding male, but the upperpart feathers are fringed with cinnamon rather than grey, the nape is nearly uniform dull light red, and the breast is deeper rusty (Cramp and Simmons 1980).

Adult breeding male. Differs from the breeding

female in being somewhat paler on the underparts, the scale-like marking on the upperpart feathers extend over the nape to the back of the head, and the upperpart feathers are mainly cinnamon and buff rather than chestnut in overall colour.

Adult non-breeding male. Like the breeding plumage, but the mantle and scapulars with less greyish and more resembling the female, the crown and upperparts black and cinnamon rather than chestnut, and the underparts less deep rusty (Cramp and Simmons 1980).

Immature. Immatures are similar to adults, but have less bright and contrasting plumages, with the feather centres of the upperparts irregularly vermiculated with cinnamon and black rather than barred, and some tertials or upperpart feathers with lateral white spots rather than streaking. Some juvenal flight feathers and tertials are retained (at least in temperate races) in the post-juvenal moult; some juvenal remiges may thus be held through the first breeding season (Cramp and Simmons 1980).

Juvenile. Similar to adults, but with a more spotted chest, white spotting on the upperparts, and less-noticeable markings on the wing-coverts (Cramp and Simmons 1980).

Identification

In the hand. Separated from other buttonquails by the combination of having the rectrices extending beyond the upper tail-coverts, the middle rectrices lengthened to a point and edged with white or buff, the greater secondary-coverts fringed with creamy buff, the back having a scaly appearance, and the central crown streak well developed. The anterior flanks and sides of the breast are typically spotted with blackish (especially in females). In Africa this breast spotting is less developed, in south-eastern Asia the dorsal markings are also weaker and streaking rather than barring predominates, and in the Philippines the birds are darker and more contrastingly patterned. This and all other buttonquails are easily distinguished in the hand from true *Coturnix* quails by the lack of a hind toe.

In the field. This species is quite variable geographically but typically the dorsal buff feather edging makes it appear rather scaly-backed, and it has conspicuous whitish superciliary and central crown stripes and a distinctly pointed tail. It has black spotting on the sides of a rufous-tinted breast, whitish eyes (except in juveniles), and flesh-coloured (not yellow) feet and toes. At least in the African race *lepurana* the female's booming advertisement call lasts about one second, and is repeated every 1–2 seconds for 30 seconds or more, whereas the call of the other African buttonquail species *hottentotta* is uttered more rapidly (Urban *et al.* 1986). Nominate *sylvatica* apparently has a similarly prolonged call lasting about a second, with equally long inter-call intervals (Cramp and Simmons 1980). This species is surprisingly similar to the widespread *Coturnix* quails, but is smaller, lacks blackish cheek and throat markings, and in flight shows darker, unpatterned flight feathers. However, its back feathers are distinctly edged with buff so as to produce a scaly pattern, which helps to separate it from the black-rumped buttonquail.

General biology and ecology

In general, striped buttonquails are birds associated with low ground cover under which they can readily run or hide, and especially favour well-drained or sandy areas having grassy, bushy, or dwarf palmetto (*Chamerops*) vegetation. Indeed, the particular type of vegetation cover is seemingly not very significant—old cultivated fields that have grown back to weedy grasses, stubble fields, plantations of various crops such as cotton, millet, maize or cassava, savanna bush with long grasses, openings in or the edges of jungle or other similar low thickety vegetation—all of these seem to be almost equally acceptable to the species. It sometimes can be found in crops that stand as high as 2 m and it greatly favours small bamboo jungle with little undergrowth but avoids dense evergreen forest. Wide stretches of grasses that are neither very tall nor very dense are particularly favoured. In Africa it ranges to at least 2000 m elevation, while in northern India it extends as high as 2400 m. Generally it is to be found on any combination of dry, warm soils and associated low vegetation, and typically avoids deserts, forests and wetlands (Cramp and Simmons 1980, Ali and Ripley 1983, Urban *et al.* 1986).

In general the species appears to be sedentary, although it may be nomadic or semi-migratory in north-western India, where it moves into this dry region only after rains (Ali and Ripley 1983).

Throughout its broad range the species feeds on seeds, especially grass seeds, and insects, including ants. The birds also particularly like to bask in the sun while stretched out in sandy or dusty sites. Pairs often dustbathe and sunbathe together, and while sunbathing may lie side by side and either spread one wing outwards while resting on one side, or stretch out on the ground while extending both wings fully.

Social behaviour

Of all the buttonquails, this species has been bred in captivity the most frequently (Butler 1905b, Mörs 1915, Sich 1927, Hoesch 1959, 1960, Trollope 1970, Flieg 1973, Bell and Bruning 1974, Werner 1975, Wintle 1975, van Praet Lucas 1976). As a result, more information is available as to its social behaviour than for any other species of hemipode.

Flieg (1973) reported that at least five adult calls are present. A 'flock' (or perhaps more realistically pair-contact) call consists of a high-pitched, almost inaudible chirping, and the alarm call is similar but much louder; presumably both sexes utter these notes. The male also utters a trumpeting buzz as a threat towards other males, and the female produces an apparently similar but 'growling' buzz as a threat. Finally, the female produces her distinctive 'booming' call (sometimes described as double-noted and at other times as single-noted) during courtship. In addition to these calls, males have been reported to produce a long and drawn-out *treee* note, which may at times be uttered in duet with the female (Hoesch 1960), and males may respond to the female's advertising call with a sharp *tuc-tuc-tuc* (Trollope 1970). Males also utter soft clucking sounds to call the young together (Butler 1905b). A few other possibly distinct calls have also been described (cf. Cramp and Simmons 1980), but at least some of these probably represent differing human interpretations of the same calls. According to Flieg (1973), female chicks may begin to 'boom' when only 2 days old, but other calls by chicks have not been described.

A practice of uncertain significance consists in individuals of either sex adopting a rigid, almost horizontal posture, and rocking back and forth in a chameleon-like manner (Hoesch 1959, Flieg 1973, Wintle 1975). It may occur at almost any time but may in general indicate a state of nervousness. Similar 'body-swaying' behaviour has been observed in the barred buttonquail (Trollope 1970), although only in females. It is similarly of unknown (perhaps concealing) function in that species. It has been suggested that in *sylvatica* it may serve as a 'cut-off' posture helping to reduce aggressive interactions; for example it was performed by both sexes of a pair when they encountered each other and after the male had left the clutch in order to feed (Hoesch 1960).

The onset of breeding is marked by the female beginning to utter her booming call. This is most frequently done at dawn, but also occurs at other times, including evening and night time hours. Calling is preceded by an expansion of the neck through an inflation of the oesophagus (see Figure 24). This is accompanied by an apparent gasping for air, as the sides of the neck are blown outwards (Butler 1905b). The call itself consists of one or two booming or *hooo* notes of remarkably steady pitch and gradually increasing volume, either sharply broken off or gradually diminishing again. Some observers (Butler 1905b, Hoesch 1960) have described the call as two-noted, the first being higher (by about an octave) and longer sustained than the second, but most descriptions decribe it as single-noted. During this call the female stands with her body at a diagonal angle, and the bill somewhat tilted downwards (Figure 24). This call probably serves mainly to attract males, rather than acting as a territorial call, since Trollope (1970) observed no response to it by other captive females. Perhaps females distinguish males from other females (which they immediately attack) through the fact that males stand their ground but do not fight back when threatened by the female.

Typically as the female utters this call the male approaches her and pecks her on the head and back. Early stages of pair formation may involve the pair walking together while performing the curious swaying movements mentioned earlier. Apparently although both sexes search for suitable nest-site, the female makes the final selection and often builds the nest herself (Wintle 1975). However, the male may also participate in nest-building to some degree (Butler 1905b, Hoesch 1960). Whilst examining nest-sites together the pair perform nest-scraping behaviour as they peck at the soil, rest on their breasts with tails raised, shuffle their wings, and generally 'try out' the site (Butler 1905b, Trollope, 1970, Wintle 1975).

After being visually separated from its mate for about 4 weeks, the male of one pair was also observed to perform a greeting ceremony by crouching with his bill tilted downward as an apparent submissive gesture towards the female (Figure 24). This in turn stimulated the female to preen the male's head (Figure 24) (Hoesch 1960). The female may also offer the male titbits of food, but males only seldom feed females (Wintle 1975). Once paired, the two birds roost together side by side and facing the same direction.

Copulation is apparently preceded by the male following or even driving the female for some time, pecking at her head and nape. When she crouches, the male mounts her and grips her with his wings during treading (Hoesch 1960). At times the female may take the initiative and mount the male (Flieg 1973).

Reproductive biology

The breeding cycle is extremely variable in different parts of this species' range. In India it generally

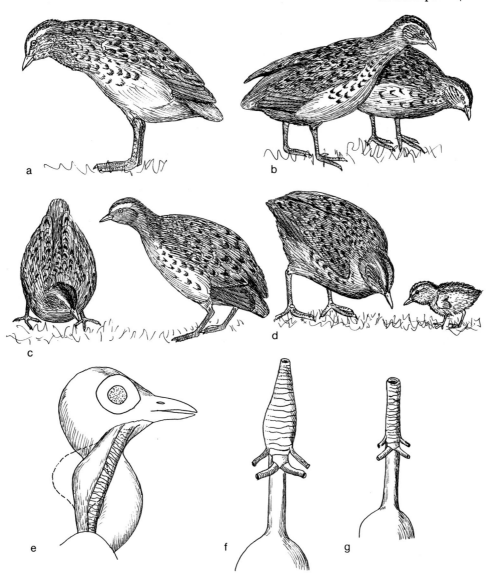

Fig. 24. Social behaviour of the striped buttonquail: (a) female advertisement calling; (b) female approaching male to allopreen him; (c) male (left) in submissive posture towards female; and (d) male titbitting behaviour toward young after photos in Hoesch 1959, 1960). Also anatomy of striped buttonquail adults: (e) female oesophagus in normal and (dotted line) inflated position; (f) female trachea and oesophagus; and (g) male trachea and oesophagus (after drawings in Neithammer 1961).

breeds from April to October, and especially during the months from June to September, which generally correspond to the rainy period (Baker 1930). In Spain the egg-laying period occurs during spring and summer, extending from April to August (Cramp and Simmons 1980). In Africa the nest records are extremely variable but extend through the entire year in South Africa, Zimbabwe and Namibia, with varied peak periods. In the more northern tropical areas the birds are often wet-season visitors, sometimes remaining to breed after the rains. In African areas somewhat to the north of the equator (Nigeria, Sudan, Ethiopia, Somalia) the breeding records are generally from April to September (plus a January record for Nigeria), but in East Africa there are roughly two breeding periods. The first of these seems associated with the January–June period and corresponds with the area's major wet season. The second generally falls between September and December and is perhaps associated with the onset of the autumn rains. There are no July breeding records for this area (Urban et al. 1986).

The nests of these species are somewhat variable but are usually well concealed under grass or other herbaceous materials. Of some forty to fifty nests found in India, no more than six were domed over (Baker 1930). The clutch-size has been reported to be almost invariably four eggs in India, with a few clutches of five (Baker 1930). In southern Africa (Malawi) the average of twenty-one clutches was 3·4, with a usual range of three to four eggs (Urban *et al.* 1986), although clutches of five have also been found (Wintle 1975). The eggs are laid on a daily schedule, and the male begins to incubate just prior to the laying of the last egg of the clutch (Butler 1905b, Hoesch 1960). Typically the female begins to remain away from the nest on the day before the last egg is laid, whilst on previous days she will often sit on the nest during the day (Hoesch 1960), thus perhaps providing the male with the proper cue to begin incubating. On the other hand, Butler (1905b) stated that during the first days of incubation the female frequently sat for some time on the clutch, but later she gave up this responsibility to the male.

Incubation lasts 12–14 days, the male often becoming restless after about 12 days. During the incubation period the male may also take 'time off' to feed with the female, or to bask in the sun, and may leave the nest eight or more times a day for periods of up to 15 minutes. In some cases the newly hatched young may leave the nest with their father after only 4 hours (Wintle 1975), but in other cases as much as 2 days may pass before the young are led out of the nest (Butler 1905b). Males feed their young by performing 'titbitting' behaviour (Figure 24), during which they offer food in the bill while standing still, uttering soft calls, and shivering their wings. In captivity it has been found that when males have been isolated from females they never refuse to accept the care of chicks, and that when a second male is introduced into a cage having a male with a large brood, both males assume care of the young (Flieg 1973). Generally females ignore or may be 'friendly' towards young chicks, but sometimes they may also attack and even kill them. Similarly, older juveniles may attack younger broods.

Introducing a second unpaired male into an aviary tends to reduce the interval between the laying of successive clutches (from about 50 days to 11 days in one case), an interval that evidently is normally controlled by the length of time required for the male to hatch the eggs and rear the young (Hoesch 1960), as was described in Chapter 4. Flieg (1973) reported that the optimum sex ratio for captive breeding is two to three males per female, which allows males to be removed to rear the young without loss of female fertility. The young are quite precocial, and can be safely removed from the parent by 5 days, as they are typically fed bill-to-bill for only the first 4 days. When frightened they close their eyes and lie perfectly still. They can fly by 7–11 days, and become fully independent of their parent by about 18–20 days.

According to Wintle (1975) only two chicks per brood are successfully reared as a rule, although the observed cases in which four young survived. He noted that hazards to the young include older buttonquails, snakes, shrews, mice, lizards and even 'inquisitive' finches. Wintle believed that theoretically a female should be able to keep five males rearing young the whole year around, the young themselves becoming able to breed when only 4 months old. However, the mortality in nature is probably very high, in spite of the fact that Wintle had a female in captivity that at the time of his writing was over 9 years of age.

Evolutionary relationships

Ogilvie-Grant (1889) included *sylvatica* in a group of taxa that also included forms now collectively included in that species and *hottentotta*, the two forms differing mainly in the fact that in *sylvatica* the neck feathers are spotted with black rather than being barred with black, as is usually but not always the case with *hottentotta*, which however lacks a buffy central crown streak and is generally darker above.

It is easy to visualize a speciation process in which *hottentotta* evolved in sub-Saharan Africa and *sylvatica* north or east of the African–Arabian deserts, followed by an invasion of Africa by *sylvatica* and the establishment of secondary contact.

Status and conservation outlook

This most widespread of the species of *Turnix* is well adapted to a variety of primary and secondary vegetation types, and is often fairly abundant over much of its range, although in Europe it occurs only peripherally and is regarded as an endangered species (Hudson 1975). There it occurs very locally in Spanish Andalusia; small populations apparently also still exist in southern Portugal, such as near Abrantes, where it once was not very uncommon (Cramp and Simmons 1980).

RED-BACKED BUTTONQUAIL (Plate 7)

Turnix maculosa (Temminck) 1815

Other vernacular names: black-backed quail, black-spotted quail, black-spotted turnix, Celestino's buttonquail (*celestinoi*), New Britain bustard-quail (*saturata*), orange-breasted quail, red-backed quail, red-collared quail, spotted buttonquail, Temminck's bustard-quail (*maculosa*), Wallace's bustard-quail;

turnix moucheté (French); Fleckenlaufhühnchen (German).

Distribution of species (Map 6)

Northern and eastern Australia, New Guinea, and various south-western Pacific islands, including the Lesser Sundas, Sulawesi (Celebes), Moluccas and Philippines (Bohol, possibly also Mindanao). Sedentary or nomadic. Mostly allopatric with and often considered (e.g. Peters 1934) as a subspecies of *sylvatica*. Regarded by Sutter (1955a) as specifically distinct, inasmuch as he believed sympatry to occur on Mindanao Island, a situation that is still unclear and 'requires further study' (C. M. N. White and Bruce 1986).

Distribution of subspecies

T. m. melanota (Gould) 1837. Australia, from eastern Queensland to eastern New South Wales.

T. m pseutes (Mathews) 1912: coastal north-western Australia.

T. m yorki Mathews 1916: Cape York area of Queensland.

T. m. horsbrughi Ingram 1909: southern and south-eastern New Guinea.

T. m. mayri Sutter 1955: Yeina and Tagula (Louisade Archipelago).

T. m. giluwensis Sims 1954: New Guinea (central highlands, 1700–2500 m).

T. m. furva Parkes 1949: Huon Peninsula (eastern New Guinea).

T. m. saturata Forbes 1882: Duke of York Island and New Britain.

T. m. salamonis Mayr 1938: Guadalcanal Island.

T. m. maculosa (Temminck) 1815: Timor, Wetar, Kisser and Moa islands, Lesser Sundas.

T. m. savuensis Sutter 1955: Savu Island, Lesser Sundas.

T. m. sumbana Sutter 1955: Sumba Island, Lesser Sundas.

T. m. floresiana Sutter 1955: Flores and Alor islands, Lesser Sundas.

T. m. beccarii Salvadori 1875: Sulawesi, Muna and Tukangbesi islands.

T. m. obiensis Sutter 1955: Obi, Tepa-Babber and Little Kei island (Moluccas).

T. m. kinneari Neumann, 1939: Peleng Island (Moluccas).

T. m. celestinoi McGregor, 1907: Bohol Island (Philippines). Also reported from Mindanao (Sutter 1955a), but of uncertain status there (White and Bruce 1986).

Measurements (mm)

Wing (all races), males 66–81, females 72–89; tail, males 21–29, females 26–33·5; tarsus, males 17–21·5, females 19–23; culmen, males 9·8–13·5

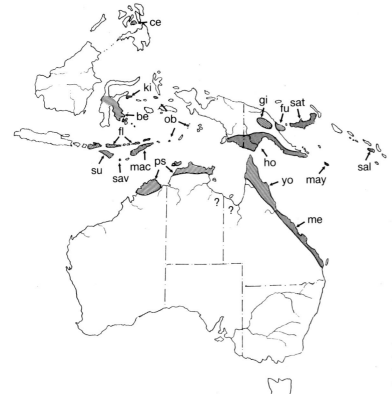

Map. 6. Distribution of the red-backed buttonquail, including the locations of its races *beccarii* (be), *celestinoi* (ce), *floresiana* (fl), *furva* (fu), *giluwensis* (gi), *horsbrughi* (ho), *kinneari* (ki), *maculosa* (mac), *mayri* (may), *melanota* (me), *obiensis* (ob), *pseutes* (ps), *salamonis* (sal), *saturata* (sat), *savuensis* (sav), *sumbana* (su) and *yorki* (yo).

(Sutter 1955a). Egg, av. 22·5 × 17 (Rand and Gilliard 1968).

Weights (g)
Females of *sumbana*, 40–51 (av. of 13, 45·3), males 31–38 (av. of 12, 33·6); females of *beccarii*, 32 and 38; males 23 and 23 (Sutter 1955a); females of *pseutes*, 38–47 (av. of 3, 42), males 27·5–34·5 (av. of 3, 31·2) (Hall 1974); female of *obiensis*, 48 (Sutter 1955a). Egg, 3·4 (estimated).

Description
Adult female. Resembles *tanki*, but distinguished by its scapulars being edged with golden buff or straw-colour. Crown feathers black, margined with brownish grey, a dull buff line down the middle of the crown, and more rufous evident in the upperparts behind the nuchal collar than in *tanki*. A distinct rufous collar is typical of some but not all subspecies. Iris pale lemon yellow to yellowish white, upper mandible dark grey, lower mandible and gape bright yellow, tarsi and toes bright yellow to yellowish green.

Adult male. Resembles the female, but no trace of a rufous nuchal collar, the crown buff and grey, and less brown on the breast and flanks. Iris yellow-white to pale yellow; bill yellow (below) to greenish gray (above); tarsi and toes greenish yellow to yellow.

Immature. Resembles the immature of *tanki*, but distinguished by the wider golden-buff margins of the scapulars. Similar to the adult male, but with a dark iris and a grey bill and tarsi (Frith 1976). In young females the rufous colour of the sides of the breast may be separated by a broad white stripe joining the white of the throat and belly (Sims 1954).

Identification
In the hand. Separated from other buttonquails by the combination of having the rectrices extending beyond the upper tail-coverts but not pointed or edged with buff, the scapulars dark chestnut to black, edged with contrasting golden-buff, the rufous-tinted upperparts (especially the nape of females) not extending to the paler breast, and the wing-coverts barred with black.

In the field. The slender, yellowish bill, pale eye, and unspotted rufous-tinted breast (becoming a bright rufous patch on the hindneck of females) helps to identify this species, as does the dark reddish brown back that is lacking in whitish scaly patterning.

General biology and ecology
Although this species is rather broadly distributed through the Pacific region, relatively little has been written on its biology. In Australia it occurs fairly widely in northern and eastern areas, but it is common only on Cape York Peninsula and along the coastal plains of Northern Territory. There it favours dense, coarse grasslands at the edges of marshes and swamps, and areas that are seasonally flooded (Frith 1976). It also occupies eucalypt woodlands and rainforests, and occurs around crops and gardens. In the Atherton region it occurs in cleared areas at elevations up to 900 m (Blakers *et al.* 1984). It has been also observed in savannah woodlands, in swamp grasslands, on a forest trail in pseudo-rainforest, and in mud and thicket areas surrounding a lake (Hall 1974). It has been suggested that the species may be quite nomadic or locally migratory in Australia, although this is still to be documented adequately (Blakers *et al.* 1984).

Three subspecies occur in New Guinea, of which two are found near sea level in grasslands bordering lowland waterways, whilst the third (*giluwensis*) occurs on mid-mountain grasslands and lightly wooded plateaux from about 1600 to 2450 m (Rand and Gilliard 1968).

The birds usually occur singly, but groups of up to five have been seen together (presumably an adult male and four dependent young). Foods include the usual buttonquail assortment of grass seeds, insects, and greens (Hall 1974).

Social behaviour
Almost nothing has been written on the social behaviour of this species, which however is almost certainly much like the other buttonquails. Frith (1976) stated that it is polyandrous, with some birds maturing at 4 months, but did not cite specific evidence for this, and the species has apparently not yet been bred in captivity. He also stated that the female utters a loud *oom* note rapidly, for a considerable period of time, which may advertise her breeding territory and serve to attract a male.

Reproductive biology
In Australia, this species breeds from October to July, when there is a maximum amount of insect life available to feed the young. The nest is a hollow depression under a grass tuft, and is lined with fine grass. Other grass stems may be bent down to form a

canopy or dome, with a side entrance. There are 'usually' four eggs, and the incubation period is reportedly 14 days (Frith, 1976).

Evolutionary relationships

Although Ogilvie-Grant (1889) placed this species in a group that also included the Asiatic form *tanki* and some now discarded taxa, and pointed out that in these the sexes are dimorphic, the middle rectrices are neither elongated nor pointed and edged with buff, and the upperpart feathers are not scaly in appearance. However, Peters (1937) later included *maculosa* as a subspecies of *sylvatica*, and this procedure was subsequently generally followed. Sutter (1955a) subsequently concluded that these two forms are parts of two well-separated phyletic groups, without close relationships, and believed them to be sympatric in Mindanao. This latter point is still uncertain, but at least there seems to be no question that they should be considered specifically distinct. He believed the nearest relative of *maculosa* to be *tanki*, and indeed Etchécopar and Hüe (1978) considered these two forms as conspecific. I have nevertheless decided, in deference to tradition, to retain *maculosa* close to *sylvatica* in taxonomic sequence although it may very well be desirable to shift it taxonomically closer to *tanki* at some future time.

Status and conservation outlook

In Australia this is apparently a rather uncommon species, and the same appears to be true in New Guinea, although the inconspicuousness of the birds in their typical long-grass cover makes any assessment of relative commonness impossible. In some parts of Australia it may have varied in abundance, but to suggest that it has declined generally is perhaps an exaggeration (Blakers *et al.* 1984). It also did not suffer noticeably from major ecological changes occurring during World War II on islands such as Guadalcanal (Pendleton 1947).

LITTLE BUTTONQUAIL (Plate 8)

Turnix velox (Gould) 1841

Other vernacular names: butterfly quail, dotterel quail, little turnix, swift bustard-quail, swift-flying quail, swift-flying turnix; petit turnix (French); Zwerglaufhühnchen (German).

Distribution of species (Map 7)

Australia, widely distributed throughout the interior except the Kimberleys, the northern parts of Northern Territory, and the York Peninsula; rare along the eastern coast. Highly nomadic and locally migratory. No subspecies recognized here; *leucogaster* North 1895 and *picturata* Mathews 1912 are considered synonyms.

Measurements (mm)

Wing, males 74–81 (av. of 21, 76·8), females 78–89 (av. of 20, 83·2); tail, males 20–34 (av. of 21, 27·9), females 28–37 (av. of 22, 31·9); tarsus, males 14·8–18·2 (av. of 23, 17·5), females 16·5–18 (av. of 12, 17·2); exposed culmen, males 10·6–13 (av. of 8, 11·1), females 10·4–11·9 (av. of 12, 11) (Museum of Victoria and Australian Museum data). Egg, 23 × 18 (Frith 1976).

Weights (g)

Females 21–64 (av. of 11, 47·6, males 25–46, av. of 17, 35·3 (Goodwin 1974, and Australian Museum data); males 28–39 (av. of 5, 37), females 47–64 (av. of 5, 56) (Morris and Kurz 1977). Egg, 3·9 (estimated).

Description

Adult female. General colour above reddish chestnut, shading into light red on the nape and the crown; otherwise the markings on the upper surface are almost the same as in *pyrrhothorax* except that in older individuals the top of the head and nape are uniform light red, the head has few or no black feathers, and the nape feathers lack white lateral margins. Sides of the head and chest dull rufous, and the median underparts from the chin backwards white. Iris white; bill light grey, becoming darker on the culmen; tarsi flesh-coloured (Frith 1976).

Adult male. Resembles the female, but the crown mottled brown with white stripes, the neck feathers with light edges, the breast mainly white at the centre and lacking the red-buff tones of females. Sides of neck and breast more mottled brown than in females (Frith 1976).

Immature. Outer primary webs mottled with rufous, those of the secondaries edged and toothed with white, and the markings on the upperparts less regular than in adults. The eyes are dark until two months old; sexual maturity occurs precocially at 3 months (Frith 1976). In young birds there are dark breast markings that disappear in the first month, the eyes are bluish, and the legs have a brownish tinge (Morris and Kurz 1977).

Bustards, Hemipodes, and Sandgrouse

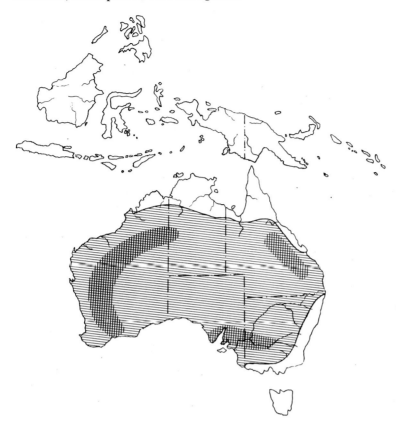

Map. 7. Distribution of the little buttonquail. Probable areas of denser populations are shown by cross-hatching.

Identification

In the hand. Separated from the other buttonquails by the combination of a pale reddish buff to off-white breast with no barring on the sides of the upper breast or neck, and a subdued dorsal pattern redominating in dull reddish-brown tones, with whitish edges and black barring. The head is almost unmarked pale reddish (females) to buff (males), grading to white on the throat and becoming somewhat darker on the crown. It is very similar to *pyrrhothorax*, but is less blackish on the head, more cinnamon-toned dorsally, and more nearly whitish below. Like most other Australian buttonquails it has a relatively short tarsus (which is no longer than the middle toe and claw).

In the field. This is the most widespread species of Australian buttonquail, and one of the commonest. It is mostly greyish brown (males) to rusty brown (females) in general tone, with little contrasting black markings evident. In flight it has an evident reddish brown rump and whitish undersides. Its female advertisement calls are a loud hooting *coo-oo* or deep *oo-ah*; squeaky chipping notes are sometimes uttered in flight. It is quite similar to various *Coturnix* species, but at close range may be distinguished from these by its whitish (rather than red to brownish) eyes.

General biology and ecology

This is perhaps the most arid-adapted of all the Australian buttonquails, and as such has the broadest distribution of any. It often ranges out into the most barren spinifex desert grasslands, but also inhabits tussock grasslands, acacia scrub, and woodland vegetation types. To a greater extent than the other Australian buttonquails it is also nomadic, moving into and out of areas according to rainfall patterns. However, it rarely moves into areas having annual rainfall patterns of more than 500 mm but instead moves into the arid interior after rains, breeds rapidly, and again disperses (Frith 1976).

Flocks of thousands of birds have been reported to occur during some irruption periods after favourable rainfall, and in some southern parts of Australia there may be a seasonal component to its movements as well. In South Australia it is a regular but highly variably abundant migrant during spring and summer, its numbers sometimes varying a hundred-fold from year to year. In the Mudgee district of New South Wales the birds are normally summer visitors, but in some years when conditions are favourable the birds may remain throughout the entire year (Morris and Kurz 1977). The availability of seeds, especially of grass seeds, seems to account for much of the species' movements (Blakers *et al.* 1984).

Social behaviour

Apparently this species has been bred in captivity on only one occasion (Lendon 1938), and few details of its behaviour are available. Like other buttonquails, the female mates with a succession of males during the breeding season, but her territory is reportedly smaller and less vigorously defended than are those of the other buttonquail species. Her advertisement call is a moaning note, and like the barred and striped buttonquails she also performs a strange front-to-back swaying or rocking movement while holding her body horizontal (Frith 1976). According to Lendon, the female displays to the male with the same kind of booming note as occurs in the painted buttonquail, only it is not so loud.

Reproductive biology

The breeding season of this species is extended throughout the entire year, and in many interior and northern areas is certainly determined by the seasonal distribution of rains. Further south it probably has a more seasonally defined breeding season. In New South Wales, chicks have been found from spring (November) to autumn (April), and a female with an egg in her oviduct was collected in March (Morris and Kurz 1977).

The clutch-size is reportedly usually four eggs, with a range of three to five, and an incubation period of 13 or 14 days has been reported (Frith 1976). Lendon (1938) observed that three eggs were laid at a rate of one every other day, the male starting incubation with the laying of the second egg. Incubation lasted 14 days, and the chicks were brooded only by the male. The female was removed the day that the young hatched, since she began attacking both the male and the chicks. The male fed the young beak-to-beak for the first 3 days, after which the chicks began to feed for themselves. After 5 weeks the single surviving chick was as large as its father, and by 7 weeks it was as large as the female and had acquired a female-like plumage. The female began another clutch of three eggs almost immediately after it was placed back with the male.

There is no good information on breeding success, but Morris and Kurz (1977) found that immatures and juveniles formed 18 per cent of fifty birds captured by them. By comparison, young birds formed 57 per cent of seventy-three red-chested buttonquail captured by the same spotlighting method.

Evolutionary relationships

Ogilvie-Grant (1889) listed this species as last in his linear sequence, immediately following *pyrrhothorax*, which he apparently regarded as its nearest relative. I would also regard these two species as fairly close relatives.

Status and conservation outlook

This is a highly opportunistic and mobile species and probably the most abundant of the Australian buttonquails. As such, it probably needs no special attention from conservationists.

RED-CHESTED BUTTONQUAIL (Plate 9)

Turnix pyrrhothorax (Gould) 1841

Other vernacular names: chestnut-breasted quail, Everett's buttonquail (*everetti*), red-chested quail, red-breasted quail, red-chested turnix, rufous-chested bustard-quail, Sumba buttonquail (*everetti*), yellow quail, Worcester's buttonquail (*worcesteri*), turnix à ventre rouge, turnix à poitrine rousse (French); Luzonlaufhühnchen (*worcesteri*), Rotbrust-Kampfwachtel, Sumbalaufhühnchen (*everetti*), (German).

Distribution of species (Map 8)

Highly nomadic, breeding widely in northern, eastern and south-eastern Australia, from north-western Northern Territory south to the Adelaide Plains (*pyrrhothorax, sensu stricto*). No subspecies recognized in Australia, but *everetti* and *worcesteri* of Sumba Island and Luzon respectively are apparently closely related forms that Sutter (1955a) considered full allospecies in the *pyrrhothorax* group but are here tentatively regarded as subspecies of this species.

Distribution of subspecies

T. p. pyrrhothorax (Gould): Australia as indicated above.

T. (p.) worcesteri McGregor 1904: Luzon Island. Considered by Peters (1934) as possibly synonymous with *sylvatica whiteheadi*; regarded by Sutter (1955a) as a full species.

T. (p.) everetti Hartert 1898: Sumba Island (Lesser Sundas). Considered tentatively by Sutter (1955a) a full species; judged by Amadon and du Pont (1970) as likely to be conspecific with *worcesteri*, with *everetti* the older name and having priority.

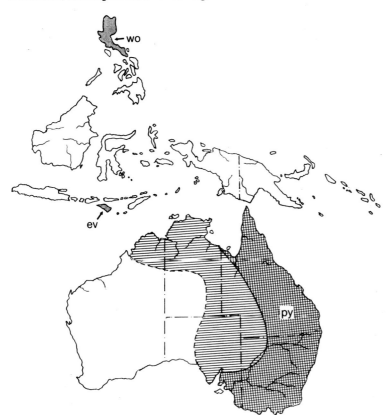

Map. 8. Distribution of the red-chested buttonquail, including its races *everetti* (ev), *pyrrhothorax* (py) and *worcesteri* (wo). These three forms are often considered to represent full species.

Measurements (of *pyrrhothorax*, mm)

Wing, males 78–79 (av. of 15, 74·8), females 64–86 (av. of 28, 78·6); tail, males 27–36 (av. of 18, 30·8), females 27–31 (av. of 18, 30·2); tarsus, males 17·2–20·9 (av. of 16, 18·5), females 17–21 (av. of 29, 18·7); exposed culmen, males 8·7–12·3 (av. of 12, 10·9), females 10·5–13·5 (av. of 21, 11·7) (Museum of Victoria and Australian Museum data). Egg, av. 22 × 19 (Frith 1976).

Weight (of *pyrrhothorax*, g)

Females 31–83 (av. of 16, 50·5), males 27–40 (av. of 12, 33·9) (Goodwin 1974, and Australian Museum data); males 27–33 (av. of 5, 33), females 44–66 (av. of 5, 50) (Morris and Kurz 1977). Egg, 4·4 (estimated).

Description (of *pyrrhothorax*)

Adult female. General colour above stone-grey, mixed with rufous and buff. Crown feathers black, with rufous margins; a whitish-buff stripe down the centre of the head; nape and nuchal region rufous grey, each feather faintly barred with darker and edged laterally with whitish buff; back, scapulars, rump, upper tail-coverts and rectrices irregularly barred and vermiculated with rufous buff and black, and margined laterally with whitish or buff; remiges and their coverts as in *varia*; rest of the wing-coverts marbled with black, buff and dull red, and somewhat widely margined with pale buff; feathers of superciliary areas, lores, sides of the face and neck white tipped with black; chin, throat, chest, sides of breast, flanks, and under tail-coverts rusty buff; the middle of the chin and throat and the remainder of the underparts whitish; some of the feathers on the sides of the chest and breast irregularly barred with black and white at the tips. Iris white to yellowish white; bill pale grey; tarsi and toes pale pink (Frith 1976).

Adult male. Resembles the female, but smaller and the rust-coloured chest not so bright. The face is also lighter, the sides of neck and breast are strongly barred with black, the throat is white, and the underparts are pale buff (Frith 1976).

Immatures. Like the adult male, but with darker breast markings (which disappear during the first month), bluish eyes, and a brownish tinge on the legs (Morris and Kurz 1977).

Measurements (of *worcesteri*, mm)

Wing, males 61·9–67·9 (av. of 8, 65·4), females 67·4–72·6 (av. of 8, 70·2); tail, males 19·8–25·2 (av. of

A. Hemipodes (Family Turnicidae)

8, 23), females 18·9–26·7 (av. of 9, 23); exposed culmen, males 8·5–9·4 (av. of 8, 8·9), females 8·7–10·7 (av. of 9, 9·6); tarsus, males 14·2–15·4 (av. of 8, 14·6), females 14·9–16·4 (av. of 9, 15·6) (data of Delaware Museum of Natural History); culmen (from base), male 10, female 11·4; depth of bill at angle of gonys, male 5, female 5·8 (Hachisuka 1932, Amadon and du Pont 1970). Egg undescribed.

Weights (of *worcesteri*)

No information.

Description (of *worcesteri*)

Adult (sexes nearly alike). General colour above black; forehead spotted with white; feathers of crown and nape tipped with pale buff and some edged with buff, producing an incomplete white line on middle of head; feathers on back and rump barred and tipped with pale buff; tertials and scapulars edged with whitish buff; feathers on sides of face mostly white with black tips; lores white; feathers on sides of neck black, each with a wide subterminal white bar; a small black spot behind ear; breast and throat rusty buff or dark clay-colour, this colour extending up each side of the white chin area, with rusty buff tips to the feathers, and bounded above by the black-tipped white feathers of the malar region; flanks, under tail-coverts, sides of abdomen and breast also rusty buff, but paler; middle of abdomen whitish; a few feathers on sides of abdomen barred with blackish brown; primaries, their coverts and secondaries all drab-grey; the four outer primaries narrowly edged with whitish and the secondaries barred with whitish on outer webs; secondary coverts blackish, mottled and edged with pale buff; rectrices blackish, edged with buff. Iris very pale yellow; bill pale bluish; legs flesh-pink, with slightly darker nails (Hachisuka 1932).

Measurements (of *everetti*, mm)

Wing, juv. male, 65, juv. female 76; tail, juv. female, 21; tarsus, juv. male 15·2, juv. female 17·5; culmen, juv. male 10·5, juv. female 12; bill depth at base, juv. male, 5·5, juv. female 6·5; middle toes and nail, juv. males, 14·8, juv. female 16·3 (Sutter 1955a). Egg undescribed.

Weights (of *everetti*, g)

Juv. male, 28, juv. female 28 (Sutter 1955a).

Description (of *everetti*)

Female (juvenile): Crown black, with wide, bright reddish brown edges; a clear bright ochre central stripe; scapulars black, the feathers with three small cinnamon brown wavy borders, and edged by bright lines, the inner part white, and the edges almost cinnamon brown; the proportions of these borders mainly white on the neck, and mostly cinnamon-brown on the scapulars, with fine black edges; greater and median wing-coverts edged with bright white and cinnamon brown, becoming cinnamon brown basally, with black tips and black-bordered cinnamon flecks; primaries brownish black, the outer three with wide and the inner ones with narrow whitish edges; lores and malar areas whitish with black flecks; superciliaries and auriculars scarcely flecked and cinnamon brown, with a small black band extending from above the ear to the neck and down the sides of the neck; sides of the neck and breast banded with black and white; the feather bases yellow-ochre, followed by a wide U-shaped black mark, a white subterminal line, and a black border; chin and throat whitish, sides mainly ochre-yellow; breast deep orange-tinted ochre-yellow; under tail-coverts reddish yellow; the rest of the undersides white. Iris whitish yellow; bill bright blue-grey; tarsi very pale flesh-coloured.

Male (juvenile): Markings of upper and under sides like the female, but the median crown stripe white, and the lengthwise markings of the back, scapulars and wing-coverts mainly white with very narrow cinnamon-brown edges; upper breast just as intensely orange-toned, the crop area feathers marked with blackish borders; rectrices covered by the upper tail-coverts; the former with whitish edge-markings and the tips progressively pointed (these lacking in the female specimen). Iris bright yellow; bill dark grey, bluish grey laterally and ventrally; tarsi pale yellowish flesh-coloured.

Identification (see Figure 25)

In the hand. Distinguished from the other buttonquails by the combination of a buffy to ochre- or rufous-coloured breast, and distinctly greyish dorsal tones mixed about equally with rufous, black and whitish markings. The bill is unusually stout and is very similar in proportion to that of *velox*, which species is less rufous-buff or ochre-toned above and whiter (less buffy) below. Similar to *Coturnix*, but lacking a hind toe. In common with most other Australian buttonquails it has a relatively short tarsus (about as long as the middle toe and claw).

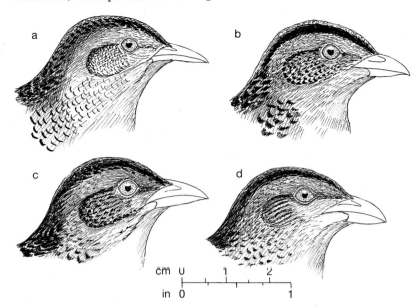

Fig. 25. Comparison of head and bill characteristics of (a) *T. sylvatica whiteheadi*. (b) *T. p. pyrrhothorax*. (c) *T. p. worcesteri* and (d) *T. p. everetti*.

In the field. In Australia, the unspotted ochre- to chestnut-coloured breast, and otherwise predominantly ochre to rufous-buff upperparts with few chestnut tones, helps to identify this relatively unpatterned species. It lacks the more cinnamon-dominated dorsal tones of the little buttonquail and the distinctly barred or spotted blackish and rufous back colour of the red-backed buttonquail. Its pale whitish (rather than brown or reddish) eyes separate it from various *Coturnix* species. On Luzon Island, *worcesteri* occurs sympatrically with the very similar *sylvatica* and might be easily confused with it, but has a distinctly heavier bill. On Sumba Island *everetti* occurs sympatrically with *maculosa*, but similarly has a considerably heavier bill.

General biology and ecology

This is a grassland-adapted species, which lives not only in native grassland areas having annual rainfalls of at least 250 mm, but also in croplands, stubble, and native pastures, although over-grazing sometimes makes this last-named habitat unsuitable. Its ecological needs overlap with those of the little buttonquail, but it is not so arid-adapted, favouring moister and more densely vegetated sites than that species. Like it, however, it is distinctly nomadic and irruptive, at which times it moves into coastal and southern areas. At times when rainfall occurs in dry areas and grassland vegetation is favourable, the birds may arrive in flocks and begin breeding, only to disperse again when the land no longer supports them (Frith 1969, 1976).

In southern Australia the species is somewhat more abundant during summer than winter, although differences are not great (Blakers *et al.* 1984). In the Mudgee district of New South Wales they are uncommon but probably year-round residents in the crop and pasture lands, but they increase rapidly when favourable conditions occur during consecutive years (Morris 1971, Morris and Kurz 1977).

Social behaviour

Relatively little has been written on the behaviour of this species but, as in the other buttonquails, females advertise their territories and solicit males with loud calling (Frith 1976). Fulgenhauer (1980) has provided some information on and illustrated several aspects of the general and social behaviour of this species, based on captive observations of two males and one female (Figure 26). He found that feeding activities tended to be evenly distributed throughout the day, but that periods of resting, preening and walking were more variable; resting usually occurred around noon, walking was most common during mid-morning and late afternoon, and self-preening was most frequent during early morning hours. Dustbathing was a very common behaviour, and typically occurred during several periods throughout the day.

Aggressive behaviour was observed between the two males, but not between sexes. It consisted of a horizontal body stance (Figure 26g), with an exaggerated back-and-forth rocking of the body and a vertical tail-flicking. This was followed by a rush towards the other bird and its retreat. A more subdued version of this backwards and forwards movement of the body was sometimes observed during normal walking. Resting was done in a relaxed sitting posture (Figure 26d), while during alarm and severe alarm the head was increasing lowered to the ground and the body flattened out (Figure 26e, f).

Allopreening, or 'courtship pecking', began between the female and one male, on the fourth day of contact between them. On that day they were dustbathing in contact, and also rested briefly in contact. At this stage the male hesitantly approached and nibbled at the female's neck (Figure 26h). By the seventh day the pair were roosting in the same scrape. Mutual bowing between the pair (Figure 26j) was observed on one occasion before the birds settled in to roost together, but was not observed at any other time. No other sexual behaviour was observed, but it was suggested by Fulgenhauer that dustbathing might serve a social function in pairing.

Reproductive biology

Although this species has been bred in captivity (Fulgenhauer 1980), almost no details of this aspect of its biology are available. It is known to breed from October to March, and it probably also breeds at other times if conditions are favourable. The nest, like those of other buttonquails, is a shallow depression at the base of a grass tuft, with an overhead dome and an entrance to one side. The female reportedly builds the nest, but before she lays her clutch it is lined by the male. There are usually four eggs, which are incubated by the male, but the incubation period is apparently still unreported. The young are fed insects at first, and later on they consume seeds (Frith 1976).

There is no direct information on breeding success, but Morris & Kurz (1977) trapped a total of seventy-three birds, of which 57 per cent were juveniles or immatures, indicating a potentially

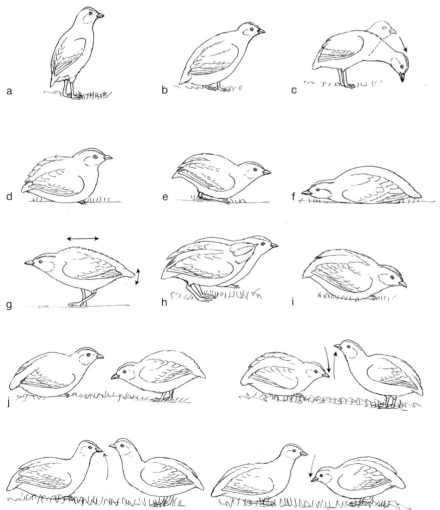

Fig. 26. Behaviour of the red-chested and barred buttonquails: (a, b) standing postures; (c) female barred buttonquail 'purring' display; (d) social roosting; (e) crouching; (f) extreme crouching; (g) threat display; (h) courtship allopreening; (i) ground-pecking; and (j) mutual bowing sequence. All but (c) are of red-chested buttonquail, after sketches by Fulgenhauer (1980); (c) after sketch by Trollope (1970).

quite high recruitment rate. Of the thirty-one adults, 65 per cent were males, suggesting that the adult sex ratio may possibly be skewed in favour of males, as would be desirable in a polyandrous breeding system.

Evolutionary relationships

This species was placed taxonomically by Ogilvie-Grant (1889) in his final group, preceded by *castanota* and followed by *velox*, to both of which it bears considerable resemblance, but it is closer in ecology and appearance to *velox*. Sutter (1955a) also suggested that these relatively thick-billed forms (including *everetti* and *worcesteri*) might be related to one another. These two grassland adapted species are substantially sympatric in Australia, and do not offer any obviously simple speciation models based on their current distributions.

The related question is that of the relationships of *everetti* and *worcesteri* to *pyrrhothorax*, for which again no easy answer exists. It is possible that these two forms should be separated and considered as conspecific (as *worcesteri*) and that other similar populations are yet to be discovered on other Pacific islands (Amadon and du Pont 1970). However, given the small number of available specimens of these two insular forms and the lack of knowledge about them, I prefer to regard them tentatively as all part of the same species.

Status and conservation outlook

Little can be said of the population status of the two insular forms (the Sumba Island form being known from only three specimens, the more recent ones collected in 1949), but at least nominate *pyrrhothorax* is probably still fairly widespread in Australia. It is often considered rare there, but its ability to move about as ecological conditions dictate tends to favour its chances for long-term survival.

PAINTED BUTTONQUAIL (Plate 11)

Turnix varia (Latham) 1801

Other vernacular names: butterfly quail, dotterel quail, New Holland partridge, painted quail, scrub quail, speckled quail, varied buttonquail, varied turnix, variegated bustard-quail; turnix bariolé, turnix varié (French); Buntkampfwachtel, Buntlaufhühnchen (German).

Distribution of species (Map 9)

Coastal and subcoastal areas of eastern, south-eastern and south-western Australia; also Tasmania

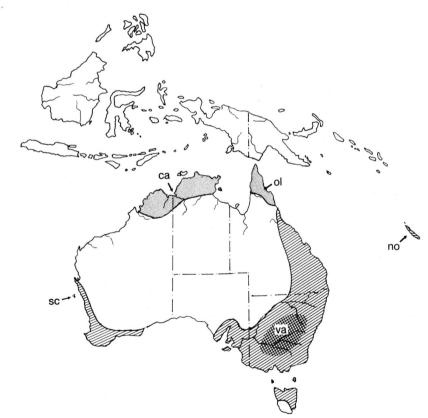

Map. 9. Distribution of the painted buttonquail (hatched), including its races *novaecaledoniae* (no), *scintillans* (sc) and *varia* (va), and the chestnut-backed buttonquail (stippled), including its races *castanota* (ca) and *olivei* (ol), which are sometimes regarded as a separate species. Probable areas of denser populations of *varia* are shown by cross-hatching.

and New Caledonia. Locally migratory or nomadic.

Distribution of subspecies

T. v. scintillans (Gould) 1845; Abrolhos Islands (off W. Australia).

T. v. varia (Latham): Australian mainland, Tasmania and Kangaroo Island (off S. Australia).

T. v. novaecaledoniae Ogilvie-Grant 1893: New Caledonia.

Measurements (of *varia*, mm)

Wing, males 66–133·3 (av. of 19, 98·3), females 103–158·7 (av. of 14, 111·4); tail, males 27–45 (av. of 18, 40·1), females 36–52 (av. of 14, 45·3); tarsus, males 15·7–21·7 (av. of 19, 20·6), females 19–24·5 (av. of 14, 22·4); culmen, males 10·5–13·7 (av. of 19, 12·6), females 12·7–14·7 (av. of 14, 13·6) (Museum of Victoria data). Egg, av. 22 × 18 (Frith 1976).

Weights (g)

Males 60–124·4 (av. of 10, 90·2), females 80–130 (av. of 14, 102·1) (Goodwin 1974 and Australian Museum and Museum of Victoria data). Egg 3·7 (estimated).

Description

Adult female. General colour above black, barred with rufous, shading into chestnut on the nuchal region; centre of the crown dark grey, the sides black, edged with grey or rufous; lores, superciliary stripes, sides of the face and throat white spotted with black; chin and centre of throat white. Nape feathers grey, with terminal black and white spots; nuchal region bright chestnut edged with grey, and dotted with black and white. Back, scapulars, rump, upper tail-coverts and rectrices grey at the base and margined or barred with black and chestnut, rufous at the tips, and with a submarginal white streak on either side; remiges and primary coverts blackish grey, the former margined with white on the outer web; secondary wing-coverts buff-grey, dotted indistinctly with black; rest of the wing-coverts grey at the base and rufous at the tip, each with two or three irregular white ocelli edged with black; chest grey, each feather with a pale buff or whitish shaft-streak widening out towards the extremity to form a black-edged spot; breast buff mixed with grey, some of the feathers on the sides chestnut, with black-edged whitish spots at the extremity; abdomen and rest of underparts pale buff. Iris red to reddish orange, bill grey-brown to bluish horn, tarsi and toes pale to deep yellow.

Adult male. Resembles the female, but there is no well-defined rufous nuchal collar, and on the chest the whitish spot-like elements predominate over the grey. Additionally there are sometimes poorly defined black splotches or bars on the back, and less evident white feather-edging dorsally.

Immature. Like the adult male, but with brown eyes and a duller plumage (Frith 1976).

Identification

In the hand. Separated from the other buttonquails by the combination of relatively short tarsi (no longer than the middle toe and claw), an olive-grey breast faintly but distinctly spotted with white, and upperparts that are predominantly brownish black, with some dull chestnut barring and buffy edging. The bill is relatively weaker than in *castanota*, the head is more distinctly spotted with black and white, and the iris is orange-red (as compared with pale yellow to whitish in *castanota* and other *Turnix* species).

In the field. This species is relatively weak-billed as compared with *castanota*, and has a greyish breast, chestnut tones that are mostly limited to the 'shoulders' and lacking from the rump area, and the upperparts patterned about equally with black and pale chestnut-brown, edged with buffy. The female's advertising call is a deep and low-pitched *oom-oom-oom*, often uttered at night.

General biology and ecology

This species is generally found in open scrublands, dry forests having a coarse grass understorey, and especially rocky hillsides having a grassy and open forest cover. Near the coast it favours dry heathlands and wooded areas, but further inland is likely to be found on timbered ridges and forests where the ground cover is sparse (Frith 1969, 1976). The birds are apparently normally quite sedentary, but at times or in certain areas they appear to be distinctly mobile, as is perhaps the case in the southeast, where the birds have been reported at a somewhat higher rate during winter than during summer, although the differences are not very substantial (Blakers *et al.* 1984).

Like the other buttonquails, this species feeds by scratching in the litter and uncovering foods from the leaf mulch, including seeds, insects and plant materials (Frith 1976).

Social behaviour (Figure 27)

This species has apparently been bred in captivity only by Seth-Smith (1905), whose descriptions are practically the only ones available for the species. He noted that the female displayed to the male by running backwards and forwards in front of him, or around him in a circle, with her tail erect and her crop puffed out like a pouter pigeon. Having run around him once or twice she would stand facing him at a distance of half a metre or less, and begin 'booming' or 'cooing' to him, at the same time stamping and scratching with her feet. Meantime the male responded with a faint clucking note. Not only was the female's crop puffed out, but also the back feathers were erected, so that the bird resembled 'a miniature balloon'.

The female's advertisement call is said to be similar to that of a bronze-winged pigeon (*Phaps chalcoptera*), and she reportedly defends her breeding territory from other females (Frith 1976).

Reproductive biology

This species breeds during the general period September to March, although nests have been found during other months as well, especially in the northern parts of the range. On the other hand, breeding in Tasmania mainly occurs between October and December. The nests tend to be fairly substantial, and are typically located among bushes, grass tussocks or fallen timbers. The scrapes are thickly lined with grasses and are often partly covered over. Most of the work on the nest is done by the male, with some help from the female, according to Frith (1976), although Seth-Smith (1905) observed that the female built the nest and if the male approached she would call almost constantly. Eggs were laid on a daily basis (five eggs in 6 days), and apparently the male did not begin incubation until the laying of the fifth egg. Incubation was completed in about 13 days, and was entirely by the male. Similarly the female took no notice of the

Fig. 27. Social behaviour of the painted buttonquail: (a) female advertising display; (b) female courtship-feeding male; and (c) male approaching nest. (a) and (b) after sketches in Seth-Smith (1905); (c) after photo in Frith (1976).

young, except to eat their food. For the first week to 10 days the young were fed bill-to-bill by the male, but later they began to feed for themselves. They were able to fly at 10 days, at 16 days they were fully feathered and were removed from their male parent, and the two parents were again reunited. Only 3 days later the female had built a nest and laid the first of a clutch of four eggs. The male, however, refused to incubate these, and the female then made a third nest and laid the first of a clutch of three eggs, which were also not incubated. A fourth nest was begun less than 2 months after the first clutch had been begun, and only about 3 weeks after the first brood had been separated from the male parent. This clutch of four eggs was incubated by the male. According to Frith (1969, 1976), the usual clutch is four eggs but ranges from three to five, and the incubation period is 14–15 days.

Evolutionary relationships

Ogilvie-Grant (1889) did not speculate on this species' closest relative, but he included it in linear sequence between *melanogaster* and *castanota*. I suspect that its nearest relative is *castanota*, and that these two currently allopatric species (separated by Australia's arid interior) are quite closely related.

Status and conservation outlook

This species has become quite rare in Tasmania since 1950, but otherwise apparently shows no marked diminution of range or numbers (Blakers *et al.* 1984).

CHESTNUT-BACKED BUTTONQUAIL
(Plate 10)

Turnix castanota (Gould) 1839 (1840)

Other vernacular names: buff-backed quail (*olivei*), buff-breasted quail (*olivei*), chestnut-backed bustard-quail, Olive's buttonquail (*olivei*), Robinson's buttonquail (*olivei*); turnix castanote (French); Robinsonlaufhühnchen (*olivei*), Rotrücken-Laufhühnchen (German).

Distribution of species (Map 9)

Disjunctively distributed in northern Australia, from the Kimberleys east to Macarthur River, and from Coen east to Cooktown, the two populations sometimes considered specifically distinct. Probably locally migratory or nomadic.

Distribution of subspecies

T. c. castanota (Gould): north-western Australia (Northern Territory; also Melville Island and Groote Eylandt). Includes *magnifica* Mathews 1912.

T. c. olivei Robinson 1900: northern Queensland (Coen to Cooktown), Australia. Larger and paler than the previous form, and often separated as a full species (e.g. Macdonald 1971).

Measurements (mm)

Of *olivei*: wing, male 93, females 99–106 (av. of 3, 102·7); tail, male 39, females 36–46 (av. of 3, 41); tarsus, male 93, females 26–30 (av. of 3, 28; culmen, male 15, females 17–19 (av. of 3, 17·8). Of *castanota*: wing, males 81·5–120 (av. of 6, 91·1), females 87–94 (av. of 4, 90·8); tail, males 28–32 (av. of 5, 29·6), females 31–39 (av. of 4, 35); tarsus, males 19–22·5 (av. of 5, 20·4), females 20–25 (av. of 4, 22·8); exposed culmen, males 11–16 (av. of 6, 14·3), females 15–17 (av. of 4, 15·9) (Museum of Victoria and Australian Museum data). Egg, av. 27 × 23 (Frith 1976).

Weight (g)

No information. Egg 7·4 (estimated).

Description (of *castanota*)

Adult female. General colour vinaceous red; a dark grey band down the middle of the head; sides of the head like the upperparts of the body. Feathers of lores, superciliary areas and sides of the face white tipped with black; nape spotted with white; most of the feathers of the upper back and scapulars edged laterally with black and white bands and some of them blotched with blackish towards the extremity; lower back, rump, upper tail-coverts and rectrices vinaceous red (in *olivei* the upperparts, including the rump and tail-coverts are predominantly reddish orange, with few black markings). Remiges and primary-coverts brownish, the outer primaries margined with whitish; rest of the wing-coverts light red, ornamented with black-edged white ocelli; chin and throat white; middle of chest and breast greyish buff (olive-buff in *olivei*), some of the feathers with whitish shaft-stripes; sides light red, with irregularly shaped black-edged white ocelli; rest of the underparts whitish buff (pale grey in *olivei*). Iris, bill, tarsi and toes all pale yellow (Frith 1976) (bill colour reported as dull blue by Pizzey 1980).

Adult male. Resembles the female, but somewhat smaller and duller.

Immature. Resembles the adult male, but duller and with darker eyes (Frith 1976).

Identification

In the hand. Separated from the other buttonquails by the combination of a short tarsus (up to the length of the middle toe and nail), a grey to olive-grey and white-spotted breast, a rather heavy bill, a mostly grey appearing head, and upperparts from the nape to the tail that are predominantly chestnut-toned, with scattered black streaking or barring and narrow whitish or buffy edging on the anterior back. Separated from the painted buttonquail by that species' more heavily black-and-white spotted face and crown, its weaker bill, and its less chestnut and more heavily black-barred upperparts, especially on the rump.

In the field. Separated geographically from the very similar but more southerly distributed previous species *varia* (the two species' ranges nearly meet between Cooktown and Cairns), *castanota* has a similar white-spotted greyish breast (but which is unspotted and more olive-toned in *olivei*) but more distinctly chestnut-toned upperparts, especially on the sides of the upper chest ('shoulder' area), rump and tail, which are distinctly bright reddish-orange, with little black present. In contrast, *varia* has a purplish grey rump, which is barred with black. The female's call, much like those of other *Turnix* species, is a repeated low, resonant *oom* sound.

General biology and ecology

The nominate race of this species is largely associated with open, dry savannah woodland having a ground cover of coarse grasses, and on sandy or rocky ridges. On the other hand, *olivei* is more associated with heath-like vegetation in wet, open areas (Frith 1976). The habitat of *castanota* has further been described as being eucalypt woodland, including that having a grassy understorey, and in areas with sandstone or lateritic substrates, whereas that of *olivei* consists of grassy clearings in rainforests, woodlands and swamps (Blakers *et al.* 1984).

The birds are said to live in groups of up to twenty, which is an unusually large number for buttonquails, and to feed on seeds, insects and herbs (Blakers *et al.* 1984, Frith 1976).

Reproductive biology

Few data are available, but nesting reportedly occurs during the wet season of December to March. The nest is a scrape at the base of a shrub or grass clump, and is sometimes domed or at least partially domed, in the common manner of buttonquails. There are usually four eggs, and the period of incubation (by the male) has been reported as 14–15 days. The species has apparently not been bred in captivity, and thus few details of breeding are available. For example, the downy young have yet to be described (Frith 1976).

Evolutionary relationships

There can be little doubt that *castanota* and *varia* are very close relatives. Macdonald (1971) reviewed the status of *olivei* and concluded that morphologically it is actually 'rather closer in affinity to *varia* than to *castanota*'. He also argued that these three forms might be regarded as semi-species of a larger super-species, and that the isolated New Caledonian form *novaecaledoniae* that is usually considered a subspecies of *varia* should be listed as a separate species or as another semi-species of this general assemblage.

Status and conservation outlook

In the course of data-gathering for the Australian atlas of birds, Blakers *et al.* (1984) noted that *olivei* was not observed at all, although there were records from as recently as the 1970s. However, they indicated that there was no reason for believing that this elusive bird no longer existed, and indeed rainforest clearing might be expected to favour this population of the species.

BLACK-BREASTED BUTTONQUAIL
(Plate 13)

Turnix melanogaster (Gould 1837)

Other vernacular names: black-breasted bustard quail, black-breasted quail, black-breasted turnix, black-fronted quail; turnix à poitrine noire (French); Schwarzbrust-Laufhühnchen (German).

Distribution of species (Map 10)

Eastern coastal Queensland, currently from about Hervey Bay (perhaps locally to Broad Sound) to extreme north-eastern New South Wales, Australia, and including Fraser Island. Historically extended further north, but the most northerly certain records are from Rockhampton and Duarina, and those from the Cooktown, Atherton Tableland area were considered equivocal by Bennett (1985). Rare, local and probably sedentary. No subspecies recognized.

Measurements (mm)

Wing, males 104–117 (av. of 9, 111·1), females 112–115 (av. of 4, 113); tail, males 36–49 (av. of 6,

A. Hemipodes (Family Turnicidae)

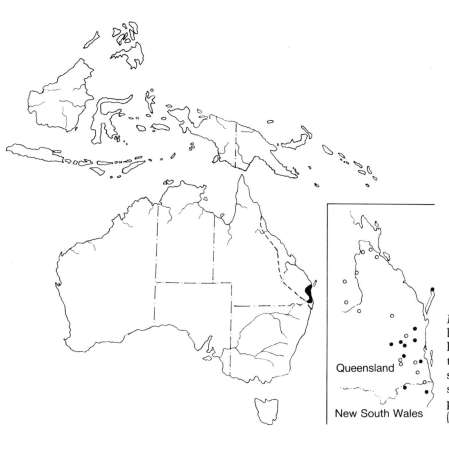

Map. 10. Distribution of the black-breasted buttonquail, including its known total historic range (dashed line), and the locations of recent sightings (solid). The inset shows the species' recent known range in southern Queensland, with the locations or pre-1970 records (○) and those since 1970 (●) (after Bennett 1985).

43·3), females 43–50 (av. of 5, 47·6); tarsus, males 23·2–25·5 (av. of 4, 24·5),two females 23·5 and 25; culmen (to skull), males 12–18·5 (av. of 7, 16·4), females 14·5–21·5 (av. of 4, 18·5) (mostly Australian Museum and Museum of Victoria data). Egg, av. 27 × 24 (Frith 1976).

Weights (g)

Males 65 and 87, females 101 and 119 (mostly Australian Museum and Museum of Victoria data). Egg 8·1 (estimated).

Description

Adult female. General colour above umber-brown. Forehead, sides of the face, chin and throat black; crown a mixture of dark chestnut and black; superciliary stripe and nape spotted black and white; most of the feathers of the upper back and the scapulars and some of the rump-feathers and upper tail-coverts with a broad chestnut patch at the tip that is barred with black, and edged laterally with white or whitish buff; primaries, inner secondaries and primary-coverts blackish brown, margined with buff on the outer web; outer secondaries and secondary-coverts paler, finely mottled with buff and black on the outer web; rest of the wing-coverts chestnut, edged with brown, and each with two or three irregularly shaped white ocelli edged with black; chest and breast black, most of the feathers widely tipped with white, producing a spotted effect, and some of those on the sides of the breast with a patch of bright chestnut near the tip; abdomen and rest of underparts dark grey, marbled with buff and black; rectrices umber-brown, with wavy transverse black bars and lateral buff spots. Iris white (tinged with yellow); bill slate-grey; tarsi and toes orange (Frith 1976).

Adult male. Differs from the female in having the crown umber-brown like the back, the lores and sides of the head white tipped with black, the chin and rest of the throat pure white, the chest and breast whitish buff, irregularly barred and marked with black, and the rest of the underparts whiter.

Immature. Resembles the adult male, but with dark eyes and grey tarsi (Frith 1976). It is generally dark brown above, marbled with rufous and heavily marked with black, and greyish below, the feathers tipped with pale-cream and crossed with black V-marks, producing a patterned appearance (Pizzey 1980).

Identification

In the hand. The extensive black and white spotting on the breast (heavier black on females, which also have blacker heads) easily identifies this species.

In the field. This is the only Australian buttonquail that is mostly to entirely black-and-white spotted on the breast, and with dark grey underparts. The female's advertisement call is a deep and low-pitched booming *oo-oom*. A rapid thumping sound and a loud crow are also reportedly produced, but the latter is probably the same as the booming call.

General biology and ecology

Most of the recent and historic reports of this species are associated with low, closed forests (with or without emergent *Araucaria* (hoop pine trees) at the drier end of the rainforest vegetational continuum. These habitats include semi-evergreen vine thicket, low microphyll vine forest and araucarian microphyll forest. These types are typically formed in areas of moderate annual rainfall (770–1200 mm), and a spring dry season that produces considerable leaf fall. Because leaf decomposition is fairly slow, a deep litter layer accumulates, but the forest floor is relatively clear of vegetation, perhaps because of the very high degree of foliage cover of the canopy (Bennett 1985).

The birds can often be located by the characteristic scrapings that they make in the litter during foraging. These scrapes are circular areas about 15–20 cm in diameter that generally do not penetrate far into the topsoil. Scrapes tend to occur in groups that are perhaps 50–100 m in diameter and about 30 cm apart. Similar scrapes are made by painted buttonquail, but these are typically found in open eucalypt forest rather than rainforest (Bennett 1985). The scrapes are made by the bird pivoting on one foot as it scratches with the other, thus rotating in a half-circle. Then it shifts feet and rotates in the other direction, forming a saucer-shaped depression. The birds feed on seeds and insects (Frith 1976).

The birds are generally found singly or in pairs, but records of as many as eight have been reported. They have at times been observed feeding on grain in company with domestic chickens or wild quails.

Social behaviour

Although the species has been maintained in captivity, and according to K. A. Muller (in Frith 1976) it has bred under such conditions, there do not appear to be any formal reports of its social behaviour and reproduction either in captivity or in the wild. Reportedly the female advertises its territory with a repeated booming call, and mates with any males it can entice into its area, while at the same time defending its territory against other females. The species was also bred during the early 1980s in the Denver and San Diego zoos, according to the *International Zoo Yearbook*, but no details of these breedings are available.

Judging from a photograph (in Frith 1976), the paired birds roost by clumping side by side in the usual manner of buttonquails.

Reproductive biology

Only a few nests have been found in the wild. These contained three or four eggs, and were found in December and probably February or March. One nest was a hole scraped under a low bush or tussock of coarse grass, and lined with blades of grass (Bennett 1985).

Breeding has apparently occurred under captive conditions in Australia. Under such conditions breeding occurred between October and March, with both sexes helping to build the nest. The male alone incubated the eggs, for a total of 16 days, and also reared the young, although females have been observed as part of a family flock. The young were fed by the male for the first few days following hatching, but grew rapidly and became sexually mature at 5 or 6 months (Frith 1976).

Evolutionary relationships

Ogilvie-Grant (1889) included this species in a group of species that consisted mostly of Australian forms, but also included the Madagascan buttonquail and listed this latter species adjacent to *melanogaster*. In all of these the tarsus is no longer than the middle toe and claw. However, I have tentatively associated the black-breasted buttonquail with the barred buttonquail, although the latter has a relatively longer tarsus. The unusually dark colour of the black-breasted buttonquail may be a reflection of its shaded environment and as such not a very good indicator of relationships. However, I believe that zoogeographically the affinities of *melanogaster* are much more likely to be with *suscitator* than with *nigricollis*.

Status and conservation outlook

Of all the species of buttonquails, this is perhaps the most threatened, to judge by available information. According to Bennett (1985) there were only ten reports of the species between 1970 and the mid-1980s. Most of these reports were from semi-evergreen vine thickets and araucarian notophyll vine forest habitats in the Upper Brisbane and South Burnett valleys, within an area bounded by the Conondale Range, Bunya Mountains, and Nanango. However, the vegetation of much of this region has been converted to plantations of araucaria and pines, from which the birds have been only rarely reported.

Reports from Fraser Island since 1968, including the finding of a possible nest, suggest a possible recent expansion of range. Furthermore, some recent observations of the birds in areas of introduced lantana and *Araucaria* plantations suggest that the species may be adapting to human-modified environments (Blakers *et al.* 1984).

BARRED BUTTONQUAIL (Plate 12)

Turnix suscitator (Gmelin) 1789

Other vernacular names: Ceylon bustard-quail *leggei*, common bustard-quail, dusky buttonquail (*nigrescens*), Indian bustard-quail (*taigoor*), island bustard-quail (*suscitator*), lower Bengal bustard-quail (*bengalensis*), Philippine bustard-quail (*fasciata*), Celebean bustard-quail (*rufilatus*), northern bustard-quail (*plumipes*), Sumbawa bustard-quail (*powelli*), Thai grey-legged hemipode (*interrumpens*); hemipode batailleur, turnix combattant (French); Bindenlaufhühnchen, Schwarzbrust-Kampfwachtel (German).

Distribution of species (Map 3)

Eastern Pakistan, peninsular India and Sri Lanka (Ceylon) east through southern Asia to the Rio Kyus, Taiwan, Philippines and Lesser Sundas. Sedentary or locally nomadic.

Distribution of subspecies

T. s. taigoor (Sykes) 1832: Eastern Pakistan (Sind and Punjab) eastwards through India, intergrading with *plumbipes* in Bihar and *bengalensis* in West Bengal.

T. s. leggei Stuart Baker 1920: Sri Lanka.

T. s. plumbipes (Hodgson) 1837: Nepal (to about 300 m), Sikkim Bangladesh and Assam east to northern Burma.

T. s. bengalensis Blyth, 1852: lower West Bengal (India).

T. s. blakistoni (Swinhoe) 1965: Shan States of Burma, and from the Tonkin area of northern Vietnam to southern China and Hainan Island.

T. s. interrumpens Robinson and Stuart Baker 1928: peninsular Burma, southern Thailand. Questionably distinct from *blakistoni* (Deignan 1945).

T. s. pallescens Robinson and Stuart Baker, 1928: Pego region of south-central Burma.

T. s. thai Deignan 1946: central Thailand.

T. s. atrogularis (Eyton) 1839: Malayan peninsula and northern Sumatra.

T. s. machetes Deignan 1946: central Sumatra.

T. s. suscitator (Gmelin): south-eastern Sumatra; also Banka Island, Java, and Bali.

T. s. kuiperi Chasen 1937: Belitung (Billiton) Island, off Sumatra.

T. s. baweanus Hoogerwerf 1962: Bawean Island (north of Java). Possibly synonymous with *suscitator* (Mees 1986).

T. s. powelli (Guillemard 1885: Lombok, Sumbawa, Sangeang, Flores, Lomblen, and Alor islands (Lesser Sundas).

T. s. rufilata Wallace 1865: Sulawesi.

T. s. nigrescens Tweedale 1877 (1878): Negros and Cebu islands (central Philippines).

T. s. fasciata (Temminck) 1815: Palawan, Calamaines, Mindoro, Luzon, Panay, Sibuyan, and Masbate islands (Philippines).

T. s. rostrata Swinhoe 1965: Taiwan, up to about 1200 m.

T. s. okinavensis Phillips 1947: Ryu Kyu islands. Also reported from southern Kyushu and on Makenoshima Island (south of Kyushu), Japan.

Measurements (of *taigoor*, mm)

Wing, males 72–85, females 77–90; tail, males c.35–37, females 33–41; tarsus, males c.22–23, females 22–25 (Ali and Ripley 1983). Egg, of *leggi*, av. 23 × 19; of *nigrescens*, av. 28·8 × 21·2 (Rabor 1977).

Weights (g)

Males (*leggi*), 35–52 (av. of 8, 46), females 47–68 (av. of 7, 60·7 (Ali and Ripley, 1983) Egg, *leggi* 4·6, *nigrescens* 7·1 (both estimated).

Description (of *taigoor*)

Adult female. General colour above rufous or greyish brown or any intermediate colour; often a more or less distinct white stripe down the middle of the head and sometimes with a rufous collar; feathers on the forehead, lores, sides of the face, throat and neck white, each with a subterminal black spot, the lores sometimes wholly black; scapulars, back, rump, upper tail-coverts and rectrices barred and vermiculated with black; all the scapulars and some of the other feathers with the lateral part of the margin edged with whitish buff; remiges and primary coverts blackish brown often narrowly margined on the outer web with buff or white, the outer web of the outer secondaries being also toothed with the same colour; rest of the wing-coverts rufous or brown on the inner web and buff on the outer, and barred with black. Middle of the throat and chest deep black; sides of the chest and the breast buff barred with black; abdomen and rest of the underparts rusty buff to pale buff, paler in the centre, the abdomen sometimes also barred. Iris pale straw-white; tarsi and toes lavender-grey or slaty to greenish; bill plumbeous to yellowish, slightly dusky towards the tip of the culmen. A duller non-breeding plumage also exists in some races (*rostrata*, *blakistoni*) of this species (Sutter 1955a).

Adult male. Differs from the female in having the chin and the centre of the throat white and the middle of the chest barred like the sides, and in the rufous collar being reduced or absent. Iris whitish straw (averaging more yellowish than in females); bill yellowish, becoming dusky toward tip; tarsi and toes bluish flesh-colour to greenish yellow.

Juvenile. Differs from the adult in having V-shaped or rounded black spots on the chest and breast instead of bars; the wing-coverts have more buff, and the outer webs of the remiges are more widely and irregularly edged or toothed with buff.

Identification

In the hand. The many subspecies of this widespread species are quite variable in appearance, but may generally be characterized by the combination of the tarsus being longer than the middle toe and claw, and with distinctive black barring present on the breast, especially medially, the blackish colour extending to the throat in females. The legs are bluish to greenish, and the iris is whitish in adults. Easily distinguished from true (galliform) quails (*Perdicula* and *Coturnix*) by the lack of a hind toe.

In the field. At least over most of its range the heavy black barring on the sides and front of the breast (becoming all black on the upper breast and throat of females), together with the distinctive bluish or greenish (rather than yellow or flesh-coloured) leg colour, should provide for identification. Some of the comparably sized bush-quails (*Perdicula*) of India have similarly barred breasts and flanks, but these lack black throats and have yellow legs, and the more widespread and otherwise rather similar *Coturnix* species also have yellow legs and streaked or spotted rather than barred chests.

General biology and ecology

Like other buttonquails, this species is adapted to a rather wide variety of habitats, the common characteristics of which probably include a dry, often sandy, substrate, with a generally bare surface having an overgrowth of scattered herbaceous or woody vegetation, under which the birds can readily walk or run in an unimpeded fashion, but which also provides visual protection from above. In the Indian subcontinent the birds extend from the plains up to about 2450 m, avoiding both deep evergreen forests and very dry areas of open wasteland without nearby cover. It regularly occurs at the edges of evergreen forests, such as where these border on cultivation or on grasslands, and often breeds in bamboo jungle, scrub, secondary growth, and especially deserted cotton-fields. Perhaps above all it favours grassy plains in which there are interspersed stretches of forest, bamboo jungle, or practically bare tracts. Similarly, in Sri Lanka it is numerous where the ground is sandy and covered with low bushes, such as in old clearings overgrown with grass and shrubs, and on open bushy lands bordering tanks. However, it is rarely found in damp soil, but instead favours dry, sandy soil. These include areas of open scrub, long grass dotted with bushes, the edges of jungle or cinnamon plantations, and similar habitats where access to good cover is combined with grass and rank vegetation (Baker 1930, 1935). Somewhat similar habitats have been reported from the Philippine Islands (Rapor 1977). Wherever the species occurs it is usually not far from water, and the birds regularly visit watering areas several times a day, according to Baker.

There is some doubt as to whether the birds are migratory in any part of their range, but it is probable that in the drier areas of the Indian plains they occur and breed only after the summer rains, and leave again during the winter drought. However, they are found throughout the year in the Himalayan foothills at elevations of more than 2200 m (Baker 1930).

Like the other buttonquails, these birds usually occur singly or in very small groups, and feed on the usual array of grass and weed seeds, insects (Rabor 1977), and probably also some green vegetation.

Social behaviour

The social behaviour of this species is best known from captive observations; it has been successfully bred in captivity on several occasions (Dharmakumarsinhji 1945, Winkel 1969, Trollope 1967, 1970). Trollope raised four broods, two each involving the same male and female in different aviaries, during two breeding seasons. He found that not only were pair-bonds formed in this situation, but the hen also fed the chicks. At night the entire family would roost together in a clumped arrangement, with each pair member brooding some of the young. An interest by the female in caring for the young was also reported in wild birds by Dharmakumarsinhji (1945). Trollope further observed that when two males were placed together in an aviary without a female present, they immediately began clumping together and allopreening. When a female was released the next day she followed each male alternately, with a combination of normal walking, a rapid walk, and an actual run, sometimes pulling the male's tail and 'pulling up the cock short'. When he pulled free the chase would be resumed. That same evening the female roosted in a clumped position with one of the two males, the other roosting on the side of the aviary. Fighting by the males the following day resulted in removal of one of the males for its safety.

Probably under normal circumstances polyandry is the usual mating system in this species, and the receptive female's advertisement calling no doubt serves to attract males to her. It also strongly attracts females, who are quickly attracted to caged females and attempt to fight with them. This would certainly suggest that the call serves as a territorial challenge to other females as well as a heterosexual attraction signal (Baker 1930).

Often during apparent display the females will perform a front-to-back body-swaying movement, moving thus about three or four times, then again proceeding forwards. This silent behaviour may serve as an actual courtship display, or perhaps simply is a kind of behavioural camouflage, such as perhaps simulates leaf movements. During calling the female typically suddenly stops her walking and 'freezes', then stretches her neck and slowly lowers her head until her bill nearly touches the ground (see Figure 26c). As her head is thus lowered, the distinctive low-pitched booming or purring call is uttered (Trollope 1970). Baker (1930) described the call as resembling a cross between a purr and a coo, and as not unlike the deep guttural purr, or grunt, of a tiger. He further stated that a calling bird gradually lowered her breast to the ground, with outstretched and gently quivering wings, and with each boom blew herself out 'until she looked like a little feather balloon'. He believed that females often climbed any convenient hillock for booming but never stood on a stump or branch.

Reproductive biology

The breeding season of this species is greatly extended, as it is in most buttonquails. For example, in Burma the birds breed more or less throughout the year, with principal periods during April–May and again in August–September (Baker 1935). Similarly, in India the season may vary somewhat locally, but probably the birds breed essentially the year around, with possible gaps during the height of the rains and during the driest months in more arid regions. In Sri Lanka the season may only last from February until May, but probably there is a second breeding period later in the year. Eggs have been found in the Philippines during March and April.

Generally the nest is hidden just inside scrub, grass, or bamboo jungle cover, alongside some open area of ground, such as beside wheeltracks or footpaths in thick grassy cover. In nearly all cases the nest consists of a thick pad of fine grasses some 10 cm in diameter, fitted into some natural hollow that has been deepened and made circular by the birds. Often the nest is wedged among the roots of a grass tuft, with the central blades of the tuft broken down and the softer parts of these blades used to help form the pad. When the nest is placed in fairly thick grasses there is often a grassy dome overhead, although it is possible that this dome is as much accidentally formed as purposely made. Among 500 to 1000 clutches observed by Baker (1930), four eggs were the usual number, three-egg clutches were rare, perhaps four clutches of five were encountered, and there was only a single clutch of six.

Both sexes take part in nest-building, the materials being gathered at the site by stem-jerking movements or being brought in from nearby by similar sharp jerking movements over the shoulder. According to Trollope (1970) the male spent more time in the nest itself than did the female. He noted twice that a canopy of stems was pulled out from a weed pile, almost obscuring the incubating male. Egg-rolling behaviour has been observed, but only by the female (Trollope 1967).

Eggs are laid on a daily basis, and incubation probably begins with the last or penultimate egg. In each of four cases observed by Trollope (1970), hatching occurred 12–14 days after the laying of the final egg, and Winkel (1969) observed a 13-day incubation period in each of four successfully incubated clutches. All the incubation is probably normally done by the male, but as noted above, Trollope observed that one of his females not only helped to feed the chicks but also helped to brood them at night, until they were about 14 days old. After that a lateral-to-lateral clumping pattern was adopted, but the female still tried to cover one or two chicks and the male did the same. The development of the hatched young is extremely rapid, and is essentially completed after about 40 days (see Figure 21) (Winkel 1969).

Evolutionary relationships

Ogilvie-Grant (1889) placed this species in a unique group of taxa now regarded as a single species, which is characterized as having the tarsus longer than the middle toe and claw, and the breast feathers distinctly barred, or in the adult female becoming wholly black. I have very tentatively included the outwardly similar Australian form *melanogaster* as a possible relative of this species (see Figure 1), but recognize that they differ from one another in several respects, including relative tarsal length.

Status and conservation outlook

This is a very broadly distributed species, although some of its races do exist as insular and perhaps intrinsically small populations. Nevertheless, the ability of buttonquails to live in close proximity to humans, and to benefit from cutting, grazing, and cropping activities, makes it unlikely that any of these populations will pose conservation problems.

B · Bustards (Family Otididae)

KEY TO THE GENERA AND SPECIES OF BUSTARDS

A. Wing 3–4.5 times tarsal length; bill usually shorter than middle toe; throat grey to white, crested or uncrested; underparts whitish, sometimes with a black breastband; mostly non-African species
  B. Smaller species (wing 238–435 mm); sides of neck black and white in breeding males; mainly Palaearctic
    C. Bill longer (culmen 28–36 mm); secondaries mostly black: *Chlamydotis*, Houbara bustard only sp.
    CC. Bill shorter (culmen 14–19 mm); secondaries almost entirely white: *Tetrax*, Little bustard only sp.
  BB. Larger species (wing 460–780 mm); no black on sides of neck
    C. Uncrested; neck feathers not vermiculated; bill short (culmen to 40 mm); Palaearctic: *Otis*, Great bustard only sp.
    CC. With a black nuchal crest; often a black band at base of neck; neck usually vermiculated grey and white; bill long (culmen over 60 mm); non-Palaearctic: *Ardeotis*, 4 spp.

D. Upper wing-coverts white, lacking black markings; black also lacking at base of neck and on breast: Arabian bustard

DD. Upper wing-coverts spotted black and white; black also present at base of neck, sometimes forming breastband

  E. Wing-coverts mostly white; African: Kori bustard

  EE. Wing-coverts mostly black; non-African

   F. Darker above, the neck much vermiculated with blackish and with a black breastband, under tail-coverts partly white: Australian bustard

   FF. Lighter above, the neck almost white and with no breastband, under tail-coverts brown: Great Indian bustard

AA. Wing 2–3.5 times tarsal length; bill sometimes longer than middle toe; adults often with black throats or underparts in males or both sexes; all African and Indian species

 B. Larger (wing 360–660 mm, culmen 44–91 mm); uncrested; crown often black-edged, with a lighter centre; throat rarely black, neck variably grey, bordered behind and/or below with rufous; abdomen, axillaries and under wing-coverts white: *Neotis*, 4 African spp.

  C. Neck all grey, becoming brownish only at base; throat often blackish; tail under 200 mm

   D. Upperparts freckled sandy; crown entirely tawny; chin and throat blackish brown with white cheeks above terminated by black eye-stripe; culmen to 56 mm: Nubian bustard

   DD. Coarser above, crown whitish posteriorly; chin and throat whitish (females) or blackish, bordered above with dull white; culmen over 60 mm: Heuglin's bustard

  CC. Neck bicoloured, grey to brownish in front, rufous to tawny behind, the throat never pure black; tail over 200 mm

   D. Sides of head, throat and foreneck dull grey to greyish brown; lower hindneck tawny; brown-edged pale superciliary stripe, primary coverts brown-tipped; tail usually under 250 mm: Ludwig's bustard

   DD. Sides of head and throat pale grey to whitish; hindneck vermiculated rufous; superciliary stripe white and black; primary coverts white-tipped; tail usually over 250 mm: Denham's bustard

 BB. Smaller (wing 180–360 mm; culmen 27–46 mm); adults (at least males) with a usually black-edged, short nuchal crest; throat and/or underparts often black in one or both sexes: *Eupodotis*, 7 African, 2 Indian spp.

  C. With black underparts in males, sometimes (in 2 of 6 spp.) also in females, which at least have black under wing-coverts and/or blackish axillaries; neck never tinged with bluish or bluish grey

   D. Wing usually under 270 mm (maximum 280 mm); primaries barred with black and brown; males never with a black nuchal crest

    E. Males with a pink crest and a black throat; both sexes with black underparts; African: ('*Lophotis*') rufous-crested bustard

    EE. Males in breeding plumage mostly black, with six long facial plumes; females buffy below; Indian: ('*Sypheotides*') lesser florican

   DD. Wing usually over 270 mm (minimum 260 mm); primaries whitish basally, increasingly barred with black towards tips; males with blackish crest, and variably black neck; females usually light-bellied (3 of 4 spp.);

    E. Entire neck black (males) or mottled black (females), females with a black or buffy abdomen

     F. Back closely barred with brown and buff; ear-coverts white (males), or with a black abdomen (females); African: ('*Afrotis*') little black bustard

     FF. Back spotted with black and brown; head all black (males), or abdomen buffy (females); Indian: ('*Houbaropsis*') Bengal florican

    EE. Forehead only black in males; whitish to buffy underparts in females ('*Lissotis*')

     F. Rump mostly blackish (males) or marked with black (females); females rich brown dorsally and on crown: Hartlaub's bustard

     FF. Rump predominantly brownish; females pale brown to grey dorsally and on crown: black-bellied bustard

  CC. Both sexes lacking black abdomen but sometimes bluish on breast; males with black on throat and usually on nape; neck often tinged with bluish; axillaries rarely black ('*Heterotetrax*' and '*Tracelotis*')

   D. Smaller (wing to 260 mm); primary coverts black-tipped white; axillaries black: little brown bustard

   DD. Larger (wing over 260 mm); primary coverts lacking white; axillaries rarely black, usually brown, grey or whitish

    E. Adults with no blue on breast but always with black throat

     F. Hindneck, breast and abdomen ashy brown, black limited to throat area: black-throated bustard

     FF. Hindneck and rear underparts buffy white, black stripe from throat to breast: Rüppell's bustard

    EE. Adults usually with bluish grey breast; if tawny then with a white throat

     F. Abdomen bluish slate; females with blackish throat and bluish grey breast: blue bustard

     FF. Abdomen whitish; females with white throat and more tawny breast: white-bellied bustard

HOUBARA BUSTARD (Plate 14)

Chlamydotis undulata (Jacquin) 1784

Other vernacular names: Macqueen's bustard (*macqueenii*); avutarda (Canary Islanders); houbara (Arabian); outarde houbara (French); Kragentrappe (German).

Distribution of species (Map 11)

Eastern Canaries (where now endangered) and the northern Sahara east through Egypt, Jordan and

Map. 11. Distribution of the houbara bustard, including its breeding (hatched), residential (solid) and wintering (stippled) ranges, and locations of its races *fuertavernturae* (fu), *macqueenii* (ma) and *undulata* (un). Approximate former USSR breeding distribution indicated by dashed line.

perhaps Syria to Afghanistan, north to the eastern Caspian Sea coast and the steppes of Kazakhstan, thence east across the southern Kirghiz steppes and foothills of the Russian Altai historically or sporadically to western Mongolia and western Xinjiang (Sinkiang), China, possibly also to western Inner Mongolia (Nei Monggol). The USSR range was much more extensive in the early 1900s, extending from the Ural River valley east locally to Tuva ASSR. Locally sedentary, migratory or nomadic, wintering south to Saudi Arabia, Iraq and northwestern India.

Distribution of subspecies

C. u. fuertaventurae (Rothschild and Hartert) 1894: endemic to the Canary Islands (Fuerteventura and Lanzarote), where it is sedentary and currently endangered. Has also been reported in the past from Gran Canaria island.

C. u. undulata (Jacquin): northern Sahara from Mauritania east to western coastal Egypt. Sedentary or locally nomadic, probably in response to rainfall.

C. u. macqueenii (J. E. Gray) 1832: eastern Egypt, the Sinai and Jordan east to Baluchistan, Kazakhstan and Mongolia; migratory to sedentary or locally nomadic, wintering on its breeding areas (Middle East population) or variably southward to northern India (USSR population).

Measurements (of *macqueenii*, mm)

Wing, males 393–431 (av. of 21, 407), females 357–377 (av. of 11, 368); tail, males 197–230 (av. of 13, 215), females 181–207 (av. of 7, 192); tarsus, males 91–106 (av. of 18, 98·8), females 83–97 (av. of 16, 89·6); culmen, males 30–36 (av. of 18, 32·9), females 28–34 (av. of 16, 30·4) (Cramp and Simmons 1980). Measurements of *undulata* are slightly smaller, and those of *fuertaventurae* average smallest (Vaurie 1965). Egg, av. 62–63 × 45.

Weights

Males in winter (India) c.1·8–2·4 kg, in summer (USSR and Mongolia) 1·5–2·22 kg (av. of 5, 19·6), females in winter c.1·2–1·7 kg, in summer 1·1–1·25 kg (av. of 5, 1·15) (Cramp and Simmons 1980). Males of *macqueenii*, 1·8–2·36 kg, females 1·18–1·68 kg (Taylor 1985). Adults of *macqueenii* in Israel typically range from 1·8 to 2·2 kg, and females from 1·1 to 1·535 kg, but considerable seasonal variation occurs (Mendelssohn *et al.* 1983). Two males of *undulata* in spring, c.2·5 and 3·2 kg (Urban *et al.* 1986); male of *macqueenii* in spring 2 kg, female 1·2 kg (Alekseev 1985). Higher spring weights (males about 3·2 kg, females about 2·5 kg) reported by Meinertzhagen (1954) seem unusually high and perhaps are questionable. Estimated egg weight 68–71 g, or 4·5 per cent of adult female (Schönwetter 1967). Average of ten eggs from ten females, 64·6 g, but a 6-year-old female produced eggs averaging 74·5 g (Mendelssohn *et al.* 1983).

Description

Adult male. General colour sandy buff (averaging palest in *macqueenii* and darkest in *fuertaventurae*), minutely freckled with blackish vermiculations, and varied with blotches of black (especially in *fuertaventurae*) where the feathers are more coarsely vermiculated or spotted with the latter; scapulars like the back; lower back, rump and upper tail-coverts a little more reddish than the back, the black vermiculations being wider apart; sides of face and ear-coverts pale sandy buff with the same blackish shafts; crown with a crest of narrow and elongated feathers that are white anteriorly and (in *macqueenii*) tipped with black posteriorly; cheeks white, streaked anteriorly with black hair-like shafts; chin and upper throat white; sides of neck black, commencing in a streak close behind the ear-coverts and extending into a ruff of stiffened feathers, the basal plumes black, the succeeding ones white, with broad black tips, succeeded by a tuft of feathery white plumes; lower throat and foreneck bluish grey, the former obscured by sandy buff, slightly freckled with black, the plumes of the foreneck elongated bluish grey, with white tips; remainder of underparts white, with some freckles on the sides of the upper breast; under tail-coverts white, with a few black cross-bars, the long ones sandy buff on the outer webs with more or less black frecklings, all the under tail-coverts with a concealed tinge of pink at the base; under wing-coverts and axillaries pure white. Rectrices cinnamon-buff, tipped with white, and banded with three (*macqueenii* and *fuertaventurae*) or four (*undulata*) grey bars. Iris varying from pale to bright yellow; bill bluish or dusky above, becoming paler on gape and lower mandible, which are usually greenish or yellowish; tarsi and toes pale yellow to creamy, usually with a dingy or greenish or plumbeous tinge.

Adult female. Differs from the male in being much smaller, and in having the crest and the ruff of the neck more feebly developed and the crest less white basally. The freckling on the foreneck and lower throat is also rather coarser than in the male.

Immature. Resembles the female in being mostly sand-soloured, but with sandy-coloured arrowhead markings that pervade the entire upper plumage. The neck ruff is very small and the crest on the head is represented only by a few elongated feathers, recognizable by their somewhat coarser black frecklings. The grey on the foreneck is almost entirely replaced by sandy frecklings, and the white of the primaries is distinctly inclined to sandy buff; sometimes there is an indication of a black band on the feathers of the foreneck. The bill is rather uniformly olive-grey, and the tarsi and toes are yellowish brown. Young birds acquire adult wing-lengths during their second summer, and reach adult weights in the second autumn. Sexual maturity in both sexes occurs the second year (Mendelssohn *et al.* 1983).

Identification

In the hand. The combination of a fairly long bill (culmen 28–36 mm) but a wing that is rather short (maximum 431 mm), and long black feathers extending down both sides of the neck, identifying this species. An ill-defined and shaggy white crest (black-tipped in *macqueenii*) is also present, at least in males.

In the field. Associated with dry steppes, this is the only bustard with a black streak extending down each side of the neck from the posterior auricular area to the sides of the upper breast. When flying it exhibits white at the bases of the outer primaries, but the wingtips are black, as is a wrist-patch. It thus closely resembles the slightly smaller little bustard in flight, which however shows more white on the secondaries and less on the outer primaries. The larger great bustard has rather uniformly greyish rather than contrasting black and white primaries. Unlike most bustards it is normally very silent, even during sexual display, but a low croaking note is produced in alarm. However, a 'peculiar high trill' has reportedly been heard during an apparent male courtship flight (Alekseev 1985); the possibility that this sound may be produced by feather vibration should not be ruled out, considering its occurrence in the little bustard.

General biology and ecology

Based on recent studies in the Canary Islands (Collar et al. 1983, Collins 1984), these birds occur almost exclusively on open and relatively flat lowlands (under 200 m elevation), especially those having a significant amount of sand in the substrate. They evidently avoid highly erodable land as well as 'badland' (*malpais*) areas in which lava fields are strewn with ash and volcanic debris, and where vascular plants are few. They also avoid both mountainous regions and those habitats located close to human settlements. Generally they appear to use areas with gently undulating terrain, and slopes of less than 10 per cent, and may gravitate to areas that are rich in herbs. It is possible that they prefer flat, open plains for foraging, but move into gently sloping hillside areas for nesting; brooding habitat has not been defined. In some transect counts in various habitat types of the Canary Islands, sandy plains held the greatest number of houbaras (0·2–0·4 per transect-hour), followed by stony plains (0·1 per transect-hour). No records of the species were obtained during transects counts in other habitat types such as *malpais* and cultivated areas.

Similarly, in breeding areas of Kyzylkum, USSR, the species favours alkaline alluvial clayey landscapes in which the vegetation consists of such halophytes as *Anabasis*, *Arthrophytum* and *Salsola*, and where adjacent sandy ridges have an early growth of spring ephemerals that provide foods for spring migrants. However, areas of pure sand are avoided, and nesting is concentrated in the *Anabasis*-dominated clayey plains (Alekseev 1985). In Morocco the species is closely associated with the arid zone as delimited by the 20 cm precipitation isohyet and as characterized by xerophytic plants such as *Artemisia*, *Haloxylon* and *Salsola*. Its southern distributional limit probably corresponds to the onset of reg-desert as characterized by grasses such as *Eremopterix* (Brosset 1961, Goriup 1983b).

More generally, the species seems to be among the most desert-adapted of the bustards. It typically inhabits little-habited steppes and semideserts, with open or scattered shrubby vegetation, and it is frequently found associated with stabilized sand-dunes. The associated shrubs at times may be 1–2 m high, and typically include xerophytic or halyophytic plants such as *Artemisia*, *Tamarix* or *Haloxylon* growing on stony or sandy substrates (Cramp and Simmons 1980). Brosset (1961) suggested that the long panoramic views associated with *Artemisia* steppe are necessary for ensuring the species' security, and this in turn affects its distribution, more so than does the availability of a specific food supply. Heavily wooded areas are therefore completely avoided, as is rocky and precipitous terrain, although the birds have been found at elevations as high as 1800 m in the Altai foothills of south-central USSR. In the USSR breeding birds are generally associated with saline clay deserts and stabilized dunes, but they avoid active dune areas (Dementiev and Gladkov 1951). The distinctly sandy-coloured plumage of the species is suggestive of its normal substrate, although the generally darker plumage colour and less sandy appearance of the Canary Islands race is probably associated with the blackish volcanic soil substrate found on these islands (Collar et al. 1983).

Except when breeding, the birds often occur in small groups that may represent families or aggregations of families. Flocking is probably especially typical of populations that undergo some kind of regular or periodic migrations; as many as fifty birds have been seen aggregated at Jordanian waterholes and up to sixty have been observed in Israel (Cramp and Simmons 1980). However, large flocks anywhere are probably now highly unlikely, owing to the great diminution of numbers throughout the species' range. Even in the past, migratory units often consisted of small parties, pairs, or even single birds, and in recent years single birds have constituted at least half the migration records during both spring and autumn. In the USSR spring migration generally occurs during late March and April, whilst autumn migration is more prolonged and may begin as early as late August and extend through October and into the first half of November (Dementiev and Gladkov 1951, Alekseev 1985). The birds begin arriving in wintering areas of Pakistan by early September, but many more will have arrived by November. They begin their return migration in February and March (Mirza 1985), leaving Pakistan in groups of twenty to thirty, or somewhat larger groups than those characteristic of autumn migrations (Mian and Surahio 1983).

During winter the birds are usually found in semi-arid to arid areas, but they often gather around dry stream channels or small valleys where rainwater collects, providing both a supply of food and some shelter. They generally avoid hilly terrain but may occupy broken plains (Surahio 1985). Although they are locally attracted to surface water and may even at times descend into wells to drink (Meinertzhagen

1954), it is probable that they can survive on moisture present in foods as well as on morning dew (Cramp and Simmons 1980).

Perhaps because of its use of desert and semi-desert habitats, densities of this species are typically very low both on wintering and breeding grounds. In wintering areas of Baluchistan, densities have been variously estimated at only 1–1.7 birds per km^2 (Mirza 1985) and about 0.3 per km^2 (Mian and Surahio 1983). These estimates compare fairly well with some rather crudely estimated densities of 0.12 to 1.25 birds per km^2 for various desert tracts used by wintering birds in Sind and Punjab (Goriup 1983a). Estimates of breeding densities include those of 0.5 and 1.0 young produced per km^2 in Israel (Lavee 1985), and a much lower estimate of 250 km^2 per breeding pair in eight breeding habitats of Kazakhstan between 1958 and 1975 (Alekseev 1985). Earlier estimates of breeding densities in these areas indicated that breeding pairs were usually distributed about 5–10 km apart, but occasionally only 1–2 km apart (Dementiev and Gladkov 1951).

The foods of this species are extremely diverse, and have been well summarized by Cramp and Simmons (1980). In brief, the species is omnivorous, with perhaps some seasonal variations. Thus in spring, plant materials including, especially, green shoots or leaves of crop plants such as barley, wheat, peas, mustard and other crops may be important, and in winter the birds also frequent cereal and oil-seed fields including mustard (*Brassica*). Plant materials known to have been consumed in various parts of the species' range include leaves and shoots of various herbs and shrubs (e.g. *Alhagi, Artemisia, Cymbopogon, Leptadenia, Triumfetta*), seeds and grains, berries, fruits of *Capparis, Grewia, Lycium* and *Ziziphus*, and drupes of *Argani*, as well as various flowers (*Senecio*) and bulbs (*Allium*).

Studies of wintering birds in Pakistan indicate that they are largely vegetarians at that period, feeding mostly nocturnally on leaves, buds, flowers and some insects, often in cultivated fields. Shoots of native halophytic plants such as *Crotolaria, Eulaliopsis, Haloxylon, Lasiurus, Tribulus* and *Zygophyllum*, and the seeds of *Fagonia* are also consumed. Other known native food plants of desert habitats in Pakistan include *Leptadenis, Cymbopogon*, and *Euphorbia* (Mirza 1971, 1985). In Pakistan's Sind province they similarly consume such native plants as *Capparis, Crotolaria, Grewia, Haloxylon, Tribulus*, and *Zizyphus*, but a favourite food plant of wintering birds there is the crop-plant mustard (Surahio 1985). Similarly, in Baluchistan a considerable variety of plant shoots and seeds are consumed, but it would appear that *Crotolaria, Haloxylon* and *Tribulus* are preferred foods, and are consumed wherever they are available (Mian and Surahio 1983).

Spring and summer foods are apparently more likely to include a variety of animal materials, which are known to include especially orthopterans (locusts, grasshoppers, mole-crickets), coleopterans (scarabeids, tenebrionids, cantharids), termites, ants, caterpillars, spiders, centipedes, and snails. Insects such as grasshoppers, locusts, and crickets are probably also extremely important foods for young chicks (Dementiev and Gladkov 1951). Some vertebrates such as lizards (agamids, lacertids) and small snakes (colubrids) are also eaten locally (Cramp and Simmons 1980). Tenebrionid beetles evidently are important spring foods for adults in Kazakhstan, USSR (Alekseev 1985). Early records from the Canary Islands suggest that there the birds used to seek out dung-beetles, and they have also been reported to eat caterpillars, lizards and land-snails, and to favour especially the nitrogen-rich pods of cultivated peas when these are ripe (Collar *et al*. 1983). More recent studies on the Canary Islands, based on faecal sample analysis, suggest that the foods include insects (mainly beetles), annual plants, the flower buds of the shrub *Launaea*, and the fruits of *Lycium* (Collins 1984).

Social behaviour

Although Cramp and Simmons (1980) reported the species as 'probably monogamous', it is perhaps more accurate to state that the mating system is still unclear, and that it is uncertain as to whether the species is monogamous or polygamous, or whether pair bonds are formed at all (Goriup 1983b). However, recent observations on the Canary Islands race suggest that the males play no further role in the reproductive cycle following fertilization, and the display and territorial behaviour of the male would suggest that it is probably polygynous or promiscuous (Collins 1984, Urban *et al.* 1986). Similarly, in the USSR Alekseev (1985) reported seeing two adults (sexes unstated) accompanying young on only a single occasion, out of a sample of at least thirteen broods. According to Dementiev and Gladkov (1951), males have never been observed on the nest, but instead after the egg-laying period they spend their time alone or in small parties on the steppes. A polygynous mating system was also suggested by Ponomareva (1982), and at least in some areas group display among males may occur, with males defending separate but nearby territories (Launay and Paillat 1990).

Males of this species often also display solitarily, their territories apparently being established shortly after spring arrival in the case of migratory populations, and migrating males may even perform displays during rest periods while on migration (Dementiev and Gladkov 1951). Breeding territories are apparently very constant from year to year, with adults and broods consistently seen at the same sites over a several-year period (Alekseev 1985).

In the Canary Islands, males defend display areas or territories that are usually about 1 km apart (or probably within the limits of visual contact), and no closer than 500 m. These display-area territories may also serve as home ranges, and include both display sites and foraging places for the males. During the breeding season both sexes tend to be quite solitary, so intersexual contacts are probably mostly limited to courtship and mating. Thus, although apparent pairs may sometimes be formed, the species is probably essentially polygynous, with the males exhibiting male-dominance rather than resource-defence polygyny. When one male intrudes on another's territory they threaten each other. Drooping of the primaries is especially important at such times. Typically agonistic 'crouching' includes a posture with the crest raised, the head held low and directed forward, the wings half opened, and the tail cocked and somewhat fanned. On one occasion this was followed by a 'dancing' behaviour, consisting of leaping into the air with the wings outstretched (Collins 1984). Such 'dancing' has not been observed as an aggressive display in other bustards, but similar crane-like behaviour does occur regularly in young great bustards and at times among adults, especially females (Grzimek 1972). Agonistic display towards smaller terrestrial animals and sometimes towards its own species may involve strong tail-cocking, wing-lowering and pulling of the head back towards the tail (Figure 28a).

Although males apparently defended quite large display areas, over which display might occur, within these larger areas there were special display sites. Three display areas occupied by different males consisted of about 0·6, 1·6, and 2·4 ha. Within these areas were special display sites, on which most displays occurred. Thus, one male performed 62 per cent of the total observed displays from his primary display site during one year, but he also had a secondary site about 400 m from the primary one. One primary display site was about 100 m in diameter and situated at the top of a low rise which was especially bare and stoneless, presumably providing both an excellent panoramic viewpoint and maximum conspicuousness.

From such sites the birds perform various self-advertisement displays which, although they have previously been described by various observers (Gérodet 1974, Mendelssohn et al. 1979), have recently been analysed in some detail by Collins (1984). Display is preceded by a quiescent phase, during which the male moves to his display site and becomes stationary. He then begins a feather-ruffling or feather-erection of the breast, neck and crest (Figure 28c). The crown feathers are raised, the neck feathers are swept back to form 'handlebars', and the breast feathers are fanned laterally. Active strutting consists of a forward walk, with the neck held vertically and the tail held at about a 45-degree angle. Feather erection reaches a climax during the strut, but the walking pace is steady and seemingly pompous. The black neck feathers are completely reversed and lie upwards, in line with the neck. Suddenly the head and neck are thrown back, the neck dropping onto the back and the chin on the foreneck. The white feathers nearly envelop the body, hiding the head and neck from lateral view, and producing a heart-shaped effect from the front. The bird then breaks into a high-stepping dance-like run, and variously runs in a series of circles about 2–20 m in diameter, in zigzags, or in straight lines. During the running phase (Figure 28d, e) the male's neck may be alternately drawn to left and right (this was observed by Mendelssohn et al. but not by Collins) and the breast bounces noticeably from side to side, suggesting that a small inflatable pouch may be present. The male abruptly stops at the end of the run, as if it had bumped into something, often with the head and neck 'kicking out' slightly and then pulling back. At the same time the neck and breast feathers flick forward like a closing fan, and then fall partway back. Gradually the male moves back into its motionless starting posture. After a few minutes it again begins its feather erection and the display sequence is repeated. Up to thirty-one displays have been seen, over a period of up to 70 minutes. The entire sequence terminates by general feather relaxation and adoption of a resting phase, sometimes with feather-ruffling or preening. Most displays were observed between 7 and 9 a.m. and between 3 and 7 p.m. Considerable individual variation in the durations of the strutting and running phase of display was observed among three different males. Aerial display has not been observed by most writers, but Alekseev (1985) stated that he saw a bird rise from the ground with a peculiar trill and, with its feathers 'quivering', fly about 300 m. The possibility that this trilled sound is caused by feather vibration rather than by vocalization should not be ruled out.

In one copulation sequence, a male interrupted his display run with short, stationary posturings, which increased as the male approached the female while circling her. When about 1 m from her, the male performed a 'kick-out', and then began throwing his

Fig. 28. Social behaviour of the houbara bustard: (a) male threat; (b) close display of male to female; (c) erection of neck and crown plumes by male before dance-run; (d) dance-run with neck drawn alternately from side to side; and (e) dance-run as viewed from opposite side, with plumes obscuring male's head. After sketches (based on photos by H. Mendelssohn) in Cramp and Simmons (1980).

head directly back on his shoulders, the head and neck held in line with the axis of the body. He also performed sideways head-throwing, turning his head left and right at about 1 s intervals, either while standing stationary or during short pursuits of the female. While standing behind the now-squatting female, the male pecked at her head and neck for about 9 s before mounting. While copulating, the male stretched his wings to maintain balance. After copulation was completed (in a separate observation), the female left the male and passed into the display area of another male (Collins 1984). A captive male has also been observed to approach a female in a head-lowered posture, with the white neck plumes lowered and the black ones spread laterally (Figure 28b) (Cramp and Simmons 1980).

Observations by Launay and Paillat (1990) and by Ponomareva (1983) are similar; the latter author additionally observed bill-clapping at the end of the display run and just prior to copulation.

Reproductive biology

The nesting season is quite variable across this species' broad range, but is concentrated during the spring months. On the Canary Islands, records extend from December to April but are probably mainly concentrated from March to the end of April. Similarly, on the mainland of North Africa the eggs are usually laid in March or April, although records extend from February to June (Cramp and Simmons 1980, Collar *et al.* 1983). There are breeding records for Syria from March to May, but birds in Israel display from the end of December to mid-June, with egg-laying starting in early March, most clutches laid in April, and continuing to the end of April. The spread of known hatching dates there is from March 29 to May 14 (Lavee 1988). In the USSR egg records extend from the first half of April to the end of June, but at least in Kazakhstan most of the young hatch during May (Alekseev 1985).

Among captive birds in Israel the egg-laying interval is usually 48 h and occasionally 72 h. The interval between successive clutches was found to be 9–12 days if the eggs were removed as soon as the last egg of the clutch had been laid, but when clutches of non-fertile eggs were allowed to be incubated, the interval ranged from 27 to 64 days. One older female produced five clutches in each of two seasons, totalling nine and eleven eggs and collectively averaging two eggs per clutch. Other females produced from one to four clutches in a single season. In captivity the usual clutch size is two eggs, but replacement clutches are often a single egg. Older captive females sometimes lay three-egg clutches (Mendelssohn et al. 1983). The clutch size under natural conditions is usually two or three eggs, but occasionally four- or one-egg clutches occur. Very rarely five-egg clutches have been mentioned, but possibly these are the work of more than one female. The mean of fifteen apparently completed clutches from wild birds was 2·8 eggs, with a range of two to four eggs (Lavee 1988).

The nest is placed on the ground, often in the shade of a shrub, but at times is located in open desert. Collins (1984) stated that nests are not placed among shrubs to provide shelter or shade, but rather are placed among low shrubs or stones for camouflage. The height of such shrubs must not obscure the female's vision, even when her head is lowered. Both of two nests found by him were on slight slopes, possibly to provide drainage during rainstorms. He noted that females incubated for long periods with the head resting on the shoulders. Foraging periods for incubating birds were irregular, and in five cases lasted from 21 to 31 minutes. These birds were never seen foraging near other birds, and the nearest male display area to one of the nests was 150 m away but visually hidden from it.

Incubation is now known to require an average of 23 days (under artificial conditions), rather than the 28 days suggested in some earlier literature. Newly hatched chicks weigh about 52 g on average, and are able to run short distances by their third day. They run quickly at 5 days, and at 6 days start to pick up live insects for themselves. Contour feathers begin to appear soon and at 11 days have covered the wing and shoulder areas. At 4 weeks the birds are fully feathered, but the wings and tail are still quite short. In spite of this they can fly short distances when a month old, and at 5 weeks can fly quite well. Two-month-old males weigh about as much as adult females, and 2 weeks later the females also weigh as much as adult females. Males acquire their full adult weight during their second autumn, and both sexes attain sexual maturity when 2 years old (Mendelssohn et al. 1983).

Lavee (1988) found that of forty-five eggs monitored in sixteen Israeli nests, 33 per cent were lost to predation, and an additional 12 per cent failed to hatch. The probability of a freshly laid egg hatching and the chick surviving through its first few hours was estimated as 34 per cent. Further chick survival was estimated at 30 per cent, and overall survival from the unincubated egg stage to an immature of nearly adult size (or to about 4–5 months of age) was judged to be 10 per cent. The average observed overall brood size was 1·76 for thirty-seven broods. It was also estimated that an average female lifespan of 9–14 years (representing an average annual adult mortality rate of slightly under 10 per cent) would be just adequate to maintain the houbara population, based on these recruitment data, with an even adult sex ratio, and with breeding starting in the second year of life. In line with this estimate, one individually tagged individual has been determined to have survived for more than 13 years in the wild.

Other data on survival of wild birds are relatively fragmentary, but Alekseev (1985) observed that thirteen Kazakhstan broods had from two to four chicks, averaging 2·5. One brood was observed to cover an area of more than 1·5 km during a period of only 3–4 hours, suggesting a high level of chick mobility.

Evolutionary relationships

The relationships of this species, which except for its unexplained placement in *Otis* by Dementiev and Gladkov (1951) has consistently been maintained as a monotypic genus, are quite obscure. It has a surprisingly long tail and relatively short legs, but the male's unusual crest and elongated and specialized neck feathers are certainly display structures and should not enter the picture in assessing the validity of the genus *Chlamydotis*. The bill shape is more like that of *Neotis* or *Eupodotis* than of *Otis* and *Tetrax*, and the downy young pattern is distinct from the young of both of the latter two forms (Fjeldså 1977). Peters (1934) placed the genus *Chlamydotis* sequentially between *Choriotis* (= *Ardeotis*) and *Lophotis* (= *Eupodotis*). Cramp and Simmons (1980), as well as Urban et al. (1986), placed the genus in linear sequence between *Neotis* and *Ardeotis*, but no reasons have been provided for any of these arrangements. I am inclined to think that the houbara is derived from a fairly isolated lineage of bustards, with no apparent nearest living relative. As a result, I have placed it at one end of the provisional phylogram that I have drawn for the group (Figure 2).

Status and conservation outlook

The world status of the houbara has been reviewed in a preliminary but comprehensive fashion by Collar (1980), as well as by Cramp and Simmons (1980). More recently an international symposium concentrating on the biology and current status of the houbara was held during 1983 in Pakistan (published as *Bustard Studies* 3, 1985). In addition to a number of papers dealing specifically with wintering houbaras in Pakistan, there were also contributions on the status of the species in the USSR (Alekseev 1985, Ponomareva 1985), in Iran (Mansoori 1985), in Israel (Lavee 1985), and in Morocco (Haddane 1985). The species' precarious situation in Morocco has also been summarized by Goriup (1983b). Additionally, the status of the endangered Canary Islands race of the houbara received special attention from the ICBP bustard group, and its report was published as *Bustard Studies* 1 (Collar *et al.* 1983). Much of these data were summarized in Chapter 4, and the worldwide situation of the houbara can be briefly stated as ranging from rather gloomy to quite dismal.

Of the species' breeding areas, the largest single component is found in the USSR, where perhaps the species once extended across the trans-Caspian region at least to the Altai Mountains and probably nearly to Lake Baikal in Tuva Autonomous Region. It was probably most common in Turkmenia, but was also probably relatively common through much of Uzbekistan and Kazakhastan (Dementiev and Gladkov 1951). More recently, the species' stronghold has become the Uzbekistan region, where the expanse of the Kyzylkum desert is one of the least exploited areas of the desert grazing zone in the USSR, and which at least once covered about 250 000 km². However, the species' numbers in the north-western Kyzylkum declined by roughly 75 per cent between 1956 and 1979, a decline that has been attributed to destruction of breeding habitats, an increase in general human disturbance, illegal shooting, and poaching. Recently there has been a large-scale development of cotton farming, with associated levels of high pesticide use, in this general region of the USSR. Added to these problems in the breeding areas are the large-scale hunting activities on the species' primary wintering areas, which include Iran, Pakistan, Palestine, Arabia, and India (Alekseev 1985).

A recent summary of the species' status in the USSR (Ilicek and Flint 1989) suggests that as of 1980 only three regions still supported more than 100 breeding pairs. These were the areas around Bukhara in south-eastern Uzbekistan, the Ustyurt area between the Caspian and Aral seas, and the Betpak-Dala area to the west of Lake Balkhash. A few dozen pairs also persist in southern Tuva ASSR and in the northern Kyzylkum steppes and scattered individual pairs elsewhere. The species is now listed in the Soviets' Red Data Book of threatened and endangered birds, and a breeding centre has been established for it at Bukhara.

South of these regions, Iran and Afghanistan may have once represented important breeding areas, especially the former. Both have suffered from massive internal strife and warfare, and little is known of the species' situation in either country. However, Iran did establish some wildlife reserves prior to its revolution, and at least for a time the species was apparently increasing. Its current situation seems much more questionable, and the same probably applies to Afghanistan, where there are very few records (Collar 1980).

Pakistan almost certainly represents the historically most important area for wintering houbara bustards from the USSR breeding population. Goriup (1983a) judged that at least 20 000 birds had been wintering in Pakistan in recent years, and that perhaps as many as 3000 or more were then being killed each year by hunters, primarily falconers. Collar (1980) observed that it was not until Arab falconers began to hunt bustards in eastern Pakistan that the numbers of wintering birds became apparent. Thus annual kills numbering at times several thousands were estimated, although in the last few years these numbers have declined, apparently because of increased hunting costs and probably also diminishing populations of bustards. However, Mian and Dasti (1985) and Mian (1986a) reported that hunting activities in western Baluchistan had been increasing recently, and that in 1982–3 probably about 5000–6000 birds were killed, of which about 1000–1500 may have been shot by local hunters, and the remainder taken by visiting Arab falconers. Additionally, habitat loss from disturbance caused by grazing, irrigation, and gathering of fuel-wood has been exacerbated by environmental changes associated with decreasing localized precipitation and increasing erosion problems. Some local breeding (in the Yakmach, Chagai and Khara districts) by houbaras probably does occur in Baluchistan (Shams 1985, Mian 1986b), but this may not be a regular event. Nevertheless, sanctuaries totalling about 40 000 km² have been established in Kharan, Chagai and Las Bela districts, and efforts are being directed towards the protection of any breeding birds there. To the east in Sind and Punjab the species has traditionally wintered in large numbers (Goriup 1983a). Circumstantial evidence points to a marked population reduction in Sind since 1960 (Surahio 1985), and in Punjab the wintering population in the early 1980s was probably more than 30 per cent lower than it was in 1971 (Mirza 1985). However, Pakistan authorities have established eleven game sanctuaries and twenty-three game reserves in Punjab. One of these in the Cholistan desert, Bahawalnagar district, was specifically established to protect houbaras. This sanctuary covers some 6600 km² and also protects the great Indian bustard. The houbara is now also legally protected in the Thai and Dera Ghazi Khan desert areas of Punjab, although in non-protected areas up to three per day can be taken (Ahmed 1985).

In India, houbaras winter in western Rajashtan in unknown numbers; unfortunately no detailed or quantitative information yet exists. However, they are known to occur in the Desert National Park of Rajasthan, which has been fenced off to exclude domestic grazing animals and improve wildlife habitat (Vardhan 1985). Hunting with falcons by Arab dignitaries has occurred in Rajasthan since the 1970s, but has been somewhat curtailed by the Indian authorities in recent years (Collar 1980).

Houbaras also breed locally in the Israel–Palestine–Jordan region, and winter from there south through much of the eastern portions of the Arabian peninsular. The species is protected fully in Jordan, but its breeding abundance is questionable; a recent statement suggests that they may still be fairly common in southern Jordan (Collar 1980). Mendelssohn (1983) suggested that probably 150–200 birds are resident in Israel, within a total maximum area of about 3000 km² of possible habitat. Of this, perhaps 1000 km² are available to houbaras for breeding. One area of natural habitat in the western Negev desert was found in August 1980 to have 142 birds present on an area of about 60 km² (or 2·4 birds per km²), within which a total of 94 birds was found on only 25 km². This is almost certainly the densest breeding population that has been reported anywhere in the species' entire range, although the survey was taken during the post-breeding period when the population was likely to be at its highest point of the year.

In Syria and Iraq, breeding certainly was regular during earlier decades, but most of the limited evidence suggests that in both countries the species has become quite rare. However, at least until the 1970s it may still have been breeding in some areas of southern Iraq (an area that has been much affected by recent warfare with Iran) (Collar 1980).

Although houbaras probably once bred regularly over considerable portions of the Arabian peninsula (Meinertzhagen 1954), there is no good information on their present breeding status there. They have generally been regarded as extirpated from the peninsula as a breeding species, but perhaps some still occur on the Summan plateau in the al-Habl area. Similarly, little or no information exists for Yemen, Oman, and the United Arab Emirates. Local wintering certainly must still occur in this region, at least in the south-eastern parts of the peninsula, and some breeding may still occur in Oman, south of Ibri (Collar 1980).

The situation across northern Africa is scarcely more encouraging, mainly because of the effects of Arab falconers over much of this region. In general, the species is believed to be widespread but patchily distributed, and although locally frequent it is generally declining, probably drastically in some areas (Urban *et al.* 1986). Little is known of its current status in Egypt, where once it bred in the near-coastal scrub regions, but where probably most of the population now represents wintering birds. This population is now being affected by Arab falconers (Collar 1980). The only known areas of probable current breeding in Egypt are along the western desert coast (between Matruh and Salum) and in the Sinai (Goodman and Meininger 1989). The Sinai population of houbaras may well include both resident and migrant birds, but its size is still unknown (Saleh 1989).

Breeding by houbaras probably locally occurs across Libya, Tunisia, Algeria, Morocco (including the western Sahara), and northern Mauritania. Hunting by local people or visiting falconers has been a serious problem for houbaras in nearly all these countries, and habitat destruction associated with agriculture or other development activities such as oil exploration have added to the species' problems. No numerical estimates are available for any of the African countries, but quite possibly only Tunisia still retains a fairly healthy population of houbaras (Collar 1980). As noted earlier, the endemic Canary Islands population is now critically low and endangered, with estimated 1980 populations of seven to twenty birds on Lanzarote Island and forty-two to one hundred on Fuerteventura (Collar *et al.* 1983).

In his summary of the world status of the houbara, Collar (1980) listed excessive hunting (eight countries), overgrazing (four countries), agricultural development (two countries), general effects of civilization (one country) and egg-collecting (in the

Canary Islands) as probable reasons for houbara population decline. Thus control of unregulated or excessive hunting would seem to be the first important step to take in preserving this species, in addition to the establishment of reserves in critical breeding and wintering areas.

LITTLE BUSTARD (Plate 15)

Tetrax tetrax (Linnaeus) 1758

Other vernacular names: pygmy bustard; outarde canapetière, outarde naine (French); Zwergtrappe (German).

Distribution of species (Maps 12, 13)

At least formerly resident in north-western Africa from Morocco (no recent proof of breeding) east to Algeria (now only a winter visitor) and Tunisia (no recent records). Now breeds locally in western Europe from Portugal and Spain (where residential) to northern France (where migratory); also residential in Sardinia and locally in south-eastern Italy; formerly also bred locally across eastern Europe from Germany, Poland and Romania south to Bulgaria and northern Greece (non-breeders last observed in Hungary in 1980, now only winters in southern Yugoslavia, where last-known breeding occurred in 1948); now local in interior Turkey (at least during winter; present breeding status uncertain), possibly still breeds locally in Syria and northern Iran (at least formerly present throughout), and more certainly from south-western USSR (the southern Ukraine and Crimea) and the northern Caspian region of Kalmykistan east disjunctively to northern Kazakhstan (formerly to north-eastern Kazakhstan and perhaps to western Siberia). Also winters or breeds locally in the USSR north to the northern Caucasus and Turkmenistan, and into western Xinjiang (Sinkiang), China. Sedentary and locally migratory, occasionally or sporadically wintering south to the coast of North Africa, Iraq, Iran,

Map. 12. Distribution of the little bustard, including its breeding (hatched), residential (solid) and wintering (stippled) ranges. Dashed lines indicate the species' approximate range at the beginning of this century.

Map. 13. European breeding or residential (solid) and wintering or unproven breeding (stippled) distribution of the little bustard, based mainly on Cramp and Simmons (1980).

Afghanistan, Pakistan, and north-western India, and rarely also to central China. Now generally reduced in range and numbers, with remaining major breeding concentrations in Iberia (mostly in Spain) and the USSR (mostly in Kazakhstan), and the wintering areas of the eastern migratory populations now mostly in Transcaucasia, the southern USSR (especially Azerbaijan), Iran and Pakistan. No subspecies recognized by Vaurie (1965); populations from central Europe eastwards have sometimes been separated racially (*orientalis* Hartert) on the basis of supposed slightly larger average measurements and generally paler upperpart colouration.

Measurements (mm)

Wing, adult males, 238–264 (av. of 29, 252), females 240–269 (av. of 15, 249); tail, adult males 95–114 (av. of 38, 104), adult females 86–112 (av. of 24, 100); tarsus, males 62–73 (av. of 40, 67·3), females 62–71 (av. of 26, 66·1); exposed culmen, males 15–19 (av. of 40, 16·9), females 14–19 (av. of 26, 15·6) (Cramp and Simmons 1980). The mean wing-length of twenty-one males of '*orientalis*' from Russia and Asia was 250, compared with 244·5 for 17 males from western Europe (Vaurie 1965). Egg, av. 52 × 38.

Weights (g)

Range of available non-juvenile weights, various locations and seasons, males 794–975; females 680–945. Two males in summer 940 and 975; males in December–January, 794–907. Juvenile male (December) 708. Females in May 740–910; other spring weights of females 680–945. Both sexes, in mid-August 630–820; in October–November 525–600 (Cramp and Simmons 1980). Range of USSR males in May, 740–910, females outside breeding season, 700–750 (Ilicek and Flint 1989). Four females in spring, 824–866, av. 842 (Schulz 1985b). Estimated egg weight 41, or 5·3 per cent of adult female (Schönwetter 1967).

Description

Adult breeding male. General colour above sandy buff, coarsely vermiculated with black, and also showing some black blotches in the centre of the feathers; rump a little greyer than the back, the feathers being freckled with whitish instead of sandy buff; upper tail-coverts white, or white mottled with a few blackish markings; wing-coverts like the back, but somewhat more sparsely vermiculated with black; lesser and median coverts white at the ends, and more or less freckled with black; the more distal coverts, alula and greater coverts white, the inner coverts slightly freckled or spotted with blackish; primary coverts blackish, narrowly tipped with white; remiges mostly white basally and blackish distally, the tips of these feathers again white; the outer primaries mainly blackish with white bases (the seventh, or fourth from the outside, unusually short and distinctively shaped, see Figure 29), the white gradually increasing towards the secondaries, which are nearly or entirely white; innermost secondaries sandy like the back; rectrices coarsely freckled with black on a white ground and crossed with four distinct bars of blackish, which are very pronounced on the basal half of the tail, which is white without any blackish frecklings, the outer feathers broadly tipped with creamy white; crown, nape, and hindneck brown, mottled with streaks and edgings of sandy buff, with a few blue-grey feathers intermixed; lores and sides of crown pale sandy buff, streaked with dark brown; feathers above and around the eye uniform creamy buff; sides of face, earcoverts, cheeks and throat light bluish grey, becoming black ventrally; this black area is bounded below by a V-shaped band of white feathers that encircle the hindneck, interrupt the black on the sides of the upper neck, and descend to the lower throat; this area is in turn bounded below by a broad band of elongated and erectile black feathers extending from the sides of the hindneck diagonally across the sides of the neck and uniting anteriorly in a broad band which runs down the centre of the neck; this broad black neck band is likewise separated from the lower neck and upper breast by a second horizontal band of white that traverses the foreneck, below which is a final narrower black band (thus two white neck bands are bounded by three alternating black ones); remainder of undersides pure white; sides of the upper breast sandy-coloured and mottled with black like the upperparts; under wing-coverts and axillaries pure white. Iris yellowish; bill horn-grey, black at the tip, dull yellowish at base of mandible; tarsi and toes dull ochre-yellow.

Adult non-breeding male. Female-like, lacking the striking facial and neck pattern, the plumage from crown to breast becoming more uniform with the upperparts, but more coarsely blotched and streaked on neck and closely barred on breast. The flanks also become barred and blotched, but remain whiter than in the female (Cramp and Simmons 1980).

Adult female. Differs from the male in being somewhat lighter rufous and more coarsely mottled with black on the upper surface; the head and neck not particoloured but resembling the back; the hindneck and mantle, as well as parts of the scapulars and back, spangled by ovate drops of sandy buff, most of these drops having a twin spot of black in the centre; sides of face sandy rufous streaked with black; throat white; lower throat, foreneck and upper breast sandy buff, the throat streaked with black and the foreneck and chest with circular bars or spots of black; remainder of undersides pure white, as are the under wing-coverts; the lower primary coverts with cross-bars of black; the wings as

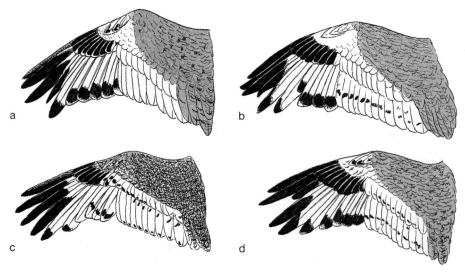

Fig. 29. Wing plumage variation in the little bustard: (a) immature male; (b) adult male; (c) immature female; and (d) mature female. After photos in Otero Muerza (1985c).

in the male, but the more distal coverts and greater coverts barred with black, the coverts and scapulars with a coarser vermiculations and heavier black spotting than in adult males, and the seventh primary not differentiated (Figure 29d).

Immature. First-year birds are generally distinguishable from adults by the greater amount of barring on the chest, by the more profuse barring on the white upper tail-coverts, and the sandy frecklings of the primary coverts; iris brownish yellow; tarsi and toes greyish. In both sexes of first-year birds the retained outermost (juvenal) primaries may be somewhat frayed and narrow, and usually have speckled margins. First-year males have a coarsely vermiculated wing-covert pattern similar to adult females, but already have modified seventh primaries (Figure 19a). First-year females have their upper wing-coverts and scapulars marbled rather than vermiculated, and have unmodified seventh primaries (Figure 29a). First-year females have their upper wing-coverts and scapulars marbled rather than vermiculated, and have unmodified seventh primaries (Figure 19c) (Otero Muerza 1985c). Males do not acquire their breeding plumage until the second year. Second-year males thus resemble older adults, but at least those of eastern (and apparently rarely western) populations may retain their outermost (tenth) juvenal primaries (Cramp and Simmons 1980).

Identification

In the hand. The combination of a very short (culmen under 20 mm) bill and mostly or entirely white secondaries identifies adults of this distinctive species.

In the field. Over its native range this is the only small (chicken-sized) species of bustard, identified by its long neck, short and stubby bill, dull yellowish grey legs and, in flight, wings that are mostly white except for contrasting black wingtips and wrist markings. Males in breeding plumage exhibit a mostly black neck that is interrupted by two white bands, the upper one V-shaped in front and the lower one nearly straight. The call uttered by territorial males is a short and crackling or snort-like *prett* note. Flying or wing-flapping males also produce a distinctive whistling *sisisi* sound produced by their specialized primaries, which is especially evident during territorial display. A low cackling is sometimes uttered by flushed birds.

General biology and ecology

This is a steppe-adapted species that at one time probably occupied large areas of Europe dominated by perennial grasses such as *Stipa*. It also used areas with scattered shrubs such as *Artemisia*, sometimes bordering woodlands or forests, or sandy, more desert-like habitats. Both stony and clay substrates are used by the birds; the former being typical breeding habitat in Iberia and the latter more frequent in drier climates. Stands of dense and tall grasses are avoided, as perhaps also are fescue-covered saline steppes. At least in the USSR the birds seem to be closely associated with *Stipa* grasslands, and they may even use more sandy or shrubby biotopes if a local growth of *Stipa* is also nearby (Dementiev and Gladkov 1951). Fallow fields are sometimes also used, and with increasing agricultural intensity the birds may adapt in varying degrees to the use of non-native vegetation in the form of pasturelands or fields of clover, lucerne, various cereal and root crops, and the like. As with other bustards, open panoramic views are needed, and level or only gently undulating terrain is favoured over steeper or rougher topography. However, compared with the great bustard it is fairly tolerant of nearby woodlands, hedges and similar vegetation that may tend to obscure the surrounding view (Cramp and Simmons 1980).

In all seasons, areas having high tree densities are avoided. Nesting sometimes occurs on fallow lands as well as on cultivated fields, but in any case small island-like areas of taller vegetation are needed to provide cover for the nest itself. Similar areas may be used by moulting birds while they are trying to avoid flying. Territorial males prefer open habitats having low vegetation, and although grazed lands may be used, areas that are fenced in and with cattle present continuously are avoided by males for use as display-sites. Nearly half of forty-five display sites studied had an average bush density of no more than one per hectare, and less than 15 per cent averaged more than nineteen bushes per hectare. Similarly, 70 per cent of seventy-nine display sites had a tree density of no more than one per hectare, and none averaged more than nineteen trees per hectare (Schulz 1986a).

Foods include a rather wide variety of materials, but are heavily shifted toward vegetation rather than animal materials. This is understandable in view of the rather grouse-like bill shape of this species compared with, for example, similar-sized bustards in the genus *Eupodotis*. Nearly all the edible parts of plants are eaten, including leaves, shoots, buds, flowers, seeds, and probably also even roots. At least on breeding areas in cultivated regions there is seemingly a concentration on leguminous plants such as clover and lucerne, although various crucifers are also eaten (Cramp and Simms 1980). The leaves of *Artemisia* are sometimes eaten during autumn in the USSR, and wintering-ground foods contain a high proportion (about three-quarters) of materials from spring and autumn ephemeral plant sources. Feeding is usually done during early morning and late afternoon hours; water sources may also be visited, but if necessary the birds can survive for some time on the moisture obtained from succulent or dew-covered plants (Dementiev and Gladkov 1951).

Animal foods are apparently limited to invertebrate sources, in contrast to the larger bustard species. Beetles (especially tenebrionids), orthopterans (field-crickets, mole-crickets and grasshoppers), dipterans, lepidopterans (probably mainly caterpillars), snails, and worms are all reportedly consumed. Certainly insects, and especially grasshoppers, are important foods for newly hatched chicks (Frisch 1976). However, even during the breeding season the majority of the foods eaten on a volumetric basis may sometimes be derived from plant materials (Glutz von Blotzheim 1973). Food materials found in stomachs of seventy-one birds collected between May and August contained from 62 per cent (in May) to 88 per cent (in July) of insect materials, mainly grasshoppers and beetles (Ilicek and Flint 1989).

According to Schulz (1985b), the birds appear to concentrate on invertebrate foods such as grasshoppers during summer, but shift to a rather wide variety of plant foods during autumn and winter, including composites, mustards, geraniums, grasses, etc. Probably relative food availability strongly influences their diet, invertebrate foods apparently being preferred whenever they are available. During the winter, beetles (carabids, chrysomelids) and their larvae are apparently taken in quantity. During late spring and summer grasshoppers are the most frequently taken foods, although beetles (especially tenebrionids) are also consumed in quantity during that season. Earthworms may be taken after periods of rain, but Schulz never observed the birds drinking.

At least in Iberia, the birds typically leave their breeding grounds after breeding has been completed, and make non-directionally orientated movements up to 30 km away, into areas that provide better foraging conditions than their breeding areas. After the rainy winter period, when feeding conditions have improved on the breeding areas, the birds again return to them. Site fidelity in the breeding areas is typical of both sexes, with males tending to establish their territories and display sites at the same places as they used in prior years (Schulz 1985b).

Migration is obviously highly variable from place to place, but in general there seems to be a good deal of vagrant wandering as well as true migration. During autumn in the USSR the birds may begin moving southward gradually as early as August, although the migration peak does not occur until late September and early October. Some birds are still seen as late as November, or rarely into December. Similarly, the spring migration may bring some birds back into the USSR by early March or even late February in the southernmost regions, but not until early May in northern Kazakhstan. Migrating flock sizes tend to enlarge as the birds move south, gradually increasing from a few birds in more northerly areas to as many as several hundreds or even thousands as they approach their winter quarters. Most birds migrate at quite low heights of less than 50 m above ground when flying over the steppe regions, but they may reach elevations as high as 3509 m above sea level when crossing mountainous areas such as the greater Caucasus (Dementiev and Gladkov 1951).

Social behaviour

During the non-breeding season this species is highly gregarious, and the birds sometimes gather in flocks of as many as several thousand of all ages and sexes. The sex ratio in such flocks is approximately equal, and the measured home ranges of radio-tagged birds of both sexes were similar, averaging approximately 400 ha. A few weeks before the start of the reproductive periods the flocks begin to break up and become segregated by sex. At that time the males become territorial and establish territories of 1–3 ha, which are not randomly distributed but rather are clustered into 'display centres'. Females visit these display centres for the purpose of fertilization, and then nest independently of the males and their territories. At this time the males have relatively small home ranges (of 27 and 83 ha in two cases with most of the time spent in areas of only 5 and 9 ha), but the females increase their home ranges (to 589 and 1206 ha in two cases) as they visit various male display centres. When they begin to nest, the females again have reduced home ranges that include the nest site and nearby foraging areas (68 ha for a female during her egg-laying period, and 21–33 ha in the case of an incubating female). Brood-rearing females also have highly restricted home ranges (of 3–5 ha in one case). With the end of the reproductive period the males' territories are dissolved and they again establish larger home ranges (of 45 and 181 ha in two cases) as they begin their post-breeding moults. Following this moult, the mixed-sex and mixed-age flocks begin to form once again and the birds begin migrating back into their 'wintering' areas (Schulz 1985b).

Recent observations by Schulz (1985b, 1986a) have confirmed the view that this species is promiscuous, the males defending display territories during the breeding season that they vigorously advertise and into which they attempt to attract females. After fertilization the females leave the male territories and nest entirely independently of them. Schulz described the mating system as being one of male-dominance polygyny, with intermediate dispersion of the males and having some similarities to that of the Eurasian capercaillie (*Tetrao urogallus*) in that males display in more or less clustered display centres, rather than being uniformly distributed over the total available habitat.

In Iberia, the breakup of winter flocks and the establishment of male territories begins about the middle of March and is nearly completed by the end of April. During that time interactions between males occur very frequently, and aggressive excitement is expressed in both sexes by a compression of the neck feathers while stretching the neck (Figure 30a). Males in spring also develop a fleshy enlargement of the neck, increasing its apparent size, although they evidently do not have an inflatable oesophagus or throat-pouch. In females the tail feathers are also cocked and somewhat folded during

Fig. 30. Social behaviour of the little bustard: (a) male aggressive posture; (b) territorial male running beside an intruder, after lateral threat behaviour; (c) territorial male during calling; (d) precopulatory posture of male towards dummy female; (e) female shock-display posture; and (f) maximum downstroke (second wing-beat) during male jumping display. After photos in Schulz (1986a) except for (e), which is after sketch in Cramp and Simmons (1980).

aggression, but tail-cocking in males is more indicative of subdominance or fear. Various aggressive signals are used by competing males, including approaching the intruder, stationary or mobile lateral threats (Figure 30b), and bill-fighting or other kinds of actual fighting, such as head-pecking or breast-to-breast jostling. Aerial chases may also occur. Calling while erecting the black and white neck feathers into a shield-like shape, in which the white V pattern is outlined against a heart-shaped black background (Figure 30c), is also commonly used by males to advertise and defend their territories. The frequency of such calling markedly increases when other males approach a resident territorial male. This 'snort-call' also involves a sharp backward neck-jerking movement that may also have visual significance as an agonistic signal, since it exposes not only the contrasting black-and-white neck but also the white underparts (Winkler 1973, Schulz 1986a).

During normal flight, as well as during territorial wing-beating and jumping displays, the specialized wing feathers of males produce a distinctive whistling *sisisi* sound that has important significance towards other birds of both sexes, sometimes stimulating territorial behaviour by other males or their fleeing. When males are flying within their territories they do so with unusually shallow wing-beats and with the neck raised at a slight angle that may serve to advertise the contrastingly coloured neck feathers. Such display flights (Figure 31b) also include gliding phases (Figure 31c). The two major territorial male displays consist of territorial wing-beating and jumping display. The former display is largely acoustic, and tends to serve to advertise and defend the territory against intruders. The jumping display is a largely visual display (Figure 30f) that serves to attract females from long distances.

Territorial wing-beating (also called 'wing-flashing') begins with an accelerating series of foot-stamping that leads to a call and, sometimes, three simultaneous fast, whistling wing-beats. These three sound sources thus produce a distinctive compound acoustic signal. Foot-stamping is performed only at specific sites, where the sound is amplified somewhat by subsurface soil characteristics, usually trapped air-pockets. This display, involving an accelerating stamping of the feet on the ground for about 2·5 s, is performed only during short periods of low light intensity, during dawn and dusk. The presence of females has no effect on display intensity, indicating that it is rather of territorial function, and since it cannot be seen over great distances it is unlikely to serve in attracting females into a male's territory.

In contrast, the jumping-display (also called 'wing-flash-and-leap') (Figure 31a) is performed only during periods of higher light intensities, and its frequency increases markedly in the presence of females. It is performed in a similar manner to wing-beating, but the preliminary foot-stamping phase is much less intensive, the wing-beats are slower, and the bird jumps off the ground a short distance during the wing-beating phase. At this time the highly contrasting wing patterns are clearly visible, and the jumping behaviour greatly helps to increase the visual conspicuousness of the male. The specific location of the jumping site also is apparently situated in such a way as to insure the male's maximum visibility. Jumping displays usually reach a height of about 65 cm above ground, and last approximately 1s (Schulz 1986a).

The direct following and chasing of females by males (Figure 30d) is an additional courtship display, which often begins as soon as a female enters a male's territory. Most males discontinue their jumping or territorial defence displays as soon as a female appears, thus often chasing the female off their territory, but some males continue to perform these territorial advertisement activities. As a result, females tend to accumulate in the territories of those males that continue to defend their territories and do not immediately begin to chase the hens. However, chasing does precede copulation. It consists of the male running rapidly behind the female, with his head retracted into the neck-collar, and repeatedly stopping and calling whilst throwing the head and body sideways. Treading is preceded by the male repeatedly pecking the hindneck of the female. Copulations normally only occur after a female has been chased for some time, and are performed very rapidly, although a captive female was observed to assume a receptive crouching position, whereupon a male immediately ran to her and mated with her (Frisch 1976, Schulz 1986a).

Like other bustards, birds of both sexes and all ages sometimes perform 'shock display' (Figure 30e), especially in the presence of potential aerial or terrestrial predators, or when confronted with almost any visually exciting stimulus (Frisch 1976). Distraction displays have also been observed being performed by brooding females (Dementiev and Gladkov 1951).

Reproductive biology

Probably May represents the major egg-laying period for little bustards in Iberia, although some laying may also occur in late April, and in Morocco there are historical records of eggs being found as early as February and as late as July. Similarly, in the USSR the egg-laying period is fairly prolonged, and extends

Fig. 31. Aerial display of male little bustard: (a) stages in the jumping display, with time points (in seconds from start of display) indicated; (b) wing-beating and (c) gliding-flight phases of display flight. After photos in Schulz (1986a).

from late April to early June. Specific sites are apparently rather varied, but in areas where the grass is quite low the birds attempt to seek out places of higher vegetation up to 40 cm tall, so as to provide visual cover for the nest. However, at least in the Crau region of southern France the species' nesting and brooding habitat is unusually sparse and generally quite stony, even the more luxuriant vegetation only having an average height of less than 20 cm. However, losses from predators and grazing sheep are very high on these open, stony plains (Schulz 1980). Of eighteen nests found by Schulz (1985b), almost half were in flat situations lacking a depressed cup, whilst the others were in footprints, in plough furrows, or associated with clods of earth. Because of the cryptic colours of both hen and eggs, the nests are extremely difficult to locate visually, and additionally the incubating female may pull vegetation over her body, thus concealing herself even more.

Eggs are usually laid at two-day intervals, either in morning or evening hours. Clutch-sizes are typically three or four eggs, although occasionally clutches of only two or even as many as six eggs have also been found. Larger clutches (of up to eleven eggs) almost certainly represent the efforts of more than one female. Of eighteen complete clutches found by Schulz (1985b), twelve had four eggs and six had three, an average of 3.7 eggs per clutch. However, a captive female laid a total of eleven eggs in a single season, over a period of slightly more than a month (Frisch 1976). Interestingly, one wild and radio-tagged female studied by Schulz (1985b) laid at least three clutches, one right after another as a result of nest destruction, before she began incubating successfully, all in slightly more than one month. All three of the nest sites were between 500 m and 1 km apart. Schulz found that in his study area the females began laying and incubating more or less synchronously, within about 2 weeks during April. Because of some renesting, some clutches were begun as late as June in each of three years of his study. Incubation began only after the last egg had been laid. Of twenty-five clutches studied by Schulz, only 4 per cent were successful, with the highest rate of nest loss (80 per cent) in pasturelands.

Incubation requires 22 days, and newly hatched chicks average about 25 g. Schulz estimated that the average size of newly hatched broods was about 3.5 young, or only slightly less than the average clutch-size. However, over the 4 weeks following hatching the average brood size dropped progressively, reaching about two young per female for broods more than 4 weeks old. He judged that on the basis of his breeding success estimates, an initial clutch-size of four eggs was statistically required for a female to produce a single successfully raised chick. Like other bustards, the young are fed bill-to-bill for their early period of growth, and only gradually begin to catch food for themselves. Fledging occurs at approximately 28–32 days after hatching, at which time the young probably weigh around 200–250 g on average. They are essentially fully grown at 50–55 days, but remain with their mother at least into their first winter and perhaps until the start of the following breeding season.

Evolutionary relationships

In general, recent taxonomists have considered the little bustard to comprise a monotypic genus (*Tetrax*); such treatment for example was used by Cramp and Simmons (1980) and Urban *et al.* (1986) in their regional monographs, but not by Ali and Ripley (1983). These authors considered the little bustard a member of the nominate bustard genus *Otis*, which otherwise includes only the great bustard, although Dementiev and Gladkov (1951) expanded the genus further to include the houbara bustard as well. Apart from their overall size differences, there are indeed some rather marked similarities between the little and great bustards. Both have distinctively characteristic short and blunt bills, which are broader than they are tall at the base. They also lack crests as adults, have rather similar bluish grey facial colouration, and have relatively short tarsi. Other similarities become apparent upon inspection of adult wing and tail plumage patterns as well as the patterns of the downy young (the great bustard's young being somewhat more buffy, and having fewer black-bordered buffy stripes present).

However, there are also some marked differences in adult male anatomy; for example, the great bustard has a complex inflatable gular pouch and oesophagus that is lacking in the little bustard, and the specialized sound-producing primaries of the little bustard are absent in the great bustard. Certainly these differences are related to sexual and species signalling, and as such cannot be considered as reliable generic traits. Thus, although I have with considerable reservations followed the recent general practice of maintaining *tetrax* as a separate monotypic genus, I have also placed *tetrax* and *tarda* relatively close to one another in my suggested phylogram of bustard affinities (see Figure 2).

Status and conservation outlook

A general review of the status of this species has been provided in Chapter 4, but some additional points may be made here as well. A considerable part of the 1980 European Bustard Symposium, held in León, Spain, was devoted to the biology and international status of the little bustard, and these materials were published as part of *Bustard Studies 2*, in 1985. An important part of that symposium was the contribution by Schulz (1985a), on the world status of the species. Based on this summary, it is apparent that the species is now either extirpated from or has become very rare in most of the countries where it was once a common breeding bird. Thus it is evidently now wholly gone from Algeria, Tunisia, Germany, Austria, Czechoslovakia, Yugoslavia, Romania, Bulgaria, Poland, and Greece. It is likewise probably gone from Morocco, Hungary, and Turkey; a recent review of the situation in Turkey (Kasparek 1989) indicates that there have been few recent sightings, and no currently known breeding places. The little bustard has undergone serious declines in France, and may well be gone as a breeding species by the turn of the century. Its 1980 population was probably between 7000 and 14 000. In Italy it has similarly declined greatly and is now confined to small populations near Foggia and in Sardinia. The Italian population is now limited to about a third of the Foggia plain's total area of about 360 000 ha. If an estimated density of one male bustard per 64 ha can be accepted (Petretti 1985), the entire plain could not have historically supported more than about 10 000 birds, and the number is now likely to be no more than about a third of this, or perhaps 3000 birds at most. The plain has been increasingly affected by ploughing, which has greatly reduced the amount of habitat remaining for steppe-adapted birds (Petretti 1988). In Sardinia, where there may have been as few as 450 little bustards in 1980 (Schenk and Aresu 1985), cereal croplands have increased in acreage about fourfold between 1958 and 1983, but human populations are still relatively low on that island (Petretti 1988).

According to data summarized by Ilicek and Flint (1989), the little bustard population in the USSR has dropped from about 8000 in 1971 to less than 6000 in 1980, with the largest remaining populations in the Ural region (3500), followed by two populations on the central and lower Volga (540 and 920), one on the north-western coast of the Caspian in Kalmyk ASSR (530), one north-east of the Caspian in Kazakh SSR (200), another in northern Kazakhstan (100), and one in the Ukraine north of the Sea of Azov (140). Major overwintering populations still occur on the steppes of Azerbaijan (6000–8000 in 1980, as compared with about 50 000 at the start of the century and about 20 000 in 1971), but only remnant wintering populations remain in other areas, such as in the Crimea and Black Sea vicinity, in Kalmyk and Dagestan ASSR (western coast of Caspian Sea), and in Uzbekistan (Aral Sea vicinity), Tadzhikistan, and Turkmenia.

Good populations of the little bustard are known to exist still in Portugal and Spain, where no significant population declines could be detected by Schulz (1985a). In Spain, the species' population may have been as high as 50 000–70 000 birds in 1980, which offers substantial hope for conservation efforts. Although country-wide estimates are unavailable for Portugal, the bustard population densities found there in recent years would suggest that good total populations must still exist (Schulz 1985b). Protection of these steppe-like habitats in Spain is needed not only for the little bustard but also for several other steppe-adapted European bird species, including pin-tailed sandgrouse, but they are increasingly being converted for use as croplands or for reafforestation with conifers (Juana et al. 1988).

Although the USSR certainly contains the largest single component of the historic breeding range, and probably still has the largest area of remaining habitat suitable for breeding, its populations have apparently plummeted in recent years (Schulz 1985a). The few data on breeding densities suggest that these are now much lower than those currently typical of Spain and Portugal. These massive population reductions are apparently the result of enormous changes in land-use, during which vast areas of steppe have been converted to cultivation, together with an increased level of grazing by domestic animals on the remaining areas of native steppe vegetation. To an additional but still unknown degree, hunting on both the breeding grounds and especially the wintering areas of Iran, Pakistan, and perhaps elsewhere had probably had a negative impact on the USSR's little bustard population. Thus it is not possible to consider that the USSR will retain a breeding population of little bustards indefinitely and might serve as a buffer against their possible extinction. Instead, Iberia must be considered as the most important single region for protecting the little bustard. This population has a management advantage over the USSR's in that it is essentially non-migratory. Thus it is not subjected to the mortality risks associated with relatively inprotected wintering grounds, as for example has been the case with the houbara.

GREAT BUSTARD (Plate 16)

Otis tarda Linnaeus, 1758

Other vernacular names: bustard (believed to be derived from a combination of the French names outarde and bistarde, both of which in turn are derived from the Latin *avis tarda*, 'slow bird'); grande outarde (French); Grosstrappe (German).

Distribution of species (Maps 14, 15)

Continental Europe (local in eastern Portugal, Spain, East Germany, western Poland, eastern Austria, southern Czechoslovakia, Hungary and Romania) east disjunctively to central Turkey, the Ukraine, the Volga and Ural valleys, the northern Caspian

Map. 14. Distribution of the great bustard, including its breeding (hatched), residential (solid), and wintering (stippled) ranges of its races *dybowskii* (dy) and *tarda* (ta). Dashed lines indicate the species' approximate range at the beginning of this century.

coastal region, and eastwardly as far north as northern Kazakhstan and south-eastern Siberia disjunctively to Lake Baikal and the Amur Valley of the USSR; southern limits at least originally extended to north-western Morocco (where now very rare, possibly extirpated), and from Iran and southern Tadzhikistan east through the Kirghiz steppes, Xingjiang (Sinkiang), Mongolia, and Nei Monggol (Inner Mongolia) to Heilongjiang (previously Manchuria), China. Mostly sedentary, but locally migratory, wintering south to Syria, Iraq, Iran, northern Afghanistan, and eastern China. Now greatly reduced, with major remaining populations in Portugal, Spain, eastern Europe (mainly Hungary and Romania), Turkey, and the USSR, and with no recent known nestings in Syria or Iraq.

Distribution of subspecies

O.t. tarda L.: northern Morocco (now extremely rare), and Portugal east through Spain and sporadically across central and southern Europe and central Asia to central Tien Shan range and the Turfan depression (Turpan Pendi) of Xinjiang (Sinkiang), China. Prior resident of Great Britain. Winters on breeding range over most of Europe; USSR component moves south to Syria, the southern Ukraine and locally elsewhere from Iraq to Afghanistan. Chinese population believed residential. Includes *korejewi* Zarudny 1905 (Vaurie 1965).

O.t. dybowskii (Taczanowski 1874: breeds from southern Siberia (Lake Baikal area) and Mongolia east to eastern Heilungjiang (Heilungking) province, north-eastern China. Winters in east-central China, from Liaoning south-west to southern Shanxi (Shensi).

Measurements (of *tarda*, mm)

Wing, adult males 598–633 (av. of 10, 617), adult females 475–497 (av. of 14, 486); tail, adult males,

222–259 (av. of 9, 243), adult females 208–219 (av. of 13, 214); tarsus, males 145–168 (av. of 15, 158), females 118–132 (av. of 19, 125); exposed culmen, males 32–40 (av. of 14, 36.8), females 27–36 (av. of 19, 30.8) (Cramp and Simmons 1980). Egg, av. 80 × 57.

Weights

Adult males (autumn to spring) 5.75–16.00 kg (av. of 11, 8.883 kg); adult females (autumn to spring) 3.26–5.25 kg (av. of 11, 4.421) (Cramp and Simmons 1980). Three zoo-raised males (one over 8 years old) weighed 7–8.8 kg, and twelve females 2.7–4.35 kg (averaging 3.683); there are some questionable reports in sporting literature of males weighing 21–24 kg (Glutz von Blotzheim 1973). Dementiev and Gladkov (1951) state that males are usually 7.5–12 kg, but exceptionally reach 21 kg, and females range from 3.8 to 6.5 kg; Ilicek and Flint (1989) give similar ranges, but suggest that average and maximum weights have declined in the USSR during the past century, with mean male weights of 7.2–11.2 kg typical in the late 1800s, compared with about 5–6 kg currently. Meinertzhagen (1954) stated that nineteen males (all wintering birds in Iraq) averaged 10.9 kg (24 lb), the heaviest being 16.8 kg (37 lb) and that 18 females averaged 7.72 kg (17 lb), the heaviest being 9.53 kg (21 lb), which also seem questionably high, especially for females. Maximum weights of males may be about 18 kg and females about 5 kg according to Osborne et al. (1984). Estimated egg weight 143 g, or about 3% of adult female (Schönwetter 1967).

Description

Adult breeding male. General colour above sandy rufous, broadly banded across with black (more heavily in *dybowskii*), the bands very strongly marked on the upper back and scapulars, less so on the lower back and rump; upper tail-coverts and rectrices light bay or vinous-chestnut, barred across with black, some of the bars broken up; the rectrices more or less distinctly tipped with white, the outer feathers white at the base, the outermost three (or more, in *dybowskii*) almost entirely white, with a broad subterminal band of black; lesser wing-coverts cinnamon with black bars like the back, but less closely arranged than on the back; remainder of wing-coverts, alula and primary-coverts white, powdered with grey towards the end of the feathers (the

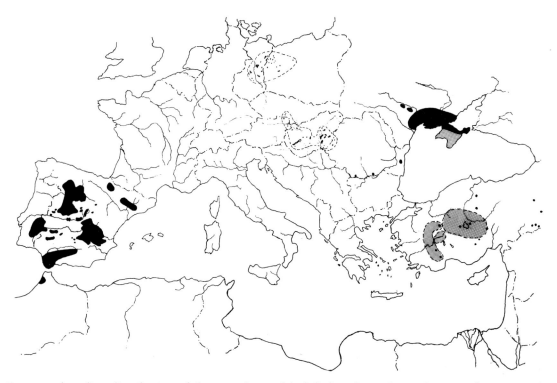

Map. 15. European breeding distribution of the great bustard (solid), based mainly on Cramp and Simmons (1980), but with updated information for Spain. Dashed lines indicate limits of remnant populations in central and northern Europe, and stippled areas show probable breeding regions in Turkey (where dots indicate areas of recent late winter or spring sightings).

median and lesser coverts mostly grey with white tips in *dybowskii*); remiges with white bases, the primaries otherwise mostly brownish, with their outer webs and tips blackish; the secondaries variably black-tipped and white basally, the white increasing in extent inwardly, the innermost secondaries being sandy rufous barred with black like the back; crown light grey, becoming tinged with rufous on the hindneck, which has numerous narrow black transverse bars; sides of face, ear-coverts, cheeks, throat and upper neck light grey, with bristly and elongated (to 190 mm) moustache-like feathers projecting from each side of the chin; lower neck mostly orange-chestnut, forming a band across the foreneck, which is washed with light grey, the sides of the neck with numerous small bars of black; sides of upper breast sandy rufous barred with black; remainder of undersides pure white. Iris dark brown; bill leaden grey, horn-black at the tip; tarsi and toes earthy or olive brown to dark grey, with a pink or blue tinge; nails horny black.

Adult non-breeding male. Like the breeding male, but with the breast blue-grey (like the neck) rather than chestnut, the moustachial feathers lost, and the thickness of the neck much reduced.

Adult female. Similar to the male, but much smaller, with poorly developed bristles of the sides of the face, and having the grey of the throat and foreneck continuous, the sides of the neck rufous, with only a few black bands, and the rufous colour grading to grey on to the sides of the neck. Females also have a double black bar crossing the tail feathers, whereas males of all ages have a single black stripe, which serves as a useful guide for field sexing (Ena *et al.* 1985).

Immature. Juveniles are generally similar to adult females, but are much paler and more freckled (less barred) on the upper surface, and have nearly all the wing-coverts barred with dusky brown, the white secondaries also freckled and barred with black, or with a large black subterminal bar; crown like the back, blackish and blotched with sandy-buff markings; hindneck greyish; a broad eyebrow and the sides of the face and throat greyish white; the lower throat and foreneck light sandy buff, obscurely freckled with dusky cross-markings; sides of the foreneck more sandy rufous, and the blackish cross-bars more distinct. The adult breeding plumage of the male is gradually acquired over a four or five-year period. The first-year non-breeding plumage resembles the adult non-breeding one, but includes several juvenal primaries and secondaries. The first-year 'breeding' plumage is similar, but the chest feathers of males may be tinged with cinnamon, and short moustche feathers may also be present. Second-year males (Figure 32b) retain juvenile outer primaries and vermiculated outer tertials, but are larger and have started to develop their inflatable display crop. Subsequent male breeding plumages become progressively more adult-like, but about 6 years may be required for the bird to attain full size and vigour (Figure 32c, d), and to develop maximally its inflatable oesophageal sacs (Figure 32e).

Identification

In the hand. The combination of a wing length of at least 475 mm, and a blunt bill that is no longer than 40 mm, identifies this distinctive species.

In the field. Over its Eurasian range this is probably the largest land bird, males being about the same size as a domestic turkey. When in flight, it is easily distinguished by white underparts and mostly white upper wing-coverts that contrast with extensive blackish areas on the primaries and tips of the secondaries. Both sexes have greyish heads and necks, and have rather long and barred brown and black tails, with white edging. Unusually silent, even during male advertisement display, which employs conspicuous visual rather than vocal signals, but in both sexes a gruff bark is used in alarm or aggression, and a grumbling snore or rattle is sometimes also uttered.

General biology and ecology

This is essentially a steppe-adapted species that inhabits a variety of relatively treeless and grass-dominated habitats, including pasturelands, but avoiding forests, parklands, savannahs, or other habitats having more than only relatively isolated trees or tall shrubs. Undulating topography is readily utilized, but steeper lands, rocky or strongly eroded terrain, and extreme desert-like regions are all generally avoided, although wintering birds have often been recorded in the deserts of Iraq and the Arabian peninsula. Nesting has also been reported on sandy and clay substrates around the Kyzylkum desert of the USSR. Probably a clear view of 1 km or more in at least three directions is needed by these bustards, as well as uninterrupted opportunities for mobility on the ground in all directions (Heinroth and Heinroth (1927–8) have suggested that the German name 'Trappe' for bustards may have been derived from 'Traben', in reference to their remarkable 'trotting' abilities). Well-drained but moist soils, supporting short to medium growths of herb-

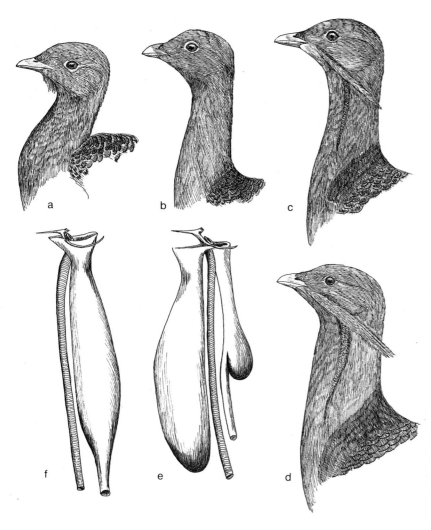

Fig. 32. Development of plumage in the great bustard: (a) adult female; (b) male at 1–2 years; (c) male at 3–6 years; and (d) male over 6 years (after sketches in Glutz von Blotzheim 1973). Also comparison of adult male tracheal and oesophageal anatomy of (e) great bustard and Australian bustard (after drawings in Garrod 1874c).

age or crops, and perhaps somewhat taller vegetation during the breeding period, are probably preferred (Cramp and Simmons 1980).

According to Isakov (1974), the great bustard was historically a species of steppe and forest-steppe (an ecotone zone between steppe and boreal forest), and was found rather seldom in semidesert habitats. With the clearing of the forests of Europe and what is now European USSR during the Middle Ages, the birds moved into the forest zone and began breeding in agricultural fields, especially those that alternated with fallow lands and pastures. Bustard numbers and their overall ranges probably reached a peak in Europe at about the end of the eighteenth century. As areas of natural and semi-natural grasslands (steppes and dry meadows) decreased with increasing agricultural activities, the range and numbers of bustards in Europe began to decline. The birds first disappeared from the subzone of mixed broadleaf–coniferous forests, then from the subzone of broadleaf forests, and finally from most of the forest-steppe areas. Isakov listed ten areas (regions, territories or republics) of the USSR where great bustards have been breeding in various densities, at least until recently. Of areas where the birds were breeding in large numbers, two were in bunch-grass steppe habitats, one was in sagebrush and bunch-grass steppe, and one in herb and bunch-grass steppe. Of areas where the birds were breeding annually in small numbers, three each were in bunch-grass steppe and in herb and bunch-grass steppe, and two each were in sagebrush and bunch-grass steppe and in forest steppe. Of areas where the birds bred only rarely, two each were in herb and bunch-grass steppe and in forest steppe, and one was in bunch-grass steppe. This would suggest that bunch-grass steppe represents ideal breeding habitat for at least the western race of the great bustard, followed in decreasing order of favourability by herb and bunch-grass steppe, sagebrush (wormwood) and bunch-grass steppe, and forest-steppe habitats. Less is known of the preferred habitats of the eastern race, which once occurred in areas of true steppes (*Aneurolepidium*, *Stipa* and shrub steppes) and also areas of meadow steppes (within wooded regions) and mountain steppes. Nesting in Soviet Central Asia has been reported at elevations as high as 3000 m.

Like the little bustard, this species is quite gregarious in the non-breeding season, and at one time gathered in flocks numbering as many as several thousands. Nowhere does that occur any longer, but in some areas maximum numbers of about 100–200 may still develop. Within such flocks there tends to be a sexual segregation, the adult males and females (including those with their still-dependent young) remaining relatively separated in distinct 'droves'. The males show a higher level of site fidelity than do females, which tend to wander more frequently. Wandering by both sexes is more common during severe winter conditions, when birds from north-eastern Europe tend to move west, sometimes reaching the North Sea region. Such flights into western Europe seem to be especially common when there is an early onset of cold weather accompanied by a large amount of snow, which apparently stimulates the species' inherent migratory tendency (Cramp and Simmons 1980, Hummel 1985).

Further east, in the USSR, the species is regularly migratory, wintering southward at varying distances. Some autumnal movement out of the western USSR begins quite early, such as in August (perhaps involving immatures or unsuccessful breeders), but the major southward movement occurs between October and December. The return flight in spring is more rapid, beginning in early March in the Ukraine and in Transcaucasia, and with birds arriving again on Soviet breeding areas during April. In Central Asia the migrations generally occur somewhat later. Autumnal flights are typically performed at fairly low altitudes, and in relatively large flocks of up to about 200. By comparison, flocks in spring more frequently comprise pairs or small parties, and migrations are sometimes performed at great heights. During both spring and autumn, flights reach substantial elevations as they pass over the Greater Caucasus mountains (Dementiev and Gladkov 1951). There is apparently also a very significant migration through south-eastern Turkey. This movement presumably represents birds that winter in the Near and Middle East and breed in the USSR. However, great bustards are present in Turkey all year and it is probable that these birds may at least in part represent an indigenous breeding population that may total several thousands (Cramp and Simmons 1980, Goriup and Parr 1985).

Foods consumed by the great bustard are extremely diverse, but they do appear to follow some seasonal patterns. During the summer animal materials, especially terrestrial insects, dominate the food intake of adults, and certainly this high animal component is also typical of juvenile birds. Thus, observations in the USSR suggest that 96 per cent of the diet (by weight) of young birds is of animal materials, and even for adult birds the animal component may account for nearly 40 per cent (Isakov 1974). Even higher proportions of animal foods taken during summer months (up to 87 per cent in August) are indicated in USSR data summarized by Ilicek and Flint (1989). Vegetation eaten during the summer is mostly cultivated plants and other green materials. As autumn progresses there is a gradual increase in the amount of plant materials, mainly seeds of cultivated plants. Plant materials continue to dominate the food intake through winter and spring. Spring volumetric measurements of Spanish birds suggest that a plant:animal ratio of 10:1 is typical (Palacios et al. 1975).

Among animal materials, insects and insect larvae predominate, orthopterans and coleopterans being especially important. Orthopteran foods include field-crickets, bush-crickets, mole-crickets, mantids and grasshoppers, with grasshoppers and field-crickets probably especially important summer foods for both young and adults. A similar wide variety of beetles are eaten (constituting an estimated 95 per cent of the total insect prey numbers in Spain), including carabids, chrysomelids, curculionids, meloids, scarabeids, silphids, and tenebrionids; other insects are consumed in smaller amounts, such as lepidopteran caterpillars and heteropteran bugs. Besides a miscellany of other non-insect invertebrates, some birds and mammals have been reported among the foods. These include such items as microtine voles and even young hares (*Lepus*), and the eggs and young of various ground-nesting birds (Gewalt 1959, Glutz von Blotzheim 1973, Cramp and Simmons 1980).

Plant foods are similarly diverse, but seem to consist of a high proportion of the green parts, flowers and seeds of composites, legumes, crucifers, and grasses, including some domesticated grain crops. Berries, fruits, rhizomes, and bulbs are also all eaten at various times. Fruit, seeds or roots of crops such as wheat, peas, beets, radishes, turnips, grapes, lentils, and olives are certainly eaten locally and some economic damage may result, especially to winter crops. However, the birds also probably help offset these losses by consuming considerable quantities of insect and rodent pests (Gewalt 1959).

Social behaviour

There has been a limited but continuing controversy as to the typical mating system of this species. This controversy largely relates to the interesting if unproven suggestion (Dementiev and Gladkov 1951, Sterbetz 1981) that a historically original 1:1 adult sex-ratio of adults has been increasingly disrupted by excessive hunting of males, especially during spring. According to this hypothesis, the once monogamous tendency of the species has necessarily progressively shifted towards polygyny or promiscuity as a result. However, even in currently protected populations the adult sex ratio ranges from about 1.5:1 to 2.5:1 in favour of females (Glutz von Blotzheim 1973). It is also hard to imagine that such extreme sexual adult dimorphism (males weighing at least three times and rarely up to seven times the females) and associated sexual bimaturism (females breeding at 3–4 years, males at 5–6 years) have evolved under social mating systems other than polygyny or promiscuity. Nevertheless, Sterbetz (1981) did observe some apparently largely monogamous sub-populations in Hungary (in two of the nine sub-populations he studied). In such monogamous matings the males would follow the fertilized female to her nesting site and keep guard nearby, sometimes remaining with her into the brood-rearing stage. However, one 'monogamous' male was observed to spend part of his time guarding the nest site, but he also spent several hours on his display site, presumably to attract additional females. In the other sub-populations Sterbetz observed a mixture of mating types apparently ranging from monogamy through harem polygyny (involving two to four females) to complete promiscuity (his 'unmated' type). He judged that the further the local adult sex ratio diverged from 1:1 in favour of females, the less active a male's defence of his territory would be. There was also less fighting observed among the adult males in areas of divergent sex ratios, the males instead tending to remain in communally displaying groups or 'rutting packs'. Sexually immature males formed smaller foraging groups that remained near or within the communal display areas of the adult males, and one old and relatively inactive male likewise did not participate in sexual display activities.

Apart from the question of occasional monogamous matings in great bustards and their possible causes, the mechanisms of polygynous or promiscuous matings in this species are highly intriguing. It is perhaps generally true that the majority of sexually active males remain in loose sex-segregated groups during the breeding season, as well as outside it. However, during the breeding period they tend to space themselves out (at distances of 50 m or more apart) over a common and traditionally utilised display ground. Within this common display ground, definite individual territorial boundaries are not maintained; rather, the males tend to move about and perhaps display randomly at various 'display stations'. Females visiting these groups may encounter and possibly mate with males at random, or perhaps they selectively choose specific males for mating on the basis of individual variations in their appearance or behaviour. A minority of males, however, perform their advertisement displays from relatively fixed territorial locations that are some distance from the common display ground. These territories are defended against other intruding males, and the resident male attempts to attract and maintain a harem of females within his own territory. The females may, however, range some distance beyond the limits of the males' defended territories (Gewalt 1959, Cramp and Simmons 1980).

A somewhat similar duality of non-monogamous mating strategies has been observed in Portugal (N.J. Collar, cited in Cramp and Simmons 1980). There most males exhibited non-territorial but otherwise 'lek-like' communal sexual behaviour, during which promiscuous mating occurred. However, a few other males exhibited definite territoriality, with an associated establishment of female harems. At least in the latter situation, females may establish a social rank order of their own, the most dominant bird of a harem sometimes pecking or even chasing other females receiving attention from the resident male. It further seems very probable that, regardless of the possible mating systems that might be used, the relative display activity and effective social dominance of individual males are likely to play important roles in the effective attraction of females and their subsequent fertilization. It would of course be of real evolutionary and ecological interest to have data on age- and experience-related behavioural variations in mating, or on other possible factors determining which of the several potential mating strategies may be used by particular males, and their relative effectiveness in fertilizing females. It might be hypothesized, for example, that monogamous males are more likely to consist of the younger age-classes, whereas older and larger birds may be more prone to polygyny or promiscuity. Similarly, information on possible female mate-choice mechanisms that might exist would be of comparable interest, as they may also have helped to shape male sexual behaviour patterns as much as have individual behavioural differences among males.

Besides these complex mating strategies, the specific male postures associated with courtship and territorial behaviour deserve special attention, inasmuch as they are among the most bizarre and spectacular of those of any bird species. Collectively, the male's advertisement display sequence can be called the 'balloon display'. This is essentially a wholly visual signal, the acoustic component associated with the inflation and deflation of the gular pouch being almost inaudible and probably insignificant as a social signal at any distance (Cramp and Simmons 1980). These displays are performed at traditional locations, much like the leks of various grouse (Johnsgard 1973, 1983a), and the sites of both have some similar physical attributes, such as providing excellent visibility and panoramic vistas, and are generally on somewhat elevated sites. One apparent difference, however, is that grouse leks are usually rather clearly subdivided by territorial boundaries, with the individual males establishing better or poorer territories (in terms of their relative size or location) on the basis of prolonged inter-male competitive interactions. Thus female grouse can perhaps rather quickly locate dominant or 'alpha' males on the basis of their territorial attributes (such as by simply moving to the more centrally located and/or larger territories). To judge from presently available information, this situation does not exist in the great bustard, and instead it may be necessary for females to make their mate-choice decisions on the basis of such possible clues as individual differences in male sizes, male display activity levels, or some other specific male variables. It is perhaps significant that, unlike grouse or other typical lek-forming birds, individual adult male great bustards vary enormously in weight (probably in direct relationship to their age, as is known to occur in Australian bustards), and perhaps these size variations are somehow positively associated with individual male mating success. In a recent review of the possible relationship of lek behaviour to avian size and plumage dimorphism, Höglund (1989) judged that although the great bustard is indeed a lekking species, it is the only member of the bustard family fitting his definition of this behaviour. He further noted that in ground-displaying lekking species such as the great bustard, there is a greater tendency to exhibit size dimorphism than in non-lekking relatives. However, he did not find the general expected correlation between plumage and size sexual dimorphism on the one hand and the presence or absence of lekking behaviour in the groups of birds that he critically analysed.

The major period of male advertisement display in Europe is concentrated from late March until the end of April, although some sporadic display activities may persist into early June. In the USSR courtship is reportedly concentrated during mid-April and May, or perhaps averages somewhat later than is typical in Europe. As with other bustards, the daily cycle of display primarily occurs during the early morning and late afternoon hours, with some nocturnal activity on moonlit nights and with occasional midday display at the peak of the display period. Additionally, at various times during the year but especially during the period of spring display, immature birds may also engage in social display activity to some degree. They often perform distinctly crane-like 'dancing' movements such as neck-stretching (Figure 33c), bowing (Figure 33d), and leaping, all being postures that in cranes (Figure 33e) apparently represent ritualized aggressive movements performed in similar conditions of generalized excitement (Johnsgard 1983b). An additional similarity between bustards and cranes is that in both groups during intense threat the birds may lie almost prostrate on the ground in a seemingly submissive posture (Figure 33f). The bowing posture may also be behaviourally related to the defensive 'shock-display' posture of both adult and young great bustards (Figure 33b), and indeed of bustards generally (Gewalt 1959).

During much of the period of spring display, adult males walk about in a distinctive 'ready' posture, with the tail variously cocked and the neck somewhat enlarged (Figure 33a). This apparent neck enlargement is partly the result of feather erection, but is largely derived from the inflation of the oesophagus. In adult males this is typically modified into an unusual double sac structure (Figure 32e), including an anterior gular pouch and a more posterior 'display crop' that evidently serves neither as a crop for storage of food nor specifically as a vocal resonating structure, but instead simply provides for a visually significant neck-enlargement mechanism. There is perhaps some individual variation in the degree to which the 'display crop' of the lower oesophagus develops, but the oesophagus itself is apparently capable of being distended in adult males (Niethammer 1937, 1940, Gewalt 1965).

The balloon-display sequence (Figure 34a–e) is a relatively rapid visual transformation from a relatively inconspicuous and generally brownish bird to an almost entirely white one, much like the unfolding of petals of a gigantic flower. By this amazing transformation, males are probably able to attract the attention of female over a distance of several kilometres. As the display begins, the male cocks his tail over the back so that it points forward towards his head. This exposes a mass of white under tail-coverts that billow out behind. Neck inflation begins almost simultaneously with the tail-cocking phase, by a series of gulping actions. The wings are at that stage suspended limply at the sides as neck inflation begins, the wrist and elbow joints both being bent at approximately a right angle (Figure 34b, c). As the neck swells up, the head,

Fig. 33. Social behaviour of the great bustard: (a) threat by male (after sketch in Glutz von Blotzheim 1973); (b) shock-display of female (after sketch in Cramp and Simmons 1980). Also (c, d) 'dancing' postures of great bustard (after sketches in Grzimek 1972), compared with (e) comparable posture of a *Grus* crane (after sketch in Johnsgard 1983b), and (f) intense threat by female (after sketch in Grzimek 1972).

which is withdrawn backwards, almost completely disappears, but the feathery 'moustache' is erected almost vertically in front of either side of the bill. A strip of bare, bluish black skin also appears along each side of the neck, the width of which increases as the neck continues to inflate until it is as large as a soccer ball. The final phase of the display involves a remarkable erection or even reversal of many of the wing-covert feathers, so that the brown upperparts are suddenly nearly hidden from view. As this begins, the wing is extended so that the humerus, radius and ulna are in a nearly straight line, but the wrist is even more strongly flexed, thus hiding the dark primaries (Figure 34d). At this time a group of white feathers (probably axillaries) is pulled upward along the anterior edge of the wing, forming a white rosette that nearly obscures the brown back and flanks. Additionally the larger brown and black banded inner wing feathers are reversed, exposing their white undersides, and the secondaries and larger wing-coverts twist forward and fan out, exposing their downy bases. This completes the visual transformation of the bird into a gigantic double ball of white feathers (Figure 34e), in which one end can

scarcely be distinguished from the other (Gewalt 1959). In this incredible posture the male may perform foot-stamping movements and wheel about, causing the gular pouch to swing from side to side (Cramp and Simmons 1980).

This full-display posture, which requires only a few seconds to assume, may be held for as little as 10–15 s but is generally maintained for about 2 min. It may rarely be held for as long as 8 min; thus presumably the maintenance of gular inflation does not interfere with normal breathing. No vocalizations accompany the posture. Periods of full display, interrupted with brief interludes of alertness, may last up to 48 min. The full display may be repeated at intervals of 1 min or more, often as the bird moves from one display site to another, at times with the gular pouch still partially inflated (Cramp and Simmons 1980).

Displaying males are approached by females, who watch and sometimes circle them cautiously. The male will then follow and attempt to circle the female. The female may at times initiate copulation by pecking at the male's cloacal area and his other white feather areas, whilst the male performs pro-

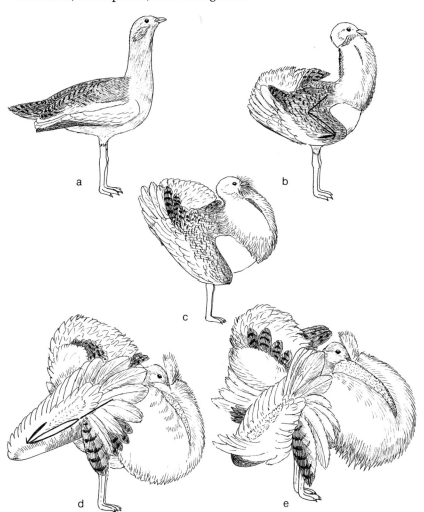

Fig. 34. Balloon-display sequence of male great bustard: (a) starting position; (b) initial neck inflation and tail-cocking; (c) later stage of neck-inflation; (d, e) maximum posturing, with extension of wing bones indicated in (d). After sketches in Gewalt (1959).

gressively more intense displays toward her, and may even attempt to push her to the ground with his wrists. Before copulation the male stands over the squatting female and pecks repeatedly at her head until she rises beneath him and insemination occurs; actual copulation lasts only 5–10 s (Glutz von Blotzheim 1973). In contrast to the situation in lek-forming grouse, most copulations reportedly occur at the edges of the display ground rather than its centre, but the two groups are probably otherwise similar in that copulations are likely to involve the most experienced and socially dominant males (Cramp and Simmons 1980).

Reproductive biology

The breeding season in central Europe closely follows the peak of male display. Full clutches may be found from the last third of April through May, and mainly occur during the first half of May. Replacement clutches are often found in central Europe during June and July but may be quite infrequent in Portugal (Gewalt 1959). In the southern USSR, eggs have been found as early as April 11, but more generally the egg records seem to be from late April through May, with the latest records for early July (Dementiev and Gladkov 1951). Clutches are usually two and sometimes three eggs, with replacement clutches not clearly any smaller than initial ones. Four-egg clutches are rarely found, and even five-egg and six-egg clutches have been reported, but thee are very rare and probably reflect the efforts of more than one female. Of twenty-four documented European clutches, twenty-one had two eggs, two were of single eggs, and only one was of three eggs, although in favourable breeding years three-egg clutches may be more frequent (Glutz von Blotzheim 1973).

Breeding densities of great bustards have generally been very low, at least in recent decades. However, one area of Portugal was judged to have at least sixteen nests in an area of 50 km² (one nest per 3.1 km², and some nests were situated as close as 58 m apart (Cramp and Simmons 1980). In an area of Germany, five females were breeding in an area of 50 km² (one breeding female per 10 km²). Both of these estimates certainly represent very high breeding densities; more typically, great bustards occur as general population densities of about one bird per 3–7 km² (Glutz von Blotzheim 1973). In Hungary the highest bustard density exists, where 1300 birds occurred on 5670 km² (one bird per 4.4 km²) (Istvan 1983). Probably nesting tends to occur in clustered 'patches'; thus in the USSR as many as three nests per km may be found near male display grounds, but large gaps of 10–15 km occur between such patches (Dementiev and Gladkov 1951).

The nest is generally only a shadow depression in low grass or similar herbaceous vegetation that provides visual protection without significantly obscuring the female's view. Thus taller grasses are only infrequently used for nest sites. Eggs are laid at one- to two-day intervals, and incubation may begin with the first, second or final egg, so that hatching tends to be staggered. Hatching in Europe and the European USSR typically begins about May 10, with a probable peak in June. Incubation period estimates range rather widely from 21 to 28 days, but the period was at least 26 days for three incubator-hatched eggs collected from a female that had begun to incubate them herself (Goriup 1985a). Like other bustards, the young are fed bill-to-bill by the female initially but gradually learn to feed themselves. Although they are extremely difficult to raise under captive conditions, largely because of the painstaking amount of hand-feeding required for the delicate chicks, Heinroth and Heinroth (1927–8) did so successfully for the first reported time. They provided some useful data on chick growth (see Figure 22) and photographs of young bustards at various ages (see Figure 23). Some additional growth curves and information on captive rearing of great bustards have been provided by Gewalt and Gewalt (1966) and Osborne (1985).

There is little information on breeding success, but probably most causes of failure are directly or indirectly caused by humans. In Hungary, where 80–90 per cent of the population nests in corn or fodder fields, about 70 per cent of the nests are destroyed by machinery (Istvan 1983). However, at least some of these losses must be balanced by renesting. In Spain it has been estimated that there is a pre-hatching loss of eggs of about 50 per cent, followed by a pre-fledging loss of 57 per cent (Ena et al. 1987). Females attempt to lure possible predators from their nests or broods by distraction displays, with the wings drooping and the tail elevated, and both males and females have rarely been known to threaten or even attack humans who venture too close to a nest or brood (Sterbetz 1981). The young birds remain with their mothers for much of their first year of life, perhaps even remaining in contact during the following breeding season. Females probably breed initially at 3–4 years, and although males may begin to show some sexual activity at that age they probably do not begin actual breeding until they are 5–6 years old. Longevity data for wild birds are still lacking, but in captivity the birds are known to be potentially very long-lived, surviving to at least 28 years and perhaps even as long as 50 years (Glutz von Blotzheim 1973). Productivity data are almost non-existent, but of a sample of 118 birds surveyed during autumn in Spain, only 5.93 per cent were chicks (Ena et al. 1985), suggesting that an annual recruitment rate of about 6 per cent may be typical. An estimated annual adult mortality rate of about 8 per cent was later estimated for this population (Ena et al. 1987). These productivity and mortality rates are quite similar to those typical of cranes, which also usually have two-egg clutches, probably mature at comparable rates, and otherwise have similar life-history characteristics (Johnsgard 1983b). Longhurst and Silvert (1985) assumed a 95 per cent annual survival rate for adults, but believed from Hungarian data that an annual population increase of 8.5 per cent per year was possible under certain hypothetical conditions (average clutch-size 2.5 eggs, fledging survival 50 per cent, survival during first, second, and third years 50, 75, and 90 per cent respectively, followed by annual 95 per cent survival rates of breeding females). This model seemingly assumes that all nesting females succeed in hatching full clutches, which is quite unlikely, given the high estimated rates of egg and nest losses cited earlier.

Evolutionary relationships

As noted in the account of the little bustard, the genus *Otis* is usually regarded as being monotypic, as for example by Ogilvie-Grant (1893) and Peters (1934), or is sometimes expanded to include the little bustard as well (e.g. Ali and Ripley 1983). An even more inclusive view is that of Dementiev and Gladkov (1951), who also included the houbara bustard in the genus *Otis*, but without special comment or justification. I am inclined to believe that these three species are all moderately closely related (see Figure 2), especially the little and great bustard, but I have followed recent tradition in maintaining their generic separation.

Status and conservation outlook

Probably few European species have suffered quite so much at the hands of humans as has the great bustard, which in the mid-eighteenth century was so common in Europe that 'bustard plagues' were common, during which schools were closed so that children could participate in collecting bustard eggs, and great damage was reportedly done to some crops. The historical chances in bustard distribution and numbers was well summarized by Isakov (1974), who judged that the great bustard was at its maximum both in its European range and population size at the end of the eighteenth century. During the nineteenth century its numbers began to decline markedly, the birds initially disappearing from the areas that they had most recently colonized, namely the original forest zone. They were not to be found in England after 1838, and shortly afterwards disappeared from France. During the first half of the twentieth century the birds had disappeared from most of the forest-steppe regions of the USSR, and now breed only in the steppe zone, their original primary habitat. Isakov estimated that in the early 1970s probably 2200–2300 pairs of the western subspecies *tarda* still remained in the USSR, and guessed that perhaps 500–600 pairs of the eastern subspecies *dybowskii* existed. A total USSR great bustard population of about 8600–8700 birds was indicated. He judged the USSR population of breeding pairs in 1971 to be as follows: Black Sea, 735; North Caucasus, 90; Volga–Don valleys, 575; Kazakhstan, 900; Tuva, 200; Trans-Baikal, 350 pairs. He also judged the European population to consist of a Baltic component (East Germany, Poland) of about 1000 birds, a middle Danube component (Austria, Hungary, Czechoslovakia) of about 3000 birds, and a lower Danube component (Romania) of about 1500 birds. Other components in Iberia, North Africa and the Near East were not estimated but regarded as probably fairly low. The Mongolian–Chinese component was not mentioned.

More recently, an attempt at a status survey of the great bustard was undertaken by Collar (1985), whose paper was summarized at greater length in Chapter 4 and who concluded that the world population about 1980 was in the region of 20 000 birds. In order of numerical strength, the most significant countries for the great bustard were judged to be Spain (6000–8000 birds), Hungary (about 3400), Turkey (3000–4000), USSR (2980 in 1978–80, or substantially less than the 1971 population estimates of Isakov), Portugal (about 1000), and probably Mongolia and China (no estimates available). Other European countries still supporting great bustards at that time included East Germany (560 in 1980), Czechoslovakia (315 in 1979), Romania (about 300 in 1976), Austria (about 150 in 1979), Yugoslavia (about 30–40 since 1965), Poland (16 in 1980), and Bulgaria (few if any). The species is now almost extirpated from Morocco, where 100 birds perhaps remained in 1980. It is possible that northern Syria still supports a few great bustards, the species having been taken by hunters there as recently as 1982. Iraq may support some breeding birds in Kurdistan, and adjacent western Iran probably likewise may have about 100–200 birds still surviving. The latter population includes perhaps as many as 100 breeding females (Cornwallis 1983), although recent warfare in both countries has certainly reduced these probabilities. No surveys of great bustards have been made in Iran since the revolution, but the breeding population has apparently undergone a tremendous decline since that time (Razdan and Mansoori 1989). Some breeding in Turkey still probably occurs in central Anatolia (upper Sakarya River plains, Malya plains) and in south-eastern Turkey (Tigris River valley), and there are a fairly good number of bustard records from eastern Turkey. Most winter records are from west-central Turkey (Kasparek 1989).

In the major world areas of great bustard occurrence identified by Collar, the bustard is probably declining in most for which any data exist. Thus in the USSR it probably declined by more than half during the period 1970–80, and in Spain it has similarly declined by perhaps as much as 50 per cent in less than a decade. In Turkey too few data are available to judge the population trend. Probably only in Hungary, where it has been fully protected since 1970, has the species increased significantly. Not only has it been provided with special 3500 ha bustard sanctuary in Devavanya, but additionally about 16–18 per cent of the 425 000 ha of the protected lands in Hungary represent bustard habitats (Istvan 1983, Dobai 1983). In Portugal it has perhaps also increased, or at least its population has been more completely inventoried in recent years. However, in East Germany it has undergone a gradual decline, and in the remaining countries of central Europe its situation has generally become critical (Dobai 1983). In Europe it is now generally protected from hunting in all countries except Portugal, Spain and Romania. All told, the 'known' world population of 14 000 (as opposed to the projected or extrapolated total of about 20 000) in the early 1980s was little more than half the maximum population estimate of 27 000 birds arrived at by similar methods in 1978 (Collar 1985).

Isakov (1974) has emphasized that not only is a complete ban on the hunting and trapping of great bustards needed to prevent their extinction, but also the preservation of steppe-like ecosystems is required, including both natural and semi-natural biotopes. An international project for the protection of such ecosystems, and their associated biota, is thus an urgent need in his view. The recent symposium on the ecology and conservation of the world's grassland birds (Goriup 1988a) has reiterated this theme. Thus Goriup (1988b) observed that most of the steppe-like habitats of Europe and the coastal belt of North Africa have either been converted into variably intensive agricultural croplands or have been seriously degraded, usually by overgrazing. As a result, at least twenty-six species of steppe-adapted birds of these regions, including both the little and great bustards, are now or might in the future be threatened with disappearance from continental Europe unless actions are taken. For example, the important great bustard population of Spain depends in part on the retention of its steppe-like habitats, which are increasingly being affected by conversion to farmlands and reafforestation (Juana *et al.* 1988). Additionally, technological changes, poaching, inadequate protection, and losses through overhead lines, pesticides, machinery, and other mortality factors have all had serious impacts on Spanish great bustard populations (Cardosa 1985). Similarly, although in China some 35 per cent of the total country, or about 260 million ha, consists of steppe-like habitats, about three-quarters of this vast area is already being utilized by humans for agriculture and grazing. People are now also being encouraged to move to such frontier regions to convert the remaining steppe habitats to farmland, and roads are now being constructed through the heart of these steppes. However, a 100 000 ha nature reserve has been established in Xianghai, where bustards are receiving special attention (Hsu 1988).

AUSTRALIAN BUSTARD (Plate 17)

Ardeotis australis (J.E. Gray) 1829

Other vernacular names: plains turkey, wild turkey; outarde d'Australie (French); australische Trappe, Wammen Trappe (German).

Distribution of species (Map 16)

Australia, now primarily limited to more northerly regions, especially of Queensland, north to the Cape York Peninsula, and west to the Barkly Tablelands and the Kimberley Plateau, with smaller populations on the Nullarbor Plain and in other arid interior habitats; also southern New Guinea (savannahs of Fly River and southward, from Pulau Kimaan of Irian Jaya to the Bulla Plains of Papua). No subspecies recognized.

Measurements (mm)

Wing, males 553–610 (av. of 4, 568), females 481–533 (av. of 5, 503.6); tail, males 235–270 (av. of

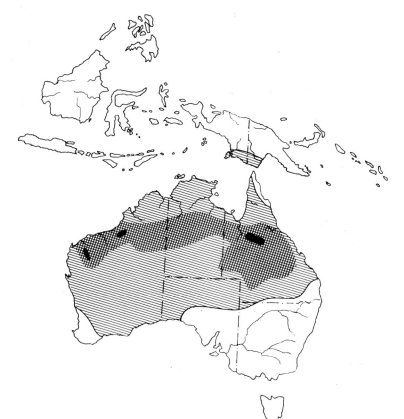

Map. 16. Distribution of the Australian bustard (hatched, with denser populations indicated cross-hatched and solid).

4, 568), females 220–292 (av. of 5, 235.6); tarsus, male 196, female 148–184 (av. of 5, 162.8); culmen, males 65–85 (av. of 6, 74), females 52–85 (av. of 5, 65) (Rand and Gilliard 1968, Museum of Victoria and Australian Museum data). Egg, av. 80 × 56 (Schönwetter 1967).

Weights

Adults c. 4.54–7.37 kg, males to c. 14.54 kg, females to c. 6.4 kg (Frith 1969, Rand and Gilliard 1968). Weights of males in captivity increased with age; those of females varied seasonally (Fitzherbert 1978). Males usually weigh 6.4–8.2 kg (14–18 lb), but sometimes reach 12.7 kg (28 lb) and rarely 14.5 kg (32 lb); females are usually 4.5–6.4 kg (10–14 lb) but may be as low as 3.4 kg (7.5 lb) (Serventy and Whittel 1962). Two females 2.1 and 3.175 kg (Museum of Victoria); three males 4.1–7.2 kg (av. 6.0 kg) (Australian Museum). Egg, 135 g (estimated).

Description

Adult male. General colour like that of *nigriceps*, but averaging darker and more heavily shaded with salty grey on the upperparts; the lower throat dull ashy white freckled with dusky vermiculations in the shape of wavy cross-lines on the rest of the throat and neck. Upperparts generally dark brown, finely marked with light brown; remiges and most of the wing-coverts brown; a black area occurs on the upper greater wing-coverts with white tips to some of these feathers; forehead, crown and elongated nape feathers black; eyelash pale olive-yellow, cheeks and neck greyish-white with transverse black bars; blackish band across chest at base of foreneck; breast and abdomen white; outer rectrices black and white, with more or less barred tips, inner ones like the back; lateral under tail-covert brown, the median coverts white. Irish greenish-white to yellowish-brown; bill straw-white to yellowish, with olive and black culmen; tarsi and toes straw-yellow to greyish or olive.

Adult female. With less black on the wings and body, and much smaller. Also generally more greyish and more vermiculated throughout, the remiges mostly bluish grey, with increasing amounts of vermiculations inwardly. The outer retrices barred with dark grey bands separated by lighter areas of vermiculated grey, and tipped with white, the median ones less heavily barred and more vermiculated like the back.

Immature. No detailed information, but probably quite similar to plumage changes described for *nigri-ceps*. Sexual maturity in both sexes may be reached in three or four years, judging from observations of hand-raised birds (D.M. White 1985).

Identification

In the hand. This species' large to very large size (wing over 475 mm) and its long, pointed bill (culmen 52–66 mm) eliminate all but the three other *Ardeotis* species. Compared with these it has almost black upper wing-coverts, with only slight white spotting. It is darker dorsally than the similar *nigriceps*, the neck is more distinctly vermiculated with greyish, and the median under-tail coverts are white.

In the field. This is the only bustard in Australia, and its large size alone separates it from all other species, except perhaps the brolga crane (*Grus rubicunda*), which has a bare red crown.

General biology and ecology

Based on recent aerial surveys, this bustard currently occurs in the highest densities in areas of Mitchell grass (*Astrebela*) plains, in or adjacent to areas where grasses dominate. It is furthermore generally confined to habitats where the upper canopy cover is less than 10 per cent, or where the upper canopy is less than 2 m high. The zone of tussock grasses and graminoids that extends across much of interior Queensland and north-eastern Northern Territory corresponds roughly to the bustard's current area of primary occurrence. Although some authors have suggested otherwise, high densities of bustards still occur in some parts of the Australian sheep pastoral regions, at least in northern areas. However, in more-southern areas heavy sheep grazing combined with the persistent vegetational effects of the rabbit plagues severely affects many of the tussock-forming perennial grasses, and may have reduced both the amount of cover and foods that are now available to bustards (Grice et al. 1986).

At one time, bustards were much more abundant and widespread in Australia, and flocks of up to a thousand or more were reported from New South Wales near the end of the last century. Currently, flocks up to 200 or so may still be seen in the Kimberleys, but usually they are found only singly, in pairs or in small groups. They have been reported to roost in tree clumps where these are available, but in treeless country they roost on the ground, usually on some high point. With dawn they move out onto the plains and spend the day foraging in the open country (Frith 1976). This daily and seasonal movement of the birds between open country and adjacent areas of heavier cover may be an important aspect of the species' habitat requirements (Downes 1982). Larger-scale migrations probably also occur, including not only regional movements in response to rainfall but perhaps also regular migrational movements between Australia and New Guinea (Blakers et al. 1984).

Although their foods have not been studied in any detail, the birds are known to eat grasses, seeds and fruits, and a wide variety of animal life, including grasshoppers, crickets, other insects, small rodents, young birds, and reptiles. Like other bustards, the birds are strongly attracted to fires, with their associated disabled or dead insects, and to plagues of mice or locusts (Frith 1969, 1976, Blakers et al. 1984).

Social behaviour

Like the great Indian bustard and the other species of *Ardeotis*, this is almost certainly a polygynous species, the much larger males displaying from territorial posts, even in the absence of females. There are a considerable number of published general accounts of the display (Murie 1868, Mattingley 1929, Minchin 1931, Frith 1969), which certainly in most respects is obviously very much like that of the great Indian bustard and the kori bustard, perhaps especially the former. However, only a single detailed description is available, that of Fitzherbert (1978), whose account is based on the close observation of several captive males.

Fitzherbert found that the onset of male display varied, the two heaviest males starting display the earliest (in June), when the males were approaching their annual average maximum weight, which in July was approximately 50 per cent greater than average weights in December and January. The two heaviest males not only started display the earliest but also performed at the highest levels of activity, whereas three lighter males continued their display longer than the heavier birds, and thus performed comparable total display activities. Regardless of these differences, males tended to perform synchronized display bouts, which suggested to Fitzherbert that perhaps similar synchronized displays may occur in the wild, possibly in dispersed lek situations.

Male display in this species is spectacular, in both intensity and duration, sometimes lasting from dawn to dark, with intervals of only about 5 or 10 min between bouts. In two birds that displayed most actively, about 95 per cent of the daylight hours were spent in full or partial display posture during the peak display period, and two birds were involved in full display for approximately 75 per cent of the recorded time. Foraging activity correspondingly greatly diminished during that period, as did pacing and preening, at least in the most actively displaying males. Females were not obviously orientated towards displaying males, and their seasonal changes in behaviour were not so well marked or predictable as in males, although fluctuations were greater.

Male 'passive displays' were typically performed at a small display site (about 1 m^2 in area) on which the male stood. On it, he assumed an erect posture, with the wings held against the body throughout and the bill variably tilted upwards. During arousal stages, the cheek pouch was expanded and the throat sac increasingly enlarged and lowered until it touched the ground or nearly so. Lastly, the tail was erected, exposing the long brown and white under tail-coverts. In this position the male was in full display or 'strut posture', and he might then make stationary stepping movements or sway from side to side, the throat sac swinging like a pendulum as the bird stepped and turned about. This grotesquely inflated posture closely corresponded to the stationary 'balloon displays' of the kori and great Indian bustards and less closely to that of the great bustard.

Calling sequences occurred at regular intervals (consistently averaging about 12 s apart from an initial period of less intense display) whilst the male was in this full display posture (Figure 35a). The call sequence began with the beak opening slightly (Figure 35b). The head was then lowered and the beak increasingly opened (Figure 35c) before being shut silently (Figure 35d). During the actual call (Figures 35e, 36) the cheek pouch and mid-neck expanded, and as the call finished, the beak opened again to expel air (Figure 35f). After the call, the bird resumed its full display posture (Figure 35g). The call was a guttural *rah* note that rose and then fell in pitch; the entire calling sequence lasted on average 2.5 s. As soon as the call was completed, the male began to turn with defined stepping movements in a partial circle, causing the pendulous throat sac to swing from side to side (Figure 35h). Thus after about eight or nine calls the male had rotated through a complete circle.

Typically, 'active display' was initiated by an approaching female, when the male left his display site and followed the female, remaining with a few metres of her, and with this throat sac swinging widely. At times he would turn broadside to the female, call, and then continue following her. Such active display sequences lasted up to 20 to 25 min, during which the calls were on average only 7.3 s apart, but the duration of individual calls remained constant. At times during active display the next call in the sequence would begin whilst the throat was still swinging (Figure 35i). During active male

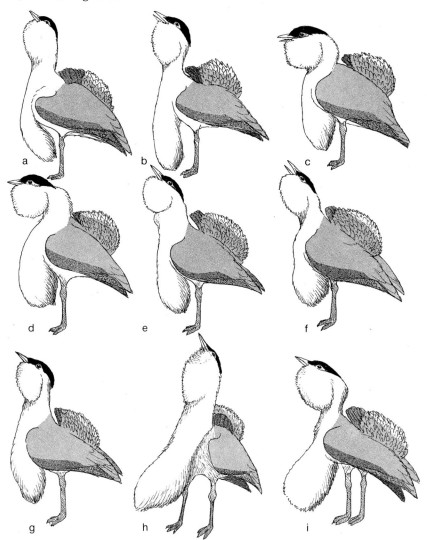

Fig. 35. Balloon-display sequence of male Australian bustard, after photos in Fitzherbert (1978).

display the inflated and pendulous throat sac was visible as far as 1 km away, and the call was audible from at least 300 m.

Copulation was observed only three times by Fitzherbert, but in each case began with a male in full display approaching a sitting or crouching female. Once beside her, he stood with his tail elevated, throat sac fully lowered, and throat pouch partly deflated. As the female remained crouching, the male moved from one side of the female to the other, always passing behind rather than in front of her. He pecked at her head, and at times pulled at her beak, while the crouching female's outstretched neck was held nearly horizontal, but her head and beak pointing upwards. This sequence lasted from 22 to 33 min, the female remaining crouched throughout. Finally the male mounted the female, the wings covering her body, the head erect, and the throat sac raised. Cloacal contact was apparently only very brief, and afterwards the female immediately ran out from under the male, who followed her and began to display again. The first observation of copulation occurred in late September, a month after egg-laying had begun in late August. Females are promiscuous, and are attracted to the dominant male (D.M. White 1985).

Reproductive biology

The breeding season is somewhat variable, but generally coincides with the period when the grass is at its maximum growth. This occurs during the spring months in southern Australia, from September to November, and during the latter wet and early dry season in northern Australia, between January and March. As with other bustards, the nest is a

Fig. 36. Balloon-display of Australian bustard, after photo by P. Goriup.

simple scrape, either in the open or near a bush. Normally only a single egg is laid, but at times there are two (Frith 1969, 1976).

Only the female incubates, for a period of 23–24 days. At least under captive conditions, second or subsequent layings usually occurs between 12 and 20 days after removal of the previous egg. Shortly after hatching, the female takes the chick away from the nest vicinity, but little is known of further development under wild conditions. In captivity the young are usually able to feed themselves after 3 or 4 days, and can be moved to outside rearing areas after 10–20 days. Brooding females will accept and foster incubator-hatched chicks, and with such methods two chicks can be reared simultaneously by a single female (D.M. White 1985).

It has been reported that males may not be sexually mature until they are 5 to 7 years old (Appayya 1982); one captive male at Serendip Wildlife Research Station was about 5 years old when it displayed for the first time, but another began displaying when only 2 years old and another 3 years after hatching. This last-mentioned male was observed copulating a year later, with a four-year-old female, and these birds bred successfully. One captive female was about 5 years old when she initially laid, although another laid an infertile egg when only 3 years old. A year later (1981), three hand-reared females obtained from eggs collected during 1976 and 1977 had begun laying (D.M. White 1985), suggesting that a four- or five-year period to sexual maturity may perhaps be typical.

Reporting further on captive breeding results, D.M. White (1985) noted that over a ten-year period 115 eggs were obtained from fifty-one females, or slightly more than two eggs per female per year. Of these, 78 were fertile, 73 chicks were hatched, and 52 young were reared. One female produced eight eggs within a period of about 2 months, half of which were laid singly and half were two-egg clutches. By the time captive-raised females were a year old they had reached their normal adult weight of 2–4 kg, averaging about 3 kg. However, males continued to increase their average weight indefinitely, with very marked seasonal fluctuations with extremes of about 5 and 10 kg. The average July weight of adult males was about 8 kg, or nearly three times the average weight of females. Over a twelve-year period captive males showed a gradual yearly increase in average weight, with peak seasonal weights attained just about at the time that sexual display began (Fitzherbert 1978).

Evolutionary relationships

Snow (1978) considered this species part of the *Ardeotis* superspecies, but I think it probable on zoogeographic grounds as well as from behavioural similarities that *nigriceps* is the nearest living relative of *australis*.

Status and conservation outlook

The current range of the Australian bustard is far smaller than its historic range (Blakers *et al.* 1984, Grice *et al.* 1986), for reasons that seem to be related to changes in vegetation structure, land use and predation. Of these three factors, vegetation seems to have overriding influence, with the birds rare in or absent from severely overgrazed grasslands. The effects of introduced foxes and other predators may be of secondary importance, but few bustards are found where foxes are common, which is mainly in southern parts of Australia. The species is also absent where land-use is intensive, but the impact of such practice may depend on the time of cultivation relative to the breeding season, the duration of the fallow period, the extent of pesticide use, and the degree of direct human disturbance (Grice *et al.* 1986). It was listed by Fitzherbert and Baker-Gabb (1988) as a threatened species primarily associated with arid tussock grassland habitats that are among the most highly prized of natural grazing lands and carry large numbers of stock.

It is possible that bustards have increased recently, at least outside the south-east region, and especially in eastern Queensland (Blakers *et al.* 1984). The species' distribution in southern New Guinea is also larger than had been previously appreciated (Hoogerwerf 1964), although little historical information on its status there is available.

GREAT INDIAN BUSTARD (Plate 18)

Ardeotis nigriceps (Vigors) 1830–1 (1831)

Other vernacular names: Indian bustard; sohan, hukna (Hindi); outarde à tête noir, outarde des Indes (French); Indische Trappe, Hindutrappe (German).

Distribution of species (Map 17)

Arid grasslands of north-western and central India, originally from Sind and Punjab (now Pakistan) east to West Bengal and south to southern Tamil Nadu (Madras). Currently local in Rajasthan, Gujarat and the Deccan plateau of Andhra Pradesh, south sporadically to Karnataka (Mysore), and perhaps occasionally west to extreme eastern Pakistan. Now classified as endangered (King 1981). No subspecies recognized.

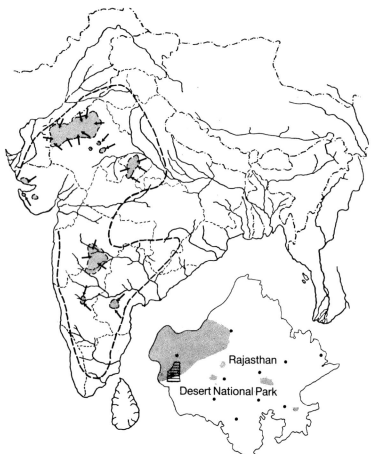

Map. 17. Distribution of the great Indian bustard, including its historic range (dashed line) and areas of recent records (stippled, with recent breeding localities indicated by arrows). The inset shows its distribution (stippled) in Rajasthan, with recent locality records indicated by dots (after Rahmani 1986).

Measurements (mm)

Wing, males 614–762, females 460–540; tail, females 229–254; tarsus, males 190–208, females 157–162; culmen, males (from feathers) c. 85–95, female (from skull) 88 (Baker 1921, Ali and Ripley 1983). Egg, av. 79 × 58.

Weight

Males 11.4–18.2 kg, females 4.5–9.1 kg (W. Elliot 1880); males c. 8–14.5 kg, females c. 3.5–6.75 kg (Ali and Ripley 1983); males 13–15 kg, females 6–8 kg (Vyas et al. 1983). A female and male aged 1.5–2 years weighed 4 and 8 kg respectively (Singh 1983). Estimated egg weight 145 g, or 3.3% of adult female (Schönwetter 1967).

Description

Adult male. General colour above dull sandy buff, everywhere finely vermiculated with blackish, some of the feathers with a rufescent tinge, but without any black or sandy-buff blotches; scapulars like the back; lower back, rump, and upper tail-coverts like the rest of the back, the latter with narrow white ends; lesser wing-coverts like the back, but with a slightly more rufescent tinge; median and greater coverts black with white tips, but the inner greater coverts mottled with white towards the base and shaded with bluish-grey, sometimes occupying the greater part of the feather; alula brown, shaded with grey and tipped with white; primary coverts slaty blue and tipped with white; the primaries mostly slaty blue, the longer ones more blackish, and with notches and bars of white on the inner ones, which are also tipped with white, these bars being irregular in shape; secondaries slaty blue, tipped with white and irregularly mottled with white towards the base of the outer web, the innermost secondaries like the back, but more greyish; rectrices like the back but somewhat more coarsely vermiculated and shaded with grey; about the middle of the tail a band of dusky blackish and two other bands of whitish obscured with dusky mottlings; crown and nuchal crest black; over the eye a few white streaks, but no distinct eyebrow; neck all round, lores, sides of face, throat, and undersides usually pure white; sides of breast with a few vermiculated sandy feathers and with a large black patch that forms a somewhat broken collar on the foreneck and extends with a few blackish feathers round the upper mantle; on the lower flanks a large patch of black feathers; under tail-coverts brownish-black with white bands at the end, some of the feathers bluish-grey freckled with brown on the margins and with a subterminal bar of black; under wing-coverts and axillaries white, but the greater primary coverts ashy, minutely freckled, and the edge of he wing black with whitish tips to the feathers. Iris pale to bright yellow; bill greyish brown to greyish white, dark at tip and near the forehead and often a little yellowish below; tarsi and toes generally yellowish creamy, a little dingy on the toes, sometimes with a light fleshy tinge or with a pale grey or plumbeous tinge.

Adult female. Very much smaller than the male, with broader white tips and more white bars on the wing-coverts and remiges, and further distinguished by the distinctive blackish vermiculations on the throat and foreneck, but the head has a contrasting broad white eyebrow stripe. The tail has a very distinct subterminal band of black, but the black band around the base of the foreneck is poorly developed or may be lacking.

Immature. The buff spots on the crown, hindneck and upper back that distinguish juveniles evidently disappear entirely at the first postjuvenal moult. Otherwise, young birds are perhaps distinguished by their coarser frecklings of the upper surface and by a greater amount of black at the base of the dorsal feathers, causing a slightly variegated appearance. Older immatures are increasingly difficult to separate from adults, and except perhaps by their weights (in the case of males) it may be impossible to distinguish young after their first year from older birds.

Juvenile. Female-like and sandy-coloured above with the usual vermiculations, but distinguished by numerous large triangular spots or bars of sandy buff, preceded by a black mark, which imparts a variegated appearance to the whole of the upper surface; the head is black, the feathers tipped with bars of pale sandy colour; there is a broad eyebrow of creamy white, as in old females; the underparts are buffy white, with indications of dusky cross-bars on the throat and chest; the markings on the wing are much more coarse and broken up into mottlings than in the adult bird, and the creamy white tips to the coverts predominate, giving a much whiter appearance to the wing.

Identification

In the hand. The combination of a large to very large size (wing 460–762 mm) and a long, pointed bill (culmen 88–95 mm) identifies this as an *Ardeotis*, and of these species it is the palest dorsally, with the neck almost pure white, with few or no vermiculations, and the under tail-coverts are almost entirely brownish black, rather than mostly or entirely white.

In the field. This is the only very large species of bustard native to the Indian subcontinent (Pakistan and India) and is thereby unmistakable.

General biology and ecology

An extensive series of important studies on the ecology of this species has now been prepared and published as Annual Reports of the Endangered Species Project, Bombay Natural History Society, including studies at Kerera, Madhya Pradesh (Ali and Rahmani 1982), at Nanaj, Maharashtra (Ali and Rahmani 1984), and at Rollapadu, Andhra Pradesh (Manakadan and Rahmani 1986). In all these areas the birds preferentially utilize wide open areas having vegetation that is below their eye-level, or about 40–50 cm high. In a study area at Nanaj, from 51 to 81 per cent (averaging 70.3 per cent) of the observations of bustards during three years were made on ungrazed grasslands of *Heteropogon*, *Chrysopogon* and *Sehimia*, that ranged in height from 5 to 70 cm. These areas were used for foraging, breeding, display, resting and roosting. Daytime resting was done in relatively tall *Sehima* grasses, and night-time roosting was in the more open and stony areas. Stony grasslands made up the second most frequently used habitat type, ranging from 2.8 to 31 per cent in three years, and averaging 13.4 per cent, but were used only for foraging and roosting. Grazed areas were rather consistently used for foraging and roosting at a consistent rate of about 11 per cent annually. Drinking is usually done on a daily basis, but the birds can go for several days without water if necessary. Preening is often done during the mid-day rest periods, and dust bathing is also an important maintenance activity.

The foods of great Indian bustards are quite diverse, but orthopteran and coleopteran insects are preferred items, especially grasshoppers. However, they also consume various plants including Bengal gram (*Cicer*), soeha (*Eruca*), sorghum (*Sorghum*), and groundnuts (*Arachis*), various fruits such as those of ber (*Zizyphus*), *Morinda* (Rubiaceae), and *Capparis*, and some small reptiles including various snakes and lizards (Bombay Natural History Society Annual Reports 1–3, Gupta 1975, Rahmani and Manakadan 1988). Studies at Nanaj indicated that maximum grasshopper densities occurred in typical grassland habitats that were heavily used by bustards. Such habitats had a relatively constant and healthy population present at the onset of breeding between August and September, and good numbers of adult insects present between September and November, the months of maximum bustard hatching. The availability of a steady adult population of grasshoppers may be critical for successful bustard breeding, which in turn may require sufficiently large grassland areas for undisturbed nesting and for supporting adequate grasshopper populations (Ali and Rahmani 1984).

Movements and flock compositions of great Indian bustards have been studied recently (Rahmani and Manakadan 1986, Manakadan and Rahmani 1986). Flock composition has been found to be sex-related for much of the year, the sexes remaining separate and gathering in fairly large groups during the early part of the monsoon season (June to August). Thus in Rollapadu both adult and subadult males were observed in flocks of ten to seventeen during July and August (approximately the peak grasshopper period). Females were usually seen in groups of up to four, with one exceptional case of a fifteen-bird group, and another sighting of twenty-two hens with juveniles in a single field. By late August in that area the dominant male began to display, and from that point on other males were seldom seen, nor were non-breeding females seen. This left only the displaying male and the nesting females in the Rollapadu study area, although foraging birds were present in harvested groundnut fields of the general area. During the peak of the dry season, from mid-February to late April, the birds were entirely absent from the study areas at Nanaj and mostly absent from Rollapadu. However, whereas the bustards breed only during the monsoon season in Nanaj, in Rollapadu there is a second minor breeding season during summer. Thus the extent and timing of migrations in this species probably varies considerably according to region and climate.

Interspecific relationships between the bustards and other species are numerous, and can be conveniently divided into 'non-associate' species that are generally feared, avoided or at most partially tolerated, and 'associate' species that are tolerated and often found in association with bustards. Non-associate avian species include scavenger vultures (*Neophron perchopterus*) and to some extent the other vultures, but the former are an especially serious potential predator of eggs. Various eagles and falcons sometimes frighten bustards, and at least females often treat eagles as potential threats, but falcons evidently pose little or no danger to adult great Indian bustards. Among other large potential avian predators, the eagle owl (*Bubo bubo*) may pose a threat, and crows (*Corvus splendens* and *C. macrorhynchus*) are feared by incubating hens and avoided by non-breeding females. Non-associate mammals include wolves (*Canis lupus*), which are potential predators of roosting or nesting birds. Foxes (*Vulpes bengalensis*), mongooses (*Herpestes edwardsi*), and monitor lizards (*Varanus benghalen-*

sis) are known to be egg predators, as perhaps are jackals (*Canis aureus*), although the last-named species is seemingly not feared by females with juveniles. Nest- and egg-trampling by domestic animals or by wild ungulates poses a potential threat in breeding areas. The four avian species most often seen as bustard associates are black drongos (*Dicrurus adsimilis*), white-eyed buzzard-eagle (*Butastur teesa*), red-headed merlin (*Falco chicquera*) and the Indian roller (*Coracias benghalensis*). All these species benefit from their association by eating insects that are flushed by the bustards (Rahmani and Manakadan 1987).

Social behaviour

During the breeding season, dominant males establish extremely large territories, into which no other adult males are allowed and within which the male displays at a prominent and traditional site. Thus at the Kerara study area, males had been displaying on a single small hillock for as long as the local villager could remember; at Rollapadu only one or rarely two males were seen displaying, and almost exactly the same display site was occupied during three breeding seasons. When the dominant male was captured and removed from this site it was quickly replaced, and in one season five males were captured at a single site, whilst in another season three males were removed. Within the 325 ha enclosure at Rollapadu twelve nests were found in one year, and nine in another, all of which apparently were fertilized by the single dominant male whose display site was at the edge of this enclosure (Manakadan and Rahmani 1986). Likewise at Nanaj only a single male was observed to display regularly during four years of study, and when a second male attempted to establish a display site about 1 km away it was not successful. Although most of the aggression of the dominant (alpha) male is directed towards other adult males, subadult males are also not tolerated near the display site at the peak of the display period, but persistent intruders may be tolerated at distances of 100–150 m. During the non-breeding season, however, males of all ages may mingle without generating aggressive interactions (Ali and Rahmani 1984).

Displays of this species have been described by a number of authors, including Dharmakumarsinhji (1957, 1966), Ali and Rahmani (1984), Manakadan and Rahmani (1986), of which the two latter references are especially valuable. These studies have shown that the males are polygynous or, more accurately, promiscuous. Thus, males gather four or five females in an apparent 'harem' that is actually unstable and varies in composition from day to day as the females move from one male's territory to another or may even leave the males' territories for a time. Promiscuity is therefore the apparent mating strategy, the male contributing only his gametes to reproduction, and the females presumably seeking out the most dominant male for fertilization.

Studies at Nanaj, Maharashtra (Ali and Rahmani 1984), indicate that both the arrival of males and the onset of sexual display are highly dependent upon the timing of significant summer rainfall. Arrival and some display sometimes occurred during pre-monsoon showers in June, but normally full display began there with the start of the full monsoon in July. During three years of observation, males always arrived in the area before the females, or remained there for a longer duration before the start of the breeding season. Peak display periods were from August to October, during which time the alpha male began display just after daybreak, and was most intense during morning and evenings, but usually did not occur midday unless a female was visible. Display during moonlit nights was not seen at Nanaj but was observed at Karera, Madhya Pradesh, where breeding occurs somewhat earlier and the peak display period occurs during May and June. Display is prolonged during cloudy weather, but is usually from 5 min to about 3 h in length, rarely exceeding 4 h.

Invariably the male selects a prominent location for display that provides a panoramic view of its territory. It spends most of its time during the display period at this site, generally strutting about with its gular pouch in a semi-pendulous state. As with other *Ardeotis* species, this pouch is capable of great distention; W. Elliot (1880) determined that the pouch of an adult male weighing about 14.5 kg could hold about 3.3 litres of water. As the male begins his display he holds his neck extended and his head directly upward, and erects his chin and throat feathers. He then begins to inflate his gular pouch, apparently by gulping air. Gradually this sac inflates until it hangs balloon-like between the legs and at times nearly touches the ground (Figure 37a). The gular pouch is especially large in older males, suggesting that the size of the inflated pouch may be an important signal denoting relative age and dominance. Tail-cocking generally coincides with the inflation of the gular pouch but can also occur independently of it.

At average intervals of 14 s the bird gulps air and utters a deep, resonant moan that can be heard as far as 1 km away under ideal conditions. Actually, three utterances are produced, including a low *chuck* as the bill is opened to gulp air, the booming moan just mentioned, and a slow and long *burrr* that is probably produced as air is expelled from the pouch and after the bird has already called ten to fifteen times. The loudest calls are produced when the gular pouch is maximally inflated, suggesting that it serves as a resonating device as well as an effective visual signal that can be seen for more than 1 km. During a display sequence (Figure 38) the male typically turns in all directions, but if a female is visible and fairly far away he turns to face her. However, at close distances of 3–6 m the male typically turns his back to her, exposing his cloaca and the surrounding feathers (Ali and Rahmani 1984).

Copulation apparently always occurs during peak display, and the female takes the initiative, by sitting down near the displaying male. After the female sits, the male begins nibbling her head and beak, as he lowers or partially lowers his tail. Both birds may catch each other's beak, and the male moves from one side of the female to the other, pecking at the female's head during each bout of nibbling, with up to twenty bouts of such nibbling. Actual copulation last only about 2 s, and thereafter the female immediately runs away, whilst the male resumes display.

Fighting among territorial males occurs occasionally, typically because of intrusion by another adult male into the territory of the dominant male. The territory of the dominant male evidently extends as far as the bird is able to see intruders, and fights are often preceded by the two males marching in parallel with tails raised, gular pouches pendulous, and crown feathers erected. Actual fights involve pecking, jumping and kicking one another (Ali and Rahmani 1984).

Reproductive biology

The breeding season of this species is seemingly rather variable, the 'normal' season being from July to September, but with at least scattered breeding

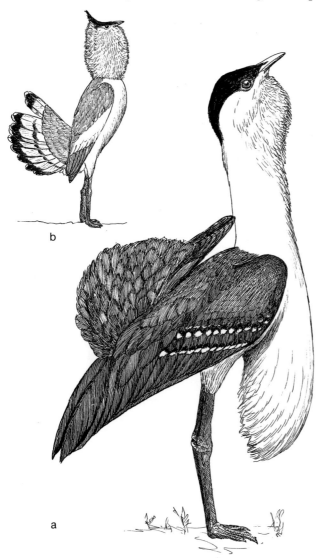

Fig. 37. Balloon-display of (a) great Indian bustard, after photo by A. Rahmani (in Ali and Rahmani 1984), and (b) Arabian bustard, after sketch by Rands (1986).

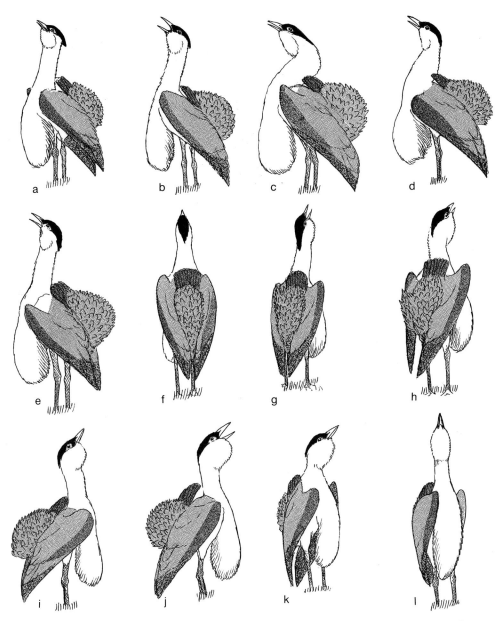

Fig. 38. Balloon-display sequence of male great Indian bustard, after a series of photos by A. Rahmani, representing a composite of several display sequences (A. Rahmani, personal communication).

records extending throughout the rest of the year from October to June (Baker 1935). Generally, in arid and semi-arid areas such as the Deccan tableland and the Indian Desert, breeding occurs during the monsoon, whilst in more mesic areas such as Madhya Pradesh breeding begins in mid-March, and reaches a peak in May, with some laying continuing until August (Rahmani and Manakadan 1988). Of three areas where recent detailed studies of this species have been conducted, the birds bred during the monsoon in one (Nanaj), during summer in another (Karera), and during both of these periods in a third (Rollapadu). In the last-named location the major breeding period was during the late summer and autumn monsoon rains, and only minor breeding occurred during early summer before the rains. Two breeding periods per year have also been reported from Ranebennur, Karnataka (Manakadan and Rahmani 1986). However, it is not known whether hens that are unsuccessful in one season attempt to breed during the other (which seems likely), or whether different males might occupy the same territory during the two seasons (which seems extremely unlikely).

During studies at Rollapadu, male display began in the autumn of 1985 within 3 days of the onset of monsoon rains in September, and within 5 days of the start of rains during April 1986. Rains during the 1986 summer monsoons were erratic, and hot weather was frequent, so although birds arrived as soon as there were good showers in early June, male display was not very regular that season. During these two years the collective monthly distribution of nests found with eggs, egg remains, or newly hatched chicks was: August 2, September 4, October 8, November 2, December 2, and January 3. Of known hatching dates, three occurred in October and four in November. Additionally there were two records of newly hatched chicks seen in May, and older chicks were seen in October and January. All of the twenty-one clutches found had a single egg (Manakadan and Rahmani 1986).

During three years of studies at Nanaj, initial dates of locating eggs or incubating females by month were: July 2, August 5, September 3, October 8, and December 1. October also represented the maximum monthly period of known or estimated hatching dates, with seven nests hatching. All of twenty-three clutches had a single egg, although a single instance of a farmer finding a two-egg clutch had been reported from the region a few years previously (Ali and Rahmani 1984).

Studies in subsequent years indicate that nest-site fidelity is common, with some of the nests found in 1986 at Rollapadu very close to 1985 nesting sites, and during earlier studies at Kerera eggs were found at the same location for three successive years at each of two different nest sites. Preferred nest sites at Rollapadu were mainly on ridges or small plateaux, where a panoramic view was visible, where the grass was not dense and usually 30–60 cm high, and where shrubs and trees were rare or absent. All of twenty-one nests were on stone-strewn or rock-bed sites, and none had surrounding vegetation higher than 1.2 m (Manakadan and Rahmani 1986). At Nanaj the birds also showed a preference for nesting on stony substrates, all but three of the twenty-one nests found being on such substrates. At only four of these sites were trees taller than 3 m located within 20 m of the nest, and at the remaining sites there were no trees or bushes within 30 m of the nest (Ali and Rahmani 1984).

Incubation is certainly by the female alone, but is of uncertain duration. The longest incubation period observed in wild birds has been 27 days, or a few days longer than that reported for the Australian bustard. Typically the female leaves the nest each morning and evening for foraging, for a variable period that in one study averaged about 24 per cent of the total daylight hours. However, during cloudy weather the female forages near the nest, returning to it as soon as it begins to rain. Similarly, during very hot weather the egg is not left unattended during the middle of the day (Ali and Rahmani 1984). Nest defence against smaller predators such as monitor lizards, foxes and jackals is strong and effective, but against larger mammals such as humans, nest-distraction behaviour may be performed (Figure 39a). In hot weather the female also drinks every day, so water availability may be an important aspect of nest-site selection where birds breed during the summer months (Rahmani and Manakadan 1988).

Hatching success data are limited, but Ali and Rahmani (1984) reported that in four years of study twenty-three eggs in as many nests produced fifteen chicks, and Manakadan and Rahmani (1986) similarly found that twenty-one eggs and nests found in two years produced fifteen chicks, the two samples collectively representing a hatching success of 68 per cent. Losses to predation were seemingly very small and insignificant, perhaps because of effective nest-defence behaviour, but some losses resulted from trampling by ungulates, nest desertion, and destruction by grass fires. Two nests were found that were believed to represent second clutches of females who had abandoned their earlier nests. In one case the second egg was laid at the same site, and in the other it was located about 100 m from the earlier nest. Because of difficulties of observation, little is known of chick growth or fledging success.

Fledging requires approximately 75 days, and this period is probably the most crucial time of the bird's life (Rahmani and Manakadan 1988). The chicks are fed bill-to-bill for a prolonged period (Figure 39b), even after fledging. The length of the post-fledging dependency period is not known with certainty, although it may last an entire year (Rahmani and Manakadan 1988). Ali and Rahmani (1984) mentioned seeing a young female of the previous year's hatch still accompanying its nesting mother, who would feed it regularly during periods of incubation recess. That incubation effort failed as a result of nest desertion (perhaps caused by undue interference from the young bird), but an apparent second nesting effort begun a month later hatched successfully after the female had driven off its yearling offspring.

Although observations from captivity may not be entirely relevant, it has been found that one hand-raised male began displaying when 6 years old, and was observed mating the following year. One captive female initially began laying eggs when she was already 21 years old, and during that year she alternately mated and laid a total of seven eggs over a period of about 75 days. During the following year she laid six eggs during a similar period, all of which were infertile (Rahmani 1986).

Evolutionary relationships

As noted elsewhere, this species is probably a very close relative of *australis*, and is perhaps only slightly less closely related to the two African species of *Ardeotis*.

Status and conservation outlook

The historical status and recent decline of the great Indian bustard has been documented in some detail by Goriup (1983c) for the species as a whole, and more recent summaries have been provided by Rahmani and Manakadan (1986, 1988). Status summaries for all of the individual states known to be inhabited by bustards in India have also been provided recently, such as those for Rajasthan (Rahmani 1986), Gujarat (Sinha 1983, Karpowicz and Goriup 1985), Madhya Pradesh (Hassan 1983), Andhra Pradesh (Kumar 1983), Maharashtra (Rego 1983), and Karnataka (Neginhal 1983) as well as for the Indian Desert region in general (Prakash 1983).

Currently, bustards are known to survive still in six Indian states: Gujarat, Rajasthan, Madhya Pradesh, Maharashtra, Karnataka, and Andhra Pradesh. Of these, Rajasthan probably holds more than half of the total species' known population, estimated in the mid-1980s at about 1500–2000. Gujarat established a bustard sanctuary in 1989, and by then all of the other states supporting bustards had already taken some conservation measures. As of 1988 there were thirteen sanctuaries or protected areas mainly established for protecting this species. The upward trends of the populations at several of these sanctuaries (Karera, Nanaj, Sonkhaliya, Desert National Parks, Rollapadu) is encouraging, and some additional sanctuaries have been proposed in Gujarat (at Banni, Abdasa, and Bhatiya-Kalyanpur), Rajasthan (Bathnoke Range), Andhra Pradesh (at Banganpalli) and Madhya Pradesh (at Pohri) (Rahmani and Manakadan 1988).

In Rajasthan, where perhaps 500–1000 bustards

Fig. 39. Behaviour of the great Indian bustard: (a) injury-feigning by incubating female (after photo by D. K. Vyas); (b) parental feeding by female (after sketch in Rahmani and Manakadan 1988); and (c) dust-bathing (after photo in Rahmani and Manakadam 1988).

exist (Rahmani 1986), several important sanctuaries are already present. Foremost of these is Desert National Park, an area of 3162 km² that supports 200–400 bustards. The Shokhaliya Closed Area (Ajmer district) supports eighty or more bustards, with an increasing bustard population. The Sorson Closed Area (Kundanpur) in Kota district supports a stable population of ten to twenty-five birds in about 40 km². In Maharashtra, the bustard sanctuary at Nanaj (Ahmednagar and Solapur districts) supports fifty to sixty birds in an area of 7818 km², with an increasing bustard population. In Madhya Pradesh the bustard sanctuary at Kerera (Shivpuri district) supports thirty or more birds in 202 km². A second sanctuary at Ghatigaon (Gwalior district) supports fifteen to eighteen birds in 512 km². In Karnataka the Rannebennur blackbuck sanctuary (Dharwad district) supports ten to fifteen birds, with infrequent breeding records, and plantations at Guttal and Bagalkot each support a few birds as well. Finally, in Andhra Pradesh, Rollapadu bustard sanctuary (Kurnool district) supports sixty or more birds by currently protecting 500 ha of core area within a much larger proposed sanctuary, and has an increasing bustard population (Rahmani and Manakadan 1988).

The increasing interest by state and federal government agencies in India favour the continuing survival of this species, which however is still in such small numbers as to be regarded as endangered in India, where it is protected in all states. It is probably now completely extirpated or essentially so from Orissa, Uttar Pradesh, Haryana and Punjab in India, as well as from Pakistan, where it once occurred in Punjab and Sind. It is on the ICBP world list of endangered bird species, and is also listed in Appendix 2 of CITES (King 1981).

ARABIAN BUSTARD (Plate 19)

Ardeotis arabs Linnaeus 1758

Other vernacular names: speckled-wing bustard; balawan, habru kharab, khrib, rawa assaid, swad (Bedouin); outarde Arabe (French); Arabische Trappe, Arabertrappe (German).

Distribution of species (Map 18)

Western Morocco east to Somalia, south of the Sahara desert; also rare and local in south-western Arabian peninsula, where now apparently limited to the Tihama region of coastal Saudi Arabia and to adjoining areas of Yemen.

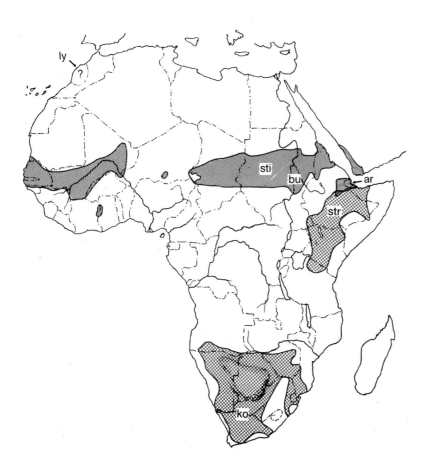

Map. 18. Distribution of the Arabian bustard (hatched), including the locations of its races *arabs* (ar), *butleri* (bu), *lynesi* (ly) and *stieberi* (sti), and the kori bustard (cross-hatched), including its races *kori* (ko) and *struthiunculus* (str).

Distribution of subspecies

A. a. lynesi Bannerman 1930: endemic to north-western Morocco (possibly now extinct; no published records since 1962).

A. a. stieberi (Neumann) 1907: Senegal, eastern Gambia and Ivory Coast east across the southern Sahara to north-eastern Sudan. Includes *geyri* Neithammer 1954, known only from two specimens collected in central Niger.

A. a. butleri Bannerman 1930: southern Sudan; vagrant to Kenya.

A. a. arabs L.: eastern Sudan, Ethiopia, Djibouti, and Somalia; also south-western Saudi Arabia and adjacent Yemen (extremely rare, formerly to Southern Yemen).

Measurements (of *stieberi***, mm)**

Wing, adult males 600–705 (av. of 8, 638), females 530–580 (av. of 6, 554); tail, adult males 290–340 (av. of 6, 315), females 270–300 (av. of 4, 281); tarsus, males 188–206 (av. of 6, 196), females 167–177 (av. of 6, 173); culmen, males 64–78 (av. of 7, 72.4), females 62–70 (av. of 5, 64.8) (Cramp and Simmons 1980). Egg, av. 75–81 × 53–56.

Weights

Males of *stieberi*, 5.68–10 kg; adult female of *butleri* 4.75 kg (Cramp and Simmons 1980); males of *stieberi*, 5.7 and 10 kg, female 4.5 kg (Urban *et al.* 1986). Estimated egg weight 115–142 g, or 3.3% of adult female (Schönwetter 1967).

Description

Adult male. General colour above sandy brown (the western and southern populations averaging somewhat darker in colour than *arabs*), minutely freckled with blackish vermiculations; upper tail-coverts like the back, but a little more coarsely vermiculated and with an indistinct edging or spot of white at the tip; wing-coverts like the back, a little more coarsely vermiculated and having triangular spots of white at the ends, the median series broadly tipped with white, the greater series also tipped with white and having the inner webs for the most part white (the white spotting much reduced in *lynesi* and *arabs*); alula like the major coverts; primary coverts brown with a bluish-grey shade, indistinctly freckled with sandy buff and tipped with white; remiges blackish, with the greater part of the inner web white on the inner primaries; the secondaries brown with a bluish-grey shade, tipped with white and crossed with indistinct bars of mottled white (this reduced in *lynesi*), the white increasing in extent towards the inner secondaries, which are coloured and freckled like the back; centre rectrices sandy brown with coarse black vermiculations, the concealed basal half of the feathers banded with white, the remainder of the underside feathers mostly partly white banded with bluish grey or brown near the base and having a broad subterminal band of sandy brown, freckled with blackish lines and sometimes shaded with greyish-blue, the ends of the feathers black and white; crown greyish-white with fine black frecklings, the sides of the crown bordered by a band of black broadening out behind the level of the eye and meeting on the nape, the posterior feathers of this band being elongated and narrow, as also are the posterior greyish feathers of the crown; lores and a distinct eyebrow white, as also the feathers below the eye; sides of face, cheeks, throat, hindneck and sides of neck, as well as the foreneck, ashy white, distinctly banded with ashy blackish everywhere except on the lower throat, where the cross markings are finer; ear-coverts ashy white with obsolete dusky bars; anterior cheeks and chin white; remainder of undersides from the foreneck downwards pure white; under wing-coverts white to ashy white, but distinctly varied with dusky cross-bars; axillaries pure white; remiges blackish below. Iris yellowish umber-brown; bill horny buff, with the culmen dusky; toes ochre-yellow.

Adult female. Similar to the male, but very much smaller and sometimes with duller cinnamon and more olive-brown, and with the wing-coverts less boldly spotted (Cramp and Simmons 1980).

First adult plumage. Like the adult, but with narrow and pointed outer primaries that are retained up to 15–18 months after hatching (Cramp and Simmons 1980).

Juvenile. Resembles the female, but with the contrasting wing pattern less distinct, the white areas tinged with buff, and pale tips to the primaries (Cramp and Simmons 1980).

Identification

In the hand. This large to very large (wing 530–705 mm) bustard has a sharply pointed bill (culmen more than 60 mm) and no black markings on the greater wing-coverts, the latter trait distinguishing it from the similar kori bustard and non-African species of *Ardeotis*.

In the field. This is the only very large bustard occurring in Arabia and over nearly all of its sub-Saharan range in Africa. In a small region of north-eastern Africa (the Awash River area of Ethiopia and in extreme south-eastern Sudan and adjacent north-western Kenya) it closely approaches and is locally in contact with the range of *kori*; there the lack of black spotting on the upper wing-coverts and at the base of the hindneck may need to be confirmed for positive identification. The smaller Denham's bustard of the Sahel zone, with which this species sometimes associates, has brown on the hindneck and lacks a definite shaggy crest. Calls include a loud, probable alarm note *bak-kak-ka*, uttered in flight and when walking, with the first two syllables drawn out and nasal. The male's display call is a rasping *pah-pah* note that carries a great distance.

General biology and ecology

This is a species associated with desert-edge or near-desert conditions that range ecologically from treeless sandy *Panicum* grasslands, shortgrass plains with scattered acacia trees and low *Commiphora* plants, and more mesic habitats grading into acacia parklands. The presence of some scattered bushes or low trees may be useful for roosting at night, providing shade during the day or, in the case of acacias, provide a source of foods in the form of seeds or gum. At least formerly the species also ranged into the well-developed but open oak woodlands of Morocco. Probably extensive grassy plains represent nearly ideal habitat, especially those with occasional wadis, and perhaps also dwarf palms (*Hyphaene*). Both wetlands and precipitous terrain are avoided, but the birds tend to tolerate the presence of visual screening by bushes or trees to a greater degree than such bustards as the little bustard (Urban *et al.* 1986). In Arabia the birds occupy undulating terrain, with gravel plains and plateaux, and long, narrow and winding valleys. These valleys are mainly dry watercourses, with xerophytic plants such as *Acacia*, *Commiphora*, *Euphorbia*, *Salvadora* and tufted grasses. Larger wadis used by the birds may also have doum palms (*Hyphaenae thebaica*) and other woody plants such as *Tamarix* and *Capparis* present (Rahmani and Shobrak 1987).

Like some of the other Sahel-zone bustards such as Denham's bustard, this species undertakes some movements of a migratory nature, and these two species have been observed migrating together (Malzy 1962). Probably both move north into the Sahel during the rainy season to breed, and return again with the onset of the dry season. This northward movement generally occurs in spring and summer, between May and August, and the return movement is during the autumn, or around October. However, some birds may remain in the Sahel zone even during the dry season (Urban *et al.* 1986).

Foods of this species are highly diverse, but primarily include larger terrestrial insects, such as swarming locusts, grasshoppers and beetles. Some other insects such as caterpillars are consumed, as well as vertebrates including reptiles, nestling birds, and small mammals such as mice. Plant materials likewise include seeds, leaves, buds and green shoots of grasses and broad-leaved plants, and seeds and fruits of plants such as *Corida*, *Grewia*, *Salvadora* and *Cucumis*, and the gum from acacia trees (Urban *et al.* 1986, Cramp and Simmons 1980).

Social behaviour

Although the Arabian bustard is much more poorly documented than the closely related kori bustard, it is believed that both species have similar, polygynous mating systems, involving display by lone territorial males. Meinertzhagen (1954) described the male's display as consisting of the bird 'inflating the neck, raising the crest and erecting the tail so that its tip touches the back of the head and displays the profuse white under tail-coverts. This is done when facing the female, who pays no attention'. However, Meinertzhagen considered *arabs* and *kori* as conspecific (along with the Australian and Indian forms of *Ardeotis*), and he may simply have been describing the relatively well-known displays of *kori* rather than those of *arabs*. Meade-Waldo (1905) observed without further elaboration a male in display, and described the species' *pah pah* call as being similar to the sounds produced by 'two big bubbles of a water-bottle'. This suggests that the calls are apparently quite different acoustically from those of *kori*, but quite possibly they are based on similar throat or oesophageal inflation and resonation mechanisms. Rands (1986) provided a sketch of the male's display posture (Figure 37b), which is certainly much like the 'balloon' display of other *Ardeotis* species, except perhaps for a more distinctly fanned tail and an apparent inflation of only the upper portion of the neck. Regrettably, however, the male's oesophageal anatomy apparently remains undescribed.

Reproductive biology

Although records are not extensive, this species apparently breeds during the wetter period, which may be quite variable locally or from year to year. Meinertzhagen (1954) stated that in Arabia the eggs were laid in late May or June, and likewise all available African breeding records are for the warmer and generally wetter months, between April and October. In Ethiopia the records are from August to October, in Sudan for April and probably also the period August to October, in Niger during July and August, in Mali from July to October, in Senegambia from July to September, in Mauritania during August and September, and in Morocco during April (Urban *et al.* 1986).

The nest is placed on bare ground, either in the open or under some scrubby cover. At least in the Sahel zone the nests are well spaced and nesting females are situated some 2–3 km apart (Urban *et al.* 1986). This would further suggest that males play no role in the incubation and rearing phases of reproduction, as is known to be the case in the other species of *Ardeotis*. Little else is known of this species' reproduction.

Evolutonary relationships

Snow (1978) regarded all four of the major *Ardeotis* forms as constituting a superspecies, the differences among the four all being of approximately equal magnitude in his opinion. Thus he thought that there was no reason to believe that *kori* and *arabs* are any closer to one another than *arabs* is to *nigriceps*, its nearest zoogeographic relative to the east. Meinertzhagen (1954) indeed considered them all as subspecies of a comprehensive species *kori*, but that is a quite questionable and certainly overly simplified interpretation. Urban *et al.* (1986) considered *arabs* and *kori* to constitute a superspecies, with no mention of the other non-African forms, and I am inclined to believe that simply on zoogeographic grounds they should be regarded as the nearest relatives among this assemblage (see Figure 2).

Status and conservation outlook

Although Urban *et al.* (1986) considered this species as generally common to abundant, they observed that it had become rare in some places such as Senegambia in the past decade, as well as having been extirpated from Morocco. It has recently been reported from Djibouti, where perhaps a healthy if widely and thinly distributed population may exist (Welch and Welch 1986, 1989), but is only a vagrant to north-western Kenya (Turner 1982). In Arabia it once occurred from Jiddah (Jedda) and the neighbourhood of Mecca south to Aden in South Yemen, with its main concentration in the Tihamah region of extreme southern coastal Saudi Arabia (Meinertzhagen 1954). However, it has declined greatly there in recent years, and is considered endangered (Gasperetti and Gasperetti 1981). For a time it was believed extirpated from the Arabian peninsula, having last been observed in the early 1970s at Wadi Hali (south of Al Qunfidhah), Saudi Arabia. However, in 1977 eggs were found and two pairs were observed between Ad Darb and Jabal Labiba, and the birds were seen again there the following year (Nader 1983). The most recent available information (Rahmani and Shobrak 1987) indicates that the species survives in Saudi Arabia only in very small, dispersed numbers, and has suffered greatly from excessive hunting and habitat deterioration. Rahmani and Shobrak found evidence suggesting that it may still occur locally between Al Qunfidhah and Baysh, and possibly occurs as far as Wadi Dahan near the Yemen border. They saw only one bird during a 3000 km survey, but also found feathers, tracks or droppings at some sites. Additionally, second-hand reports of birds seen in the past five years came from Assarb, Wadi Wusa, Al Khabra, Assadha, and Al Qunfidhah. Its current status in Yemen also should be further investigated, where seven great bustards were seen south-east of Al Hudaydah in 1979, and seventeen were counted during a 1985 survey. It is possible that Djibouti supports a healthy if widely distributed population on the basis of recent observations by G. and H. Welch, and further studies in that region are certainly warranted.

KORI BUSTARD (Plate 20)

Ardeotis kori (Burchell) 1822

Other vernacular names: Burchell's bustard, giant bustard, greater bustard, northern kori bustard (*struthiunculus*); kori (Bantu dialect); Gompou (Afrikaans); outarde de kori (French); Riesentrappe (German).

Distribution of species (Map 18)

Eastern Ethiopia and western Somalia south to central Tanzania, and most of southern Africa south of the Zambezi.

Distribution of subspecies

A. k. struthiunculus (Neumann) 1907: Ethiopia south to Lake Victoria and central Tanzania.

A. k. kori (Burchell): Zimbabwe, Botswana, southern Angola, Namibia, South Africa, and southern Mozambique. Listed as 'vulnerable' in the *South African Red Data Book* (Brooke 1984).

Measurements (mm, of *struthiunculus*; kori averages slightly smaller)

Wing, males 752–767 (av. of 3, 761), females 600–655 (av. of 13, 629); tail, males 370–387 (av. of 3, 378), females 280–342 (av. of 13, 312); tarsus, males 230–247 (av. of 3, 241), females 181–205 (av. of 15, 190); culmen, males 95–120 (av. of 3, 109), females 81–95 (av. of 15, 88.5) (Urban *et al.* 1986). Egg, av. 80–81 × 58–59 (Schönwetter 1967).

Weights

Males (of *struthiunculus*), c. 11.4–13.5 kg, females c. 5–6.4 kg (Someren 1933); male 10.9 kg, two females 5.9 kg; males (of *kori*) 13.5–19 kg (Urban *et al.* 1986). Males to at least 18 kg (Clancey 1967). Egg, 149–158 g (estimated).

Description (of *struthiunculus*)

Adult male. General colour above dark sandy brown, with blackish vermiculations and with a slight greyish shade, some of the feathers of the mantle and upper back rather more blackish; lower back, rump and upper tail-coverts like the back, the latter rather more coarsely freckled; lesser wing-coverts like the back; median coverts mostly white, coarsely mottled with black or grey frecklings, and having a broad black subterminal bar and a white tip; greater coverts also like the median, but more thickly mottled with black or grey vermiculations; alula like the median or greater coverts, but the subterminal bar not so strongly indicated; primary coverts ashy brown, the inner ones mottled and broadly tipped with white; remiges brown, the two outer ones scarcely freckled with white on the outer web, but the inner ones becoming more white on the inner web, barred with bluish-grey and tipped with white, some of the inner primaries chequered with sandy buff on the outer webs; secondaries bluish-grey, everywhere mottled with white, the feathers tipped white and a subterminal bar of blackish-brown, the innermost secondaries like the back; rectrices ashy brown at the base, crossed by two broad bands of white, which are separated from each other by black bands, one broad and one narrow, the latter followed by an indistinct white band which merges into the sandy-brown ending of the tail, this portion having a narrow band of black, a much broader subterminal band of black, and a white tip; crown strongly crested, black, with a greyish band of feathers down the centre, and a black post-ocular stripe (the latter reduced or lacking in *kori*); the nape and sides of posterior crown greyish white barred with black, exactly like the whole of the neck; sides of face, throat, a streak over the eye, a patch in front of the eye, anterior cheeks and chin all white; on the foreneck, partly concealed by the long barred feathers of the lower throat, is a crescentic band of black, the sides of the upper breast also marked with black; remainder of undersides white, including the axillaries and most under wing-coverts, the lower primary-coverts ashy freckled with white. Iris lemon-yellow to orange-brown; bill light horn-colour, darker brown above and yellowish below; tarsi and toes light yellowish.

Adult female. Smaller, with the black on the crown and eye-stripe somewhat reduced (Urban *et al.* 1986).

Juvenile. With shorter neck-ruff feathers (no filamentous plumes), the crown paler, and more heavily marked over the mantle, scapulars and tertials (Clancey 1967, Urban et al. 1986). Young birds have more brownish and less greyish lesser upper wing-coverts, and have more white on the inner webs of the outer remiges (Friedmann 1930).

Identification

In the hand. This large to very large (wing 600–767 mm) bustard has a long, pointed bill (culmen 81–120 mm), and a shaggy black crest like that of *arabs*. Unlike that species it has black on the upper wing-coverts and at the base of the hindneck.

In the field. Except in a very small area of north-eastern Africa (upper Nile drainage of western Ethiopia and extreme south-eastern Sudan) this species and the very previous one do not come into close proximity, and thus in most parts of its range its very large size and combination of a grey neck and black, shaggy crest provide ample basis for field identification. It is in broader contact with the similar but smaller Denham's bustard, which species lacks a shaggy crest and has a distinctive rufous hindneck. Displaying males utter very low-pitched *wum* sounds, in groups of three or six notes, with greatly expanded necks. They produce barking or snoring *bar-kah-ka* notes both when alarmed and when undisturbed.

General biology and ecology

This is a species of open grassland plains and open thornbush at generally rather low elevations. In Kenya the race *struthiunculus* usually does not range above about 2000 m elevation, although it has occurred as high as about 2700 m around Laikipia (Pitman 1957). In southern Africa the race *kori* occurs in such habitats as karoo, acacia- or open-bushveld, light woodlands, and park-like environments in non-montane country (Clancey 1967).

In a summary of records from Zimbabwe, Rockingham-Gill (1983) observed that nineteen out of twenty occurrence records were for flat landscapes, fourteen out of nineteen records were associated with clay or loam soils, seven out of ten locations had intermittent streamflows, and ten out of twenty records were from open mopane woodland habitats with medium to heavy grass cover, with trees and shrubs distributed singly or in discontinuous clumps. All of the records were from savannah-woodland habitat types. An unpublished recent summary of kori distribution records from southern Africa indicates that the species is centred in arid savannah habitats, where only scattered trees occur in grasslands. This habitat type covers most of Botswana, and extends locally into central Zimbabwe along the Central Dyke and perhaps also where woodland clearing has occurred. Similarly, in South Africa most records are from the far southern and south-eastern Cape Karoo, where scattered trees are typical, but there are also records from the north-central Karoo where tree-lined dry watercourses exist.

The birds typically favour locations where the grass is not too long, and where stony outcrops are present. They also are attracted to ridges and hilltops, but often take shelter in valleys, or rest in the shade of thorn-trees or tall grasses. They are strongly attracted to areas where a fire has passed through recently, for not only may new grass be sprouting there but insects and other animals exposed or burnt by the fire can readily be found. They likewise often associate with herds of large ungulates such as wildebeest (*Connochaetes*), perhaps feeding on disturbed insects or on insects attracted to the dung piles. They also are attracted to waterholes to drink, and often find shaded sites in which to rest during the hottest parts of the day. The birds are usually found solitarily or in pairs, but as many as twenty birds have been seen in favoured areas within a radius of about 3 km (Someren 1933), and up to a dozen have been seen in a single small caterpillar-infested field (Pitman 1957).

The foods of this species are evidently quite diverse, although they are still rather poorly documented. They especially include locusts or grasshoppers and beetles (particularly dung-beetles), plus a variety of rodents, lizards, small snakes, and the like. Plant materials that are consumed include seeds, roots, and also the gum-like exudate of acacia trees (the basis for the Afrikaans name Gompou, or 'rubber peacock') (Urban *et al.* 1978).

Migrations to wintering areas at lower elevations and more easterly sites in some regions of southern Africa are apparently well documented, at least for previous times (Snow 1978, Clancey 1967). Similar migrations, generally in response of rainfall or food supplies, have also been reported in East Africa, and at least some of these migrations may be on foot by these large and highly mobile birds (Urban *et al.* 1986).

Social behaviour

Although some writers have suggested that this may be a monogamous species, mainly because of some observations of seemingly paired birds and also because males have been observed to courtship-feed females, most of the evidence would favour instead the view that males display solitarily in dispersed territories or possibly more localized and lek-like congregations (Urban *et al.* 1986). Up to as many as four males displaying within 100 m of one another have been observed in Nairobi National Park, Kenya.

Good evidence that the kori bustard is polygynous comes from the fact that males display almost continuously during the height of the breeding season; thus Niethammer (1940) noted that during November (the start of the rainy season) in Namibia the birds could be seen displaying all day long, standing with their necks inflated to about four times their natural size, their tails cocked, and their folded wings almost touching the ground. Like Murie (1868), Niethammer described the anatomy of the male's gular pouch, an expanded area of the anterior third of the oesophagus that is capable of substantial inflation, and which can be held in an inflated state for an indefinite period. In conjunction with the long ruffled neck feathers, the whole neck is seemingly expanded during display and somewhat resembles an oversized stovepipe (Figure 40). During this period the male stands still or may parade about, with the crest raised, the tail cocked so strongly as to touch the ruffled neck feathers, and the billowing white under tail-coverts forming a white 'blinker' that is almost as large and perhaps even more conspicuous than are the greyish-white neck feathers, and which can be seen for great distances (Hoesch 1959). The strutting display as described and illustrated by Urban *et al.* (1986) may also include a drooping of the wings so that the tips of the primaries touch the ground, bill snapping, and neck vibration. During direct courtship of a female, the male walks slowly around her or stands within 10 m of her, bowing with the body tilted forward, the neck inflated, and the head never reaching below the level of the shoulders.

The booming display is performed with maximally inflated neck, the wings drooping, and the tail lowered so as to form a straight line with the wings, hiding the white under tail-coverts. However, the whitish neck feathers are strongly ruffled and the black crest is raised. Hanby (1982) illustrated the very erect stance (Figure 41a) adopted during the call, as has also Hellmich (1988). Hellmich described the call as consisting of three pairs of drum-like sounds, *ump-ump, ump-ump, ump-ump*. Hoesch stated that the associated call is a six-note *wum-wum-wum-wum-wumwum*, and Hanby likewise observed that the call consisted of six deep, rhythmic notes, as if the bird were saying 'I'm...a...ko-

156 *Bustards, Hemipodes, and Sandgrouse*

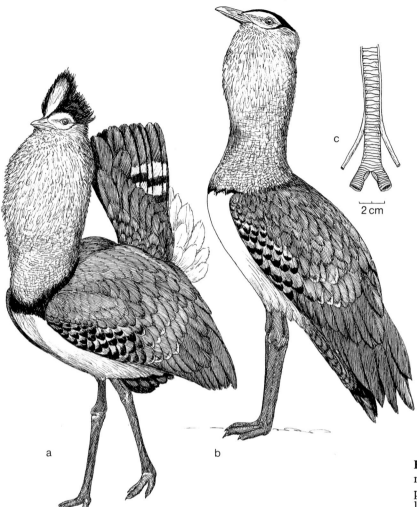

Fig. 40. Male Kori bustard: (a, b) stages of neck-inflation and balloon-display (after photos by P. Johnsgard); (c) ventral view of lower trachea and syrinx (from a specimen).

ri...bus-tard!'. Others have described the vocalization as consisting of three-note or four-note sequences, or in various other ways, so perhaps there is also some individual or regional variations in the note sequence or cadence. Besides six-note sequences, Hellmich (1988) occasionally heard five- and seven-note sequences. He noted that the intervals between call sequences averaged about 29–37 s and gradually increased as calling continued. The overall period of male display by a captive male (in Germany) lasted about 2 months.

During aggressive encounters between displaying males they may stand breast-to-breast, with tails raised, and hold on to each other by their bills, all the while pushing each other forwards, backwards or in circles (Figure 41b) (Schaller 1973, Allen and Clifton 1972). Such direct physical encounters related to achieving individual male dominance, together with the probable heterosexual attraction of large males to females, probably help to account for the evolution of the large male size and strong adult sexual dimorphism (an approximately 2:1 adult weight ratio in the sexes) that exists in this and closely related species.

Reproductive biology

The breeding season in South Africa is during the spring and summer from September to February, and in Namibia is similarly from November to January, during the wetter summer period (Clancey 1967, Niethammer 1940). Records for Zimbabwe are from September to December, plus April, and in East Africa the breeding records are rather varied, apparently depending upon the local rainfall pattern. The general breeding period is from January to June, but there are also records for August and November. Spring breeding between March and June has also been reported for Ethiopia and Somalia (Someren 1933, Urban *et al.* 1986).

Nests are well scattered in locations well away from the displaying males and sometimes are partly

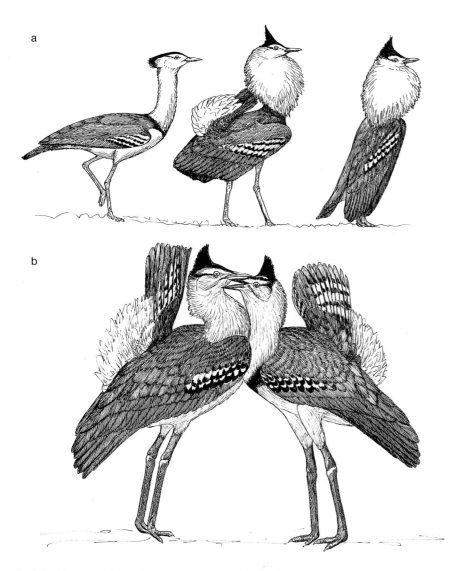

Fig. 41. Display sequence of male Kori bustard (a) after sketches in Hanby 1982), and fighting behaviour of two male Kori bustards (b) (after photo by G. Schaller, in P. Osborne *et al.* 1984).

hidden by or at least located near a tuft of tall grass, a shrub, or a rock. The clutch consists of one or two eggs, incubated only by the female (although some early literature suggests that both sexes incubate) (Osborne *et al.* 1984). Although there are earlier estimates of the incubation period of about 31 days, unpublished research at Karoo National Park in November 1987 established an incubation period of 23–25 days for a single nest of two eggs that were laid at least 2 days apart. Only the female incubated, and no male frequented the nesting area, although one was heard several times at about 500 m from the nest, where a display site may have been present. During incubation the female spent very little time off the eggs for foraging, and probably utilised fat reserves during this period. This condition may prevail because inasmuch as no male is present, the female must remain near the nest to protect it from possible predators. Small predators such as the small grey mongoose (*Galerella pulverulenta*) were observed to be aggressively driven off by the female, but when a martial eagle (*Polemaetus bellicosus*) flew over, the female ran and took shelter under nearby acacia trees.

The female tends the young alone, although according to some reports a male may at times attend her. The fledging period is unreported but, as with the great Indian bustard, the young chicks continue to be fed by the female until they are well grown, and follow very closely behind her when the female is walking. The large size of the adult birds probably deters many predators, although adults have reportedly been killed by both jackals and martial eagles (Urban *et al.* 1986).

Evolutionary relationships

As noted in the *arabs* account, this species is part of a large superspecies that includes all the *Ardeotis* forms, and there is some question as to whether *arabs* and *kori* are the most closely related forms within this assemblage. I am of the opinion that this is most likely the case (see Figure 2), although additional behavioural and morphological data would be highly desirable.

Status and conservation outlook

This very large and conspicuous species is still fairly common in some areas of southern Africa such as Botswana, Namibia and Zimbabwe. An estimated 10 700 birds were judged to be present in Zimbabwe in 1980, when the species was observed in twenty of Zimbabwe's fifty districts, and where in some areas it is not harmed by the natives inasmuch as it is regarded as possessing magical attributes. However, the kori bustard is probably generally declining in southern Africa, and it is considered 'vulnerable' in South Africa (Brooke 1984). Allan (1988) observed that the kori had declined in the Transvaal, Orange Free State and possibly in parts of Cape Province, as a result of intensive agriculture, illegal hunting, and possibly poisoning by strychnine baits put out for small predators. The same deteriorating situation exists in at least some parts of eastern Africa. This includes Somalia, for example, from which there are no recent records (Ash and Miskell 1983). In the 1987 Karoo research mentioned above, of twenty recorded 'unnatural' kori bustard deaths, twelve were caused by overhead lines, five were the result of road kills, and three were caused by fences.

DENHAM'S BUSTARD (Plate 22)

Neotis denhami (Children) 1826

Other vernacular names: Barrow's bustard, Burchell's bustard, Jackson's rufous-necked bustard (*jacksoni*), Stanley's bustard (*stanleyi*); Veldpou (Afrikaans); grande outarde d'Afrique, outarde de Denham (French); Kafferntrappe, Stanleytrappe (German).

Distribution of species (Map 19)

Guinea and southern Mauritania east across the Sahel zone to western Ethiopia and south through the interior of East Africa to Cape Province, South Africa; local in the Congo Basin of Zaïre and elsewhere in western Africa. Migratory in northern portions of range.

Distribution of subspecies

N. d. denhami (Children) 1826: Guinea and Mauritania to Ethiopia and northern Zaïre. Includes

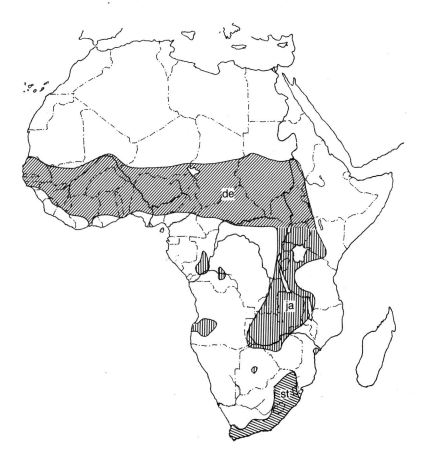

Map. 19. Distribution of Denham's bustard, indicating the locations of its races *denhami* (de), *jacksoni* (ja) and *stanleyi* (st).

burchellii Heuglin 1867 (Mackworth-Praed and Grant 1952).

N. d. jacksoni Bannerman 1930: Angola and southern Zaïre east to Kenya, south to Angola, Botswana and Malawi.

N. d. stanleyi (Gray) 1831: Botswana (local), Lesotho, and South Africa. Listed as 'vulnerable' in the *South African Red Data Book* (Brooke 1984).

Measurements (of *denhami*, mm)

Wing, males 575–660 (av. of 7, 612), females 480–530 (av. of 7, 507); tail, males 300–330 (av. of 6, 315), females 255–290 (av. of 5, 275); tarsus, males 170–183 (av. of 7, 176), females 130–153 (av. of 7, 142); culmen, males 70–76 (av. of 6, 74.6), females 55–66 (av. of 5, 60.8) (Cramp and Simmons 1980). Egg (of *stanleyi*, av. 76 × 55. Additional measurements and inter-racial comparisons have been provided by Benson and Irwin (1972), who reported that although body size generally decreases southwardly, the tarsus is longest in nominate *denhami* and non-montane populations of *jacksoni*, whereas it is shorter in *stanleyi* and shortest in the populations of *jacksoni* that occur in high plateau areas.

Weights

Both sexes generally 4.08–10.4 kg, with males reaching 14.3 kg; breeding male 4.5, pre-migrant male 6.35; female 3.2 (Cramp and Simmons 1980); males 6.8–8.2 kg, females 4.5–5.5 kg (Someren 1933); males 9–10, female 3 kg (Urban *et al.* 1986); males 5–7.3 kg (Pitman 1957). Estimated egg weight 126 g, or 9.1% of adult female (Schönwetter 1967).

Description

Adult male. General colour above greyish to dark sandy brown finely varied with sandy and black vermiculations, these markings somewhat coarser and more pronounced on the scapulars and inner secondaries, where there are also occasional blotches of black; lower back and rump more obscurely vermiculated, the grey shade of the upper back not so distinct; the upper tail-coverts more distinctly vermiculated, like the two central tail-feathers, the latter having three distinct bars of black and one broad one of white; lesser wing-covers like the back, but somewhat browner and less distinctly vermiculated; median coverts black, with a good deal of white at the base and a small white tip; greater coverts black, or white with black bars near the tip, and occasionally some other black markings on the outer web, the inner web freckled with brown; alula and primary coverts black, with small white tips; remiges black, white at the base of inner web, this increasing greatly on the inner webs of the inner primaries, which are also chequered with white on the outer web, the secondaries black with white tips; all but the central rectrices barred with black and white, the base being uniform black, with other bars of black across the feathers, the subterminal white bar somewhat mottled with black spots, the ends of the feathers mottled with ashy; crown black, with a white mesial band, flanked on each side by a broad whitish eyebrow; nape greyish-white, extending down the sides of the neck for a short distance; hindneck rufous (darker and richer in *stanleyi* and *jacksoni* than in nominate *denhami*), this colour extending to the mantle, which is very finely vermiculated with black; lores, sides of face and ear-coverts white, this white area sometimes extending back along the sides of the neck and separating the bluish foreneck from the rufous hindneck; cheeks white, coarsely freckled with black vermiculations; throat white; foreneck and chest light blue-grey, the lower part with black vermiculations that extend down to the chest, bordered on each side by the rufous on the hindneck, which extends on to the sides of the chest, where it is also varied by black markings, forming an indistinct crescent; remainder of the undersides pure white, including the under wing-coverts and axillaries. Iris light hazel (sometimes yellowish-white during breeding); upper mandible greyish, with a darker culmen, lower mandible yellowish; tarsi and toes dingy yellowish-white to pale yellow or greyish-yellow (Cramp and Simmons 1980).

Adult female. Somewhat like the male but smaller. Crown black with central white streak expanding towards nape but not extending over the black; lores freckled black on sandy ground, this colour extending over the front of eye, followed by a wide white supercilium that extends over the ear-coverts; cheeks and ear-coverts whitish freckled on the former and streaked on the latter, with blackish; chin and throat white; the entire foreneck finely and closely vermiculated ashy grey, the grey shading to lavender at the sides, encircling the nape, and extending to the upper breast where it is sharply defined from the white breast; hindneck rufescent; abdomen flanks, thighs, under tail-coverts white; flanks buffy; mantle, back, scapulars, rump, long inner secondaries and lesser coverts freckled and vermiculated sandy buff and black; median and greater coverts black and white; rectrices and remiges as in the male. Iris hazel; bill mostly lead-grey with darker culmen, lower mandible whitish; tarsi and toes whitish.

Immature. Like the adult, but narrow and more

pointed juvenile outer primaries retained for at least a year, possibly 1½ years in males, and short and narrow outer tail-feathers for a year. Some males have slight dusky vermiculations on sides of head and upper foreneck (Cramp and Simmons 1980).

Juvenile. Very like the female, but with freckled streaks to crown, coarser freckling on the throat, less rufescent on the back of the neck, scapulars and long inner secondaries more coarsely vermiculated longitudinally, and with a considerable rusty wash on the lesser coverts; iris brown; bill blackish above, dirty white below; tarsi and toes dirty whitish (Someren 1933). There is a post-juvenile moult that begins about 2 months after hatching, involving the body, inner rectrices and inner primaries. It is soon completed on the body, but the outer rectrices are not lost until about a year later, and the outer primaries of females may be lost at about the same time. The outer three or four juvenal primaries of males may not be shed until the bird is about 18 months old (Cramp and Simmons 1980).

Identification

In the hand. This is a large-sized bustard (wing 480–660 mm), with a long bill (culmen 55–76 mm), a strongly black-and-white striped head, a neck that is vermiculated rufous behind and bluish grey in front, the two areas often separated on each side by a white stripe, and primary-coverts that are white-tipped.

In the field. The strong contrasting colours (white underparts, black-and-white wings, bluish-grey foreneck, and rufous hindneck) and large size (only slightly smaller than the kori and Arabian bustards) all provide for rather easy field identification in most areas. In South Africa the very similar Ludwig's bustard is sympatric in a few parts of Cape Province and Orange Free State, but this species shows little if any white on the folded wing. Generally silent, but the male Denham's bustard utters guttural barking notes *kaa-kaa, kia, kia, kia,* perhaps as an alarm call, and clucking notes are sometimes produced while in flight. A resonant, far-carrying call note is uttered by the male during display.

General biology and ecology

This is generally a species associated with open grasslands, usually at moderately high elevations or temperate rolling uplands. In East Africa it occupies similar habitats to those used by the kori bustard, but ranges to higher elevations. In northern Malawi and eastern Zambia the birds occur in high rolling grasslands of the Nyika Plateau, at an altitude of about 1800–2500 m above sea level, and in a region averaging about 110 cm of annual rainfall (Wilson 1972).

Similarly, in Orange Free State the birds breed in temperate montane 'sour' grassveld, in a region having an annual rainfall of more than 70 cm. Birds using these sour grasslands for breeding tend to move eastwardly to lower altitudes during winter, and some also move west into the bushveld of central Transvaal. There is also a small isolated population on the Waterberg Plateau of the central Transvaal, and a population on the coastal grasslands of northern Natal. In Cape Province the species has a more coastally oriented distribution, and is mostly found in the coastal Macchia vegetation or coastal fynbos belt, but it also wanders into the Karoo Midlands (Herholdt 1987).

Flock sizes are typically small in this species; Wilson (1972) noted that of sixty-five sightings, the maximum number of birds seen together was seven, and only ten sightings were of groups larger than two. The largest number (twenty-six) of sightings was of pairs, but there were nineteen sightings of single males or single females.

Like many other bustards, these birds are strongly attracted to burnt fields, where they forage on insects, small mammals and lizards that have been rendered helpless or left exposed (Someren 1933). They also often examine the droppings of animals such as wildebeest (*Connochaetes*) and zebra (*Equus*), pecking them open and looking for fly larvae, dung beetles, and the like. They also feed on termites, including the winged forms as well as the wingless ones. Small frogs are apparently taken by wading in shallow water. Additionally, grass, flower heads, and other vegetable matter may be eaten at times (Howells and Fynn 1979). Three specimens collected in Malawi were analysed carefully by Wilson (1972), and all contained some plant matter of flower heads, shoots, seeds, or grass materials. One whole skink (*Mabuya varia*) was present, as well as the head and neck of a *Duberria lutrix* lizard. Two of the specimens had millipede remains, two had hymenopterans (bees and pompilid wasps), one had orthopterans (cricket nymphs), one had lepidopterans (small moth larvae), one had coreid hemipterans, and all had beetles present in some number. These beetles included tenebrionids (in all), curculionids (in all), staphilinids, scarabeids, and histerids. Tarboton (1989) observed a male eating the contents of a pipit's nest, and noted that a dropping contained hundreds of fragments of beetle exoskeleton, and a few mandibles of termites. Flower-heads of various plants (*Hypochoeris* and *Senecio*) were also eaten by adult birds.

There is evidently a substantial amount of migratory seasonal movement in this species in various parts of its range, although this is not yet well documented. Thus in the Sahel zone of Mali and Nigeria the birds move north between April and June, or sometimes as late as August, in response to rainfall, and breed during the rainy period in the Sahel. They usually return south between September and December, but some may remain in the Sahel throughout the year (Elgood et al. 1973, Lamarche 1980). Birds arriving in the Sahel as late as July or August are probably non-breeding immatures. Similarly in Senegambia, non-breeding birds are present from mid-July to mid-November. In East Africa the movements are less extensive and may be locally variable, but in some cases there is a movement to lower elevations during winter, and in others there may be a seasonal northward shift (Urban et al. 1986). It has been suggested that the species may be migratory in Natal, with seasonal movements to coastal areas, and in Cape Province it is subject to movements that are not as yet understood, but those, as noted above, may include occasional wanderings north to the Karoo (Herholdt 1988). Such wanderings would place the species in competitive contact with Ludwig's bustard, which is otherwise ecologically rather well separated from Denham's bustard, although in some areas such as the Albany district they might come into limited contact (Skead 1965).

Social behaviour

Although the evidence is still slightly contradictory as to this species' mating system, it is very probable that males are polygynous, and occupy individual territories that may be well separated but still organized into a lek-like arrangement. Tarboton (1989) suggested that 'dispersed lek' may best describe the spacing characteristics of the males' display areas, in which three or four males may be seen displaying simultaneously, but at least 700 m apart. He suggested that male-dominance polygyny may be the mating-system model that best applies here, inasmuch as the male has apparently been emancipated from the nesting phase of the reproductive cycle, and the environmental resources (food) are widely dispersed rather than clumped and easily defensible. There has been some speculation that monogamy might occur at least at times in this species. This idea has at least in part been perpetuated by Wilson (1972), who reported seing both sexes sitting at active nests in two cases, and speculated that they might have been waiting to exchange incubation duties. With this exception, all the available information, including such physical evidence as the marked sexual dimorphism of the species, would suggest that a polygynous mating system exists as in the other large bustards.

In line with the probability that the species is normally polygynous, Tarboton (1989) reported that in South Africa breeding males establish territories at least 700 m apart in a dispersed-lek pattern. They display in response to one another or to any available female, and play no further role in nesting after fertilization has occurred. According to Someren (1933) the species' display is even more elaborate than that of the kori bustard, partly because 'the neck ruffle is very much longer and fanned out into an immense ball in front and the feathers are kept quivering all the time'. Additionally, the neck skin is greatly inflated (the male's oesophageal anatomy is apparently still undescribed), the white under tail-coverts are fanned, and a far-carrying call is produced. According to some authorities this call is very similar to the 'booming' of a Eurasian bittern (*Botaurus stellaris*) (Clancey 1967), although Tarboton (personal communication) stated that he found no such similarity. With its neck feathers thus billowed out, and its neck bent back so that the head is brought close to the back, the displaying bird stands or slowly walks about, sometimes with a bouncing motion, for as long as an hour or more. Such a displaying bird may be visible for as much as 2 km. One male under observation for 13 hours displayed for 20 per cent of the day. Display occurred intermittently throughout the day, but was most frequent during early morning and late afternoon hours. The birds frequently displayed when other adult-plumaged males were in view, and almost always displayed when grey-breasted birds (females or immature males) were visible. Some indication of population density and territorial size is provided by the fact that five adult males (plus nine other birds, including two breeding females) occurred on an area of about 1800 ha. The males were spaced 700–2000 m apart, and each of these adult-plumaged males occupied areas of about 60 ha (Tarboton 1989).

Smith (1987) has provided some useful descriptions of male displays based on observations in Kenya, where four displaying males occurred on a study area of about 24 km^2, as well as at least fourteen other birds. Males performed two displays, including a 'balloon' display (Figure 42d) that appeared to advertise territory and establish male dominance. A 'boundary' display was observed twice, and involved parallel pacing by two males at the edges of their territories, with raised napes but not with inflated necks. Neck inflation is a major part of the balloon display, together with a fanning of the orange nape feathers. In this posture the male walks slowly about, occasionally with more intense

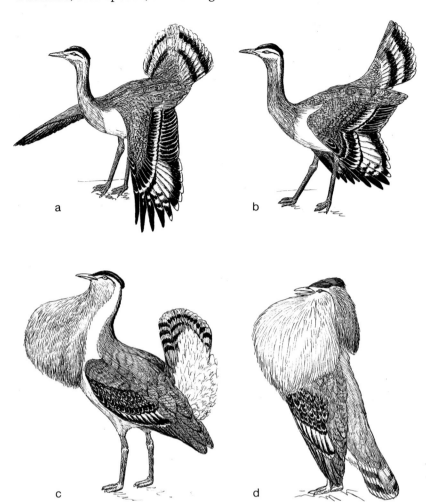

Fig. 42. Social behaviour of male Denham's bustard: (a) defensive display; (b) 'strutting' posture; (c) tail-cocking and neck inflation; and (d) extreme balloon display. (a, b) after sketches in Howells and Fynn (1979); (c, d) based on various descriptions including Smith (1987).

'display struts' that have even greater neck enlargement. At the end of such struts the male often turns, and then rises so that the body is held nearly vertically, the white neck 'balloon' and orange nape almost hiding the head. There is then a shudder, the bill is opened, and a vocalization is probably uttered, although wind conditions prevented hearing this.

Some additional observations on display have been provided by Howells and Fynn (1979). When defending a termite mound a bustard threatened black kites (*Milvus migrans*) that were on the ground by adopting a wings-lowered and tail-cocked posture (Figure 42a). Its more typical 'strutting' posture consisted of a turkey-like posture, with a fanned tail. This posture was sometimes seen during preening periods, and often terminated them. Three possible flight displays were observed. One consisted of jumping vertically to about 3 m, stalling, and then spiralling back to the ground. A second version consisted of rising to about 7 m, then closing the wings and falling back to the earth, snapping the wings open to break the fall only at the last moment.

In a third type the male spread its wings at the top of its climb and flew strongly for about 40 m before landing and adopting a strutting posture. It is possible, perhaps probable, that one or more of these displays corresponds to the 'rocket flights' of the smaller *Eupodotis* species, but because of the limited description available I have not identified any as such in my phylogram of possible bustard affinities (see Figure 2).

Reproductive biology

The breeding period of this species is quite varied, depending upon geographical location, and is evidently somewhat opportunistic, at least in the northern parts of its range. In South Africa the major breeding period is the moist summer months between October and March (Clancey 1967). Of twenty-five egg records listed by Herholdt (1988), the largest number (nine) was for November, followed closely by December (seven), with extremes of September and April. Of six records of chicks, four were in November and December, also suggesting that these two months represent the peak breeding period in South Africa.

Outside South Africa, Malawi records extend from October to January, plus March and August, and Zambian records are from November to February, plus July and August. Records from western Tanzania are for March and July, and for the northern Uganda–western Kenya region January to March. Further west, in northern Zaïre and the Central African Republic the records are from December to February, and in Nigeria breeding has been reported in May. Further north in the drier latitudes of Chad and Mali the records are from June to August and July to October respectively, during the rainy period.

The nest is a scrape on the ground, often being hidden by grass or in the shade of woody vegetation. Of twenty-five clutches listed by Herholdt (1988), fourteen were of at least two eggs (two of the clutches were simply identified as having 'eggs'), and eleven were of a single egg. Furthermore, of six broods, three were two-chick broods, suggesting that two-egg clutches are at least as common as single-egg clutches, if not more so. Almost certainly only the female incubates (Uys 1963), although as noted earlier, males have been seen accompanying females at the nest (Wilson 1972). Uys (1963) observed a female perform a male-like display when she approached her nest for the first time after a hide had been erected nearby and was occupied. She extended her wings by her sides so that the primaries dragged along the ground, the tail was fanned above her back, and the neck and chest feathers were also erected. She walked around for a few minutes in this posture before walking up to the nest and sitting down on the egg. At no time was a male seen in the immediate vicinity of the nest. One incubating female observed by Tarboton (1989) was nesting about 900 m from the nearest male display site, and about equidistant between two such males. Neither of these males approached the nest or showed any interest in it.

Evolutionary relationships

As observed by Snow (1978), this species seems to provide a sort of central 'core' to the genus *Neotis* from which the other three extant species all seem to have sprung in different parts of the African continent. The genus *Neotis* also seems to be a more or less intermediate evolutionary grade between those represented by species here placed in the genera *Ardeotis* and *Eupodotis*, in terms of both their physical structure and probably also their mating systems. At least in an ecological and behavioural sense there are also some similarities between *Neotis* and *Otis*, since both represent large, essentially grassland-adapted bustard types with high levels of sexual dimorphism and apparently similar variably dispersed lek mating systems.

Status and conservation outlook

The large size of this species has unfortunately made it an attractive target for sportsmen, who have traditionally hunted it with rifles (Someren 1933). As a result it has become rare or perhaps extirpated from some of the northern parts of its range (Darfur, Sudan), and is distinctly rare in Kenya, where perhaps only 100–150 birds remained in the mid-1980s (Urban *et al.* 1986). It is similarly now essentially limited to the Nyika Plateau and probably the Luangwa Valley in Malawi (Wilson 1972), and has only rarely been seen in north-western Zimbabwe (Howells and Fynn 1979).

In South Africa it has been judged that as of the late 1980s less than 100 breeding pairs remained in the Transvaal, and 100–200 pairs may then have existed in eastern Cape Province. In Orange Free State the species is rare on the plains but more frequent in temperate grasslands at higher elevations. It is apparently holding its ground in Natal, where it is still widespread in suitable habitat. Although it is legally protected in South Africa, inadvertent snaring is probably responsible for some mortality, and some deaths are also caused by the birds flying into telegraph wires (Herholdt 1988).

In line with this last comment, it has been reported (unpublished work in the Karoo National Park) that five out of ten reported cases of 'unnatural' mortality were caused by collisions with overhead lines, four by snares put out for mammals, and one by collision with a fence. The incident of mortality through illegal hunting could not be determined. Allan (1988) judged that the species' habitat in South Africa was gradually disappearing as a result of agriculture and commercial afforestation. Like that of the two other large bustards, the Ludwig's and kori bustards, its population status is considered 'vulnerable' in South Africa.

NUBIAN BUSTARD (Plate 21)

Neotis nuba (Cretzschmar) 1826
Other vernacular names: maqur (Arabic); outarde Nubienne (French); Nubiertrappe (German).

Distribution of species (Map 20)

The sub-Saharan Sahel zone from Mauritania, Mali, Niger and Chad east to northern Sudan.

Map. 20. Distribution of Heuglin's (he), Ludwig's (lu) and Nubian bustards, inlcuding the locations of the Nubian bustard's races *agazi* (ag) and *nuba* (nu).

Distribution of subspecies

N. n. nuba: Endemic to Sudan and eastern Chad.
N. n. agaze Vaurie 1961: Mauritania to Chad.

Measurements (mm)

Wing, males 423–475, females 340–418 (various sources); tail, males 249–266 (av. of 6, 260); females 210–236 (av. of 4, 223); tarsus, males 117–130 (av. of 6, 123), females 95–107 (av. of 4, 102); culmen, males 48–55.5 (av. of 8, 50.8), females 49–53 (av. of 4, 50.6) (Urban *et al.* 1986). Egg, av. 70 × 47 (Schönwetter 1967).

Weights

Males to at least 5.4 kg (Urban *et al.* 1986). Egg, 84 g (estimated).

Description

Adult male. General colour above very light tawny with coarse but sparsely distinct black frecklings, generally in the form of cross-bars with a few irregular arrowhead marks on the scapulars and mantle; lower back and rump also tawny with indistinct dusky cross-lines not so pronounced as on the back; upper tail-coverts rather more ashy than the back, with somewhat coarser frecklings and indications of narrow blackish bars like the tail; wing-coverts like the back, with very few cross-markings; median and greater coverts with white tips, the latter almost uniform tawny colour and with scarcely any vermiculations; alula and primary-coverts almost entirely white with a subterminal shade of black; remiges blackish, all but the outermost primary with white tips, the greater part of the bases white or tinged with tawny, the innermost secondaries like the back; rectrices ashy freckled with black mottlings and crossed by two narrow black bands, the bases of the feathers being whitish; crown of head also light tawny, with minute cross-lines of blackish, and with black bands on the sides of the crown that meet on the nape; lores, eyebrow and sides of the face white; ear-coverts slightly washed with ashy blue like the hindneck, lower throat, and foreneck; cheeks and chin white, as also the rest of the sides of the throat; in the centre of the throat a broad black streak; foreneck and chest deep tawny rufous, this colour extending backward on the sides of the neck and reaching upwards so as to separate the grey on the hindneck from the mantle; breast, abdomen, under wing-coverts and axillaries white, with a few dusky cross-markings on the upper breast. Iris ochraceous; bill horny yellow, more dusky towards the tip, sometimes pale green with a dusky base; tarsi and toes pale yellow to cream-white (Cramp and Simmons 1980).

Adult female. Like the male, but with some colours less intense and the mantle faintly streaked pale (Cramp and Simmons 1980); the black on the throat is confined to the centre (Mackworth-Praed and Grant 1952).

Immature. Like the adult, but some juvenile outer primaries retained. Birds with some coarse dark grey bars on neck are probably not fully adult (Cramp and Simmons 1980).

Juvenile. Like the female, but the chest collar indistinct and with vermiculations more widespread below and over the white areas of the wing (Cramp and Simmons 1980).

Identification

In the hand. This medium to large (wing 340–475 mm) bustard is distinctive in having a nearly all-grey neck (becoming brownish on the basal hindneck) and a tricoloured head of rufous (crown), white (cheeks), and black (throat and postocular stripe).

In the field. This species of the southern fringes of the Sahara is pale brown above and white below, with more white on the underwing than in the considerably larger *arabs* and similar to the slightly larger *denhami* but lacking rufous on the hindneck and mostly brown rather than mostly black-and-white upper wing-coverts. A shrill *maqur* call (the basis for its Arabic name) of uncertain significance has been noted.

General biology

This species occupies desert scrub and desert-fringe or arid savannah areas, extending further into the Sahara than the other species of *Neotis*. The birds probably move out of these extremely arid areas during the dry season, but almost nothing is known in detail of the species' ecology.

Foods include large insects such as locusts and beetles (tenebrionids and elaterids reported), plus ants, other small hymenopterans and bugs, vegetable matter including leaves, shoots and seeds, fruits of *Salvadora*, and the exuded gum of *Acacia* trees (Urban *et al.* 1986).

Social behaviour

The mating system of this entirely unstudied species is completely unknown, although is unlikely to differ from the other *Neotis* forms. Not even its oesophageal anatomy, which might provide a clue to male displays, has been described. The strong sexual dimorphism would suggest that polygyny prevails.

Reproductive biology

There are few breeding records, but those from Mali are from July to October, breeding in Niger is reported for August, and Chad egg-laying likewise probably ocurs in this July–August period (Urban *et al.* 1986). Two-egg clutches have been found, but it would not be surprising if single-egg clutches also occur at times.

Evolutionary relationships

Snow (1978) stated that *nuba* might be the result of an early offshoot of early *denhami* stock in north-eastern Africa, which seems a likely possibility. I have judged (see Figure 2) that these two forms are probably quite closely related, and they are nearly parapatric in distribution, with both species perhaps occurring together or nearly so in the vicinity of Lake Chad.

Status and conservation outlook

Although information is quite limited, this species is evidently threatened by hunting throughout its range. Unfortunately, at present it is apparently not under study by any conservation agency as to its status or biology.

HEUGLIN'S BUSTARD (Plate 23)

Neotis heuglinii (Hartlaub) 1859

Other vernacular names: Heuglin's red-breasted bustard; outarde de Heuglin (French); Heuglintrappe (German).

Distribution of species (Map 20)

Eastern Somalia and Ethiopia south to the vicinity of Lake Turkana (Rudolf), northern Kenya. No subspecies recognized.

Measurements (mm)

Wing, males 489–505 (av. of 3, 495), females 405–432 (av. of 4, 423); tail, males 185–190 (av. of 3, 188), females 160–180 (av. of 4, 171); tarsus, males 151–162 (av. of 3, 157), females 126–136 (av. of 4,

131); culmen, males 72–79 (av. of 3, 76), females 62–65 (av. of 4, 64) (Urban *et al.* 1986). Egg, av. 73 × 53 (Schönwetter 1967).

Weights

Two males 4.0 and 8.0 kg, two females 2.6 and 3.0 kg (Urban *et al.* 1986). Male, 2.7 kg (Someren 1933). Egg 110 g (estimated).

Description

Adult male (after Someren 1933). Crown black, tapering to a point at the nape, with a central irregular white streak slightly expanded towards the nape, followed by a white superciliary line, narrow at the base of the mandible, expanding in the region of the eye and extending back over the ear-coverts; lores white, speckled with black; a black streak runs from the gape to below the eye and ends above the ear-coverts; chin and cheeks white, the latter black-speckled; throat white with a large black patch; foreneck and upper hindneck lavender grey, shading into vinous-grey and to chestnut on the base of the foreneck and upper breast, these areas indistinctly barred with wavy dark-grey lines; feathers on back of neck sandy buff with large buffy central area freckled and vermiculated with black at margins and tip; mantle, back and lesser wing-coverts sandy buff with buff arrowhead marks, vermiculated and freckled at margins with blackish; the upper mantle slightly washed with rusty; scapulars and long inner secondaries coloured as the mantle, but the buffy arrow marks more broken up by wavy black lines; median coverts black and white with terminal white spot; greater coverts black with white tips; secondaries brown-black with white tips and varying amount of white on inner webs; primaries mostly black with slight freckling on inner webs, remainder black-brown with white on inner webs; rump and upper tail-coverts buff-sandy distally with gradually increasing distinct blackish vermiculations, where there is a distinct broad black bar followed by clear blackish vermiculations and a pure white tip; breast, flanks, abdomen and under tail-coverts white. Iris hazel; bill olive at base, horn at tip, lower mandible whitish; tarsi and toes ivory with greenish tinge posteriorly.

Adult female (after Someren 1933). Very similar to the male above, but top of head less black and the pale central line buffy; very much less black on throat; chestnut of breast merely a wash; size smaller.

Immature. Like the female, but with a grey nape spot (Mackworth-Praed and Grant 1952). The black parts of the head are browner, and the black throat is reduced to a stripe (Urban *et al.* 1986).

Identification

In the hand. This large (wing 405–505 mm) bustard has a nearly all-grey neck (becoming rufous on the base of the hindneck) and is coarsely barred above, with the face less distinctly tricoloured than in the previous species, the crown being whitish rather than chestnut, at least posteriorly, and the cheeks less pure white, especially in males.

In the field. This is the only *Neotis* species found in the deserts of the eastern horn of Africa, and is not only smaller than the kori bustard but lacks its black crown and crest, the black instead occurring as a superciliary stripe and elsewhere on the face or chin, forming in the male a black triangular patch on the anterior throat. The lower breast is also separated by black and rufous bands from the white underparts. Vocalizations of the species are still undescribed.

General biology and ecology

Like the other *Neotis* species, this is an arid-adapted species, in this case occupying the lowland Chambli desert area of the African horn, where it ranges from extreme nearly vegetation-free rock desert through open deserts having only annual grasses, semi-desert savannahs, and tussocky grasslands. The birds are apparently somewhat nomadic, probably moving as necessary with local or seasonal rainfall patterns. They are usually found only singly or in pairs, but small groups have been seen feeding on berries, leaping into the air to reach them if necessary. Only a few specimens have been examined as to their foods, which are known to include grasshoppers and vegetable matter such as berries. A mouse and a lizard have also been reported as foods (Urban *et al.* 1986).

Social behaviour

There is no information on the social behaviour or mating system of this species.

Reproductive biology

The nest is a scrape on nearly bare ground, with little surrounding vegetation, the female relying on her concealing sand-coloured plumage pattern for protection from visual predators. Breeding records for Somalia and Ethiopia are from April to June, during the wetter period, and for northern Kenya in

January and June, or later in the rainy season when the grasses are tallest. The clutch is reportedly two eggs (Urban *et al.* 1986).

Evolutionary relationships

This species is presumably a fairly close relative of *denhami* (see Figure 2), and might be regarded as an early offshoot from this basic stock (Snow 1978). The two forms have essentially parapatric distributions in north-eastern Africa, perhaps coming into contact or nearly so in the vicinity of Lake Turkana.

Status and conservation outlook

Apparently this species is generally uncommon to fairly frequent over much of its range in Somalia, Djibouti and Ethiopia. It occupies only a very small range in northern Kenya, but regularly occurs east to Marsabit (Urban *et al.* 1986).

LUDWIG'S BUSTARD (Plate 24)

Neotis ludwigi (Rüppell) 1837

Other vernacular names: Ludwigse Pau (Afrikaans); outarde de Ludwig (French); Ludwigtrappe (German).

Distribution of species (Map 20)

Extreme south-western Angola, western Namibia and west-central South Africa east to Lesotho and the Drakensberg foothills of Orange Free State. Not yet definitely reported from either Botswana or the Transvaal, but probably of rare or casual occurrence in the south-western portions of both regions. Also not authentically reported from coastal Transkei or western Natal. No subspecies recognized. Listed as 'vulnerable' in the *South African Red Data Book* (Brooke 1984).

Measurements (mm)

Wing, males 495–561 (av. of 6, 536), females 433–470 (av. of 5, 452); tail, males 235–263 (av. of 4, 255), females 205–256 (av. of 3, 231); tarsus, males 133–149 (av. of 4, 137), females 114–122 (av. of 3, 117); culmen, males 53–64 (av. of 4, 58), females 43.5–63.7 (av. of 3, 52.2) (Urban *et al.* 1986). Egg, av. 74 × 54.

Weights

Males 4.2–4.7 kg (av. of 4, 4.525); females 2.2–2.5 kg (av. of 4, 2.35) (Earlé *et al.* 1989). Estimated egg weight 118 g, or 3.9% adult female, reported as 3 kg (Schönwetter 1967).

Description

Adult male. General colour above brown, profusely variegated with black wavy lines and blotched with large arrowhead marks of uniform sandy buff; scapulars and inner secondaries like the back, but the blotches taking the form of bands; lower back, rump, and upper tail-coverts sandy buff, rather more finely vermiculated; the upper tail-coverts a little more coarsely lined with indications of narrow bands; lesser wing-coverts like the back, those on the bend of the wing white with blackish spots; median coverts for the most part blackish, with broad sandy-coloured bands, which are freckled with blackish; greater coverts almost entirely black, more or less vermiculated on the inner web; primary coverts also black, with white bands near the base and tipped with white, like the greater coverts; remiges black, the inner primaries white at the base and especially along the inner web; the secondaries black, with more or less white tips; rectrices broadly banded with four black and three lighter bands, and tipped with whitish or ashy; crown and nape brown to black with a white coronal patch before the occiput; upper hindneck and the sides of the neck white, the rest of the hindneck sandy rufous, this colour spreading on to the upper mantle, which is finely freckled with blackish; the entire sides of face, throat and foreneck blackish-brown; chest mottled with bars of white and brown, and separated from the rufous hindneck by a line of white extending down the sides of the neck and joining the sides of the breast, which are white like the rest of the undersides and under wing-coverts. Iris greyish-brown; bill dark horn-colour, grading to flesh-white basally, tarsi and toes greyish or greenish-white (Clancey 1967).

Adult female. Similar to the male in colour, but smaller, and distinguished by the colour of the lower throat and chest, which is brown, mottled all over with freckles and bars of white; chin and upper throat white; eyebrow white, mottled with blackish markings; sides of face brown, also mottled with white; wings as in the male, but the primary coverts for the most part white with blackish ends.

Immature. Like the female but more heavily barred with blackish-brown and white over the entire forethroat and upper breast (Clancey 1967).

Identification

In the hand. This large (wing 430–560 mm) bustard is distinctive in having a brown or brownish-black foreneck, becoming paler on the sides and more rufous toward the base of the hindneck, and wings with brown-tipped primary coverts. The head is mostly striped with greyish brown and white, lacks black on the throat and the pale superciliary stripe is edged with brown rather than black.

In the field. This species is found on open plains, and is similar in size to *denhami* but smaller than *kori*, the other two large bustards of the region. It is the only one of the three to lack black-and-white striping on the head, and the only one with a brownish forehead. It also shows less white on the folded wings than either of these species, although white wing markings are visible in flight. Silent except during display, when a deep, resonant croaking is produced by males, and loud foot-stamping is also performed.

General biology and ecology

This is a more arid-adapted species than the larger and more contrastingly patterned Denham's bustard, although the two species overlap in range and to some degree in habitats. It generally occurs in open plains of the Karoo, along the edges of the Namib desert, in light thornbush or 'degenerate' acaciaveld, and in rolling upland areas having short grasses, rock outcrops and scattered termite mounds (Clancey 1967).

There are some seasonal movements evident in this species, the birds becoming more common in northern Cape Province and Orange Free State after the first spring rains, and probably being mainly a winter (May to October) visitor to Namaqualand, Namibia, which is a winter-rainfall area. Thus they tend to move from Namaqualand east into the Karoo, a summer-rainfall area, during the summer (November to April), where they concentrate in areas that have received rain and have a subsequent outbreak of locusts or caterpillars. Seasonal records for the species have shown that indeed the birds are present in the drier winter-rainfall area mainly during winter, and that they are present throughout the year in the summer-rainfall area, but with increased numbers during the rainy period.

The extent of the Karoo has increased in recent years, and thus additional habitat has been created for this species, which there occurs primarily on sheep-farming areas (Herholdt 1987, 1988). The birds are fairly gregarious, foraging together in groups of up to about twenty at times. Besides caterpillars and grasshoppers, the species' foods reportedly include crickets, beetles, lizards, small mammals, and some vegetable matter, including seeds. A recent study by Earlé *et al.* (1989) on the stomach contents of seven birds shot during January and February indicates that although this species is essentially omnivorous, orthopterans may constitute the single most important taxonomic group in its diet, forming 89 per cent of estimated dry mass and 30 per cent of the numbers of food items found, grasshoppers being taken in large quantity. A wide variety of beetles had also been eaten, especially curculionids but also including carabids, scarabaeids, buprestids, tenebrionids, and meloids. Most of the remaining insect materials consisted of hymenopterans, especially ants. Vegetative plant materials were in most of the stomachs, and one contained the tail of a reptile.

Social behaviour

Little is known of the mating system of this species, but it is believed to be polygynous, the territorial males having a rather static advertisement display. Like Denham's bustard, displaying males inflate the neck, cock the tail so as to expose their white under tail-coverts, and utter deep, resonant *klump* or *wup* calls, at intervals of about 5 s (Urban *et al.* 1986). There is scarcely a true crop present, but the oesophagus is somewhat spindle-shaped, with the middle region somewhat larger than that of either end, especially in males (Niethammer 1940). This oesophageal enlargement no doubt accounts for the neck inflation and perhaps helps to resonate the male's territorial vocalizations. Loud foot-stamping also apparently occurs.

Reproductive biology

South African breeding records are generally spread over a wide period from July to March, whereas the available records from Namibia are for the summer from December and January (Urban *et al.* 1986). Egg and chick records summarized by Herholdt (1987) include four for August, two for September, two for October, four for November and one for December. This would generally suggest that breeding may be more closely timed to local rainfall patterns than to a strict seasonal schedule.

The nest is a simple scrape on bare ground, often hidden among stones on a hill slope or ridge crest. Records summarized by Herholdt include four single-egg clutches and seven two-egg clutches, representing a mean of 1.6 eggs. There is no additional information on the species' breeding biology, which is probably much like that of Denham's bustard.

Evolutionary relationships

Snow (1978) suggested that this species and *heuglini* are probably both closely related to *denhami* and the result of isolation of early *denhami* stock in the north-eastern and south-western parts of Africa. He regarded all of the *Neotis* forms as constituting a species-group, and did not recognize any superspecies. I have suggested (see Figure 2) that *ludwigi* might be the most isolated of these four forms, and a very early offshoot of early *denhami* stock.

Status and conservation outlook

Herholdt (1988) has thoroughly reviewed the status of this species and has concluded that it has probably not declined as much as some prior authors have believed, in part because of the spreading of suitable semi-desert habitat in the Karoo. He said that there were no valid recent records from Natal, and judged that the occurrence of this species in the Transvaal required confirmation, although it might wander into the south-western parts of this province from Orange Free State. In that region breeding does occur, and has been reported recently from the southern portion of Orange Free State. It occurs more often in more arid regions there than in the eastern parts of the province. Similarly, in Cape Province it occurs in the central and western Karoo, and other arid areas of the interior. It has also been reported from some south-eastern areas of Cape Province near the coast, such as Adelaide, Grahamstown and King William's Town. It has also been reported from several localities in Lesotho. In Namibia the species is generally distributed over the hotter and drier areas of the west. Thus it mainly occurs in the Namib and along its edges north to the Angola border, extending a short distance into south-western Angola.

According to Allan (1988), the long-term degradation of the Karoo ecosystem from grazing may have had subtle but still unmeasured effects on its status there, and in spite of protection, hunters continue to take a toll on the species. Finally, considerable numbers of birds are killed by collisions with power lines during their long-distance movements, one farmer reporting twenty birds killed in this way during a ten-year period. Its population status in South Africa is considered 'vulnerable'. An unpublished study by the same author revealed eighty-nine 'unnatural' deaths of Ludwig's bustards, of which sixty-nine (77 per cent) were caused by collisions with overhead lines. Fences (twelve cases) and road kills (two) accounted for most of the remainder, although six were reportedly killed by traps, trap guns, and coyote-getter devices. Illegal hunting kills were not listed, and their importance was difficult or impossible to assess.

BLACK-BELLIED BUSTARD (Plate 25)

Eupodotis melanogaster (Rüppell) 1835

Other vernacular names: black-bellied korhaan, Longbeen-korhaan (Afrikaans); outarde à ventre noir (French); Schwarzbauchtrappe (German).

Distribution of species (Map 21)

From Senegal east to Ethiopia and Somalia, and south to Angola, Namibia, northern Botswana, Zimbabwe and eastern South Africa.

Distribution of subspecies

E. m. melanogaster (Rüppell): Senegal to Ethiopia, Angola and Mozambique, north of the Zambezi.

E. m. notophila (Oberholser) 1905: south-eastern Africa, south of the Zambezi, to western Zimbabwe and the vicinity of Port Elizabeth, South Africa.

Measurements (of *melanogaster*, **mm)**

Wing, males 325–358 (av. of 26, 346), females 302–345 (av. of 23, 319); tail, males 158–201 (av. of 26, 178), females 146–166 (av. of 23, 157); tarsus, males 120–137 (av. of 26, 129), females 119–140 (av. of 23, 130); culmen, males 37–46 (av. of 26, 40.5), females 36.5–42.5 (av. of 23, 40.1) (Urban *et al.* 1986). Egg, av. 58 × 48 (Schönwetter 1967).

Weights

Two males of *melanogaster* 1.8 and 2.7 kg, female 1.4 kg (Urban *et al.* 1986). The average weight of males is about 1.8 kg, with some males reaching 3.2 kg; females rarely exceed 1.35 kg (Pitman 1957). Egg, 73 g (estimated).

Description

Adult male. Generally dark brown dorsally, this colour predominating over the sandy-coloured portions of the upperparts, the feathers being vermiculated rather coarsely with blackish and having large median ovate or sagittate streaks of black; lower back and rump nearly uniform brown with very few transverse frecklings of sandy buff; wing-coverts lighter and more tawny, with the same kind of black frecklings and with very distinct arrowhead blotches of black; the marginal and median coverts white, as are the alula feathers, which have slight indications of black spots, the outer alula feather having black bars and mottlings and tinged with buff; greater coverts white with a slight fulvous tinge, and regularly barred with three black bands, the inner webs somewhat coarsely mottled with the same;

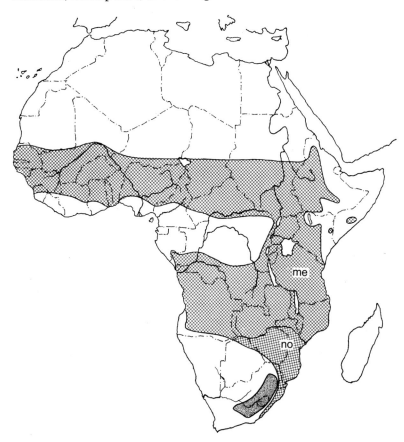

Map. 21. Distribution of the black-bellied bustard (cross-hatched), including the locations of its races *melanogaster* (me) and *notophila* (no), and of the blue bustard (stippled).

primary coverts black tipped with white and an additional bar of white at the base, the latter tinged with tawny; most primaries almost entirely white with black tips, the inner ones white with a subterminal band of black, the secondaries black with two large spots of white on the outer web, the second one almost concealed by black mottling, the innermost secondaries dark brown like the back; upper tail-coverts and centre rectrices rather more ashy brown, the remainder either tawny or blackish-brown, thickly freckled with blackish-brown and crossed by five black bands, the outer feather almost entirely black; crown and hindneck sandy buff minutely freckled with black and ornamented with some tiny spots of lighter buff; lores and eyebrow dark slaty grey, continued in a black line which skirts the sides of the head and unites on the nape, separated from the crown by a less distinct band of white somewhat tinged with ashy; feathers below the eye, cheeks, chin and upper throat hoary grey shading into black in the lower throat and continued broadly down the centre of the neck, this stripe bordered on each side by a band of white that passes over the posterior cheek and joins the ear-coverts, which are also white; side of neck and of upper breast sandy buff vermiculated with black, and separated from the upperparts by a line of white; foreneck, chest, breast and underparts black, including the axillaries and under wing-coverts; thighs black with a white ring just above the bare portion of the tibia; long under tail-coverts sandy buff, barred and freckled with black towards the ends. Iris light brown, gradually darkening inwardly; bill mostly pale yellowish, the culmen dark brown; tarsi and toes yellowish brown.

Adult female. Similar to the male above, but more profusely spotted with black arrowhead markings, relieved by a creamy white line that borders each black mark and brings it into strong relief; head blackish spotted with sandy buff; eyebrow, sides of face and ear-coverts also sandy buff, with no black on the face or sides of the crown; chin and upper throat white; lower throat, foreneck and chest sandy buff minutely freckled or barred with blackish; the feathers of the breast sandy buff with lines and spear-shaped markings of black; remainder of underparts white tinged with sandy buff, sides of body with large blotches of black; axillaries black; under wing-coverts barred black and white; the greater coverts black barred with white, some of the bars broken up into mottlings of sandy buff; rectrices sandy buff mottled and barred with blackish; under tail-coverts also sandy buff barred with dark brown.

Immature male (after Someren 1933). Very like the female but with some indication of a black throat patch and central black line down the front of the neck; breast and abdomen sepia to black blotched with white; thighs black spotted with white and with a white circlet distally; mantle and back generally like the female, but the arrow-like marks more distinct and the wing-coverts paler and inclining to white medially.

Juvenile (after Someren 1933). Crown dark brown heavily spotted with ochraceous; neck ochraceous with black bases; mantle, back and wing-coverts sepia with black central area to each feather vermiculated with sandy buff and with a terminal buffy spot; upper breast spotted like the neck; remainder of underside dirty whitish, freckled and vermiculated with blackish on the sides and on the under tail-coverts.

Identification

In the hand. This small to medium (wing 302–358 mm) bustard has black underparts and foreneck (in males), the sides of the neck grey to whitish, or (in females and young) a sandy buff to ochraceous neck and a similarly pale greyish crown and rather light brown dorsal body colouration, by which the species can usually be told from the very similar Hartlaub's bustard. The latter has darker upperparts (especially on the rump) and a more brownish crown.

In the field. Except in a few areas of eastern Africa (parts of Ethiopia, Kenya, Tanzania and Somalia), this species can be recognized by the males' black underparts and the contrasting black and grey neck, with but little black reaching the face. The very similar Hartlaub's in the areas just mentioned has more extensively black underparts and the black on the face is more intense. Females of the two species are extremely similar, but those of *melanogaster* average slightly lighter in tone. During ground display, males of this species utter a dull *waak* followed by a throaty grunt and, after a pause, a sharp whip-like sound (Newman 1983). Another description of the male's advertisement call is that of a short *quick* note, followed about 4–5 s later by a low *quok*, the latter sounding like a healthy burp, and both very frog-like in quality (Kemp and Tarboton 1976). Aerial display, involving a wing-beating phase followed by long glide back to earth with the mostly white wings held conspicuously and stiffly upward in a rather steep dihedral, may also occur.

General biology and ecology

This is primarily a tall grassland- and savannah-adapted species that extends into areas that may have moderate tree cover and perhaps prefers lightly wooded areas over pure grasslands. It is ecologically quite adaptable, and also occurs in cultivated areas, fallow fields, and somewhat damp sites such as around wetlands, and in coastal flats with scrub cover. In Uganda it reportedly occurs at all altitudes below 3600 m having suitable grassland habitat (Pitman 1957).

Schulz and Schulz (1986) studied the species in Senegal during the breeding season. There the birds occurred in dry savannah, in cultivated fields having a vegetation height of up to 30 cm, or other areas such as fallow lands that at least during the rainy period had a vegetation height of at least 50 cm. Areas with trees and bushes were not avoided, but only rarely did the birds land among such tall vegetation.

For courting, the males preferred areas covered with only sparse vegetation, and with few elevations, but decayed termite hills are often present. Typical courtship territories were about 200 by 200 m in width, and were often closed in by taller bushes or trees. Now and then bustards were seen in very open territories, such as on freshly sown millet or peanut fields, and one probable breeding female used such an area. Only once were the birds ever seen in a damp, thickly wooded area.

The birds are probably sedentary in most regions, but may move north into the Sahel zone of Mali and Sudan during the rainy period. Similar wanderings or other possible local migratory movements have been observed elsewhere as well (Urban *et al.* 1986).

Foods consist of the usual diverse array of plant and animal materials, especially larger terrestrial insects. Thus in Zaïre, beetles were present in seven out of eight stomachs, grasshoppers in four, bugs and caterpillars in three, and crickets, termites, mantids, centipedes and seeds in two (Chapin 1939).

Social behaviour

This species is almost certainly polygynous, based on its male display behaviour. Schulz and Schulz (1986) found no evidence of a pair-bond between the sexes, and believed that aerial male displays function to attract females to the vicinity of a breeding male, which are then supplemented by terrestrial displays that help lead the female directly into the male's territory. That territory is visited by the female only for copulation during the first weeks of the display season; thereafter the female departs to

172 *Bustards, Hemipodes, and Sandgrouse*

lay her eggs and tends to the nest without further male participation.

Two types of male advertisement display occur, flight display and ground display, the latter being performed much more frequently than the former. Both involve optical as well as acoustic signals, the acoustic components being rather soft and nonmelodic, and the optical signals being very contrasting and conspicuous, especially on the head, neck, breast and wing regions. Schulz and Schulz (1986) observed males displaying during mid-July in Senegal. At that time the males remained in closely defined territories, and defended them against other males. At the height of their activity the birds displayed at all times of the day, but displays were of lower intensity during the hottest part of the day.

Courtship on the ground was usually done on flat, vegetation-free areas, such as decayed termite hills, the birds moving from one site to another within the territory. Each ritual was regularly performed on a regular schedule, with intervals between the displays lasting about a minute. During these intervals walks between display sites were sometimes taken, often with foraging during such walks. In the case of one male that was watched continuously by Tarboton (personal communication), the average interval display between calls was 54 s (range 45–70 s, total of forty displays).

Ground display begins with the male standing in a tall, alert posture, stretching his neck first diagonally toward the front, then moving it straight up, and then opening his bill slightly. At this point the bird is motionless for about a second, and emits a moaning 'ooohh' sound that lasts about half a second (Figure 43a). From the front at this point the black throat and neck stripe is clearly outlined with white, but the neck is not significantly inflated except in the throat area itself. Shortly after this call the head is brought quickly back until the nape almost rests on the back (Figure 43b). The head and bill are then held almost horizontally, a position held for up to several seconds as a snoring or growl-

Fig. 43. Social behaviour of male black-bellied bustard: (a, c) advertisement calling sequence (after photos by P. Johnsgard and in Kemp and Tarboton 1976); (d) wing-drooping and tail-cocking (after painting in Newman 1983); (e) throat pattern during first phase of calling sequence; and (f) body posture during gliding phase of display flight (e, f after sketches in Schulz and Schulz 1986).

ing sound lasting about half a second is produced. Then there is a muffled plopping sound, like that of a cork being removed from a bottle, and simultaneously the wings are jerkily lifted a few centimetres, exposing the white wing areas momentarily. This last call and wing-jerk may be repeated while the body position remains unchanged (Figure 43c). Then the neck and head are lifted back to an upright position similar to the initial alert position. The bird may also preen at this time. All the vocalizations associated with display are quite soft but nevertheless carry well, especially the plopping sound, which may be heard for at least 100 m. Kemp and Tarboton (1976) described these sounds as being 'frog-like', but mentioned only the last two of the three notes described by the Schulzes. Little (1964) described the call as sounding like *ke-waa-aa-k*, which likewise apparently refers only to the last two notes. Newman (1983) illustrated an apparent general display posture, in which the tail is strongly cocked and the wings are somewhat drooped (Figure 43d).

Display flights take place independently of ground display, and are done quite irregularly and less frequently. However, flight displays sometimes occur without apparent direct cause. The male takes off and with deep wingstrokes delineates an area of 50–200 m, reaching a height of 7–15 m. The body and neck are held in a straight line during this flight, and after attaining this height the male stretches out his wings diagonally upwards in a V-shape and, with his neck and head held in this rigid position, gradually glides back to the ground (Figure 43f). There are no interruptions by wingbeats during the gliding phase, which may cover 50–150 m, and during the entire flight the male may cover a distance of up to 1000 m. At the point of landing the male will often begin ground display.

Kemp and Tarboton (1976) observed that when directly courting a female, a male approached her with his neck fully stretched, his throat puffed out, and his neck and throat feathers erected to form a ruff. He approached the female with short steps, and when near her he repeatedly withdrew his neck, stretched it out horizontally towards her, and then raised it back into the vertical position, all in a reptilian-like sinuous manner. This stimulated the female to crouch, whereupon the male straddled her, pecked and pulled at her head, trod on her, and attempted to copulate.

Reproductive biology

The breeding season of this widely distributed species is of course quite variable, occurring in South Africa and Zimbabwe during the spring and summer period of October to February. In Malawi it similarly occurs between November and February, and in Angola from January to March. The records from Zambia range from September to March, but thirteen out of seventeen breeding records are for the period October to January (Benson *et al.* 1972). In East Africa the records are quite varied, namely between February and April in Uganda and adjacent Kenya, and during January, May and September over much of Tanzania, the birds apparently preferring to breed during the latter part of the dry season. In West Africa (Mali, Senegambia and Nigeria) the records are concentrated between June and September, and in Ethiopia there are breeding records for April and September (Urban *et al.* 1986).

The nests are scrapes on bare ground, often in the protective cover of vegetation or a termite hill. The clutch size is one or two eggs, probably more often one than two, since Pitman (1957) noted that only two out of twelve Uganda nests that he observed had two eggs present. Tarboton (personal communication) reported that seven out of twelve South African nests had single eggs, and five had two. They are incubated by the female without any assistance from the male and indeed the nests are probably situated well away from the male territories (Schulz and Schulz 1986).

Evolutionary relationships

Snow (1978) judged that this species and *hartlaubi* are a very closely related species pair that comprise a superspecies. Where the two come into contact in north-eastern Africa, *melanogaster* is found in more mesic habitats, such as in broad-leaved tall-grass savannah, and occurs over a considerably broader altitudinal range, extending about 1400 m higher in elevation (or to 2400 m) in Ethiopia.

Status and conservation outlook

This is one of the most widely distributed of the African bustards, and in many areas is still relatively common. However, it is declining in some regions, such as in Senegal and Gambia (Schulz and Schulz 1986), in East Africa, in Zambia, and in Zimbabwe, often because of agricultural effects, overgrazing, or hunting (Urban *et al.* 1986). In southern Africa there are perhaps still good populations in Mozambique and South Africa's eastern Transvaal, and it is present locally or marginally in Cape Province, Swaziland, coastal Natal and Zululand, but at least most of these areas the population has suffered from the effects of sugar-cane cultivation (Clancey 1972–3).

HARTLAUB'S BUSTARD (Plate 26)

Eupodotis hartlaubii (Heuglin) 1863

Other vernacular names: Hartlaub's black-bellied bustard; outarde Hartlaub (French); Hartlaubtrappe (German).

Distribution of species (Map 22)

Eastern Sudan to Uganda and central Tanzania. No subspecies recognized.

Measurements (mm)

Wing, males 328–359 (av. of 10, 338), females 229–319 (av. of 8, 309); tail, males 151–183 (av. of 10, 164), females 100–145 (av. of 8, 127); tarsus, males 111–131 (av. of 10, 124), females 113–128 (av. of 8, 120); culmen, males 38–45.5 (av. of 10, 42.9), females 40.4–45.5 (av. of 9, 43.6) (Urban *et al.* 1986). Egg, no information.

Weights (kg)

Male, *c.* 1.135 (Someren 1933); two males 1.15, 1.6 (Urban *et al.* 1986).

Description

Adult male. Centre of forehead to mid-line of crown blackish with creamy tips, paler posteriorly, with fine whitish vermiculations; this area to the nape surrounded by a white band; lores, supercilium, cheeks, chin and throat shading to black, this black zone extending from the posterior angle of the eye through the posterior part of the supercilium and back to the nape where it meets the line of the opposite side; it is continued down to the lower throat and extends down the mid-line of the fore-neck to meet the black of the upper breast, which colour extends over the whole of the underside to the under tail-coverts. Beyond the black on the head is a narrow white zone that expands over the ear-coverts, passes down on either side of the median black line of the throat and expands out into two white patches on either side of the upper breast. Remainder of neck very finely vermiculated black and greyish-white, the vermiculations broader and widening out at the base of the neck. Mantle, scapulars, back and long innermost secondaries sepia, with bold creamy speckling and vermiculations, each feather with an irregular central arrow-mark accentuated by a creamy outline. Lesser wing-coverts similarly coloured by paler; the marginal,

Map. 22. Distribution of Hartlaub's bustard.

median and greater coverts white; primary coverts black and white; outermost primary black, the remainder mostly pure white with increasing black tips outwardly, the second and third with black outer webs; secondaries white with black inner webs and an increasing amount of black extending to the outer web inwardly, the innermost secondaries sepia as described above. Rump and upper tail-coverts black with fine sparse white speckling; rectrices black with narrow bars of white speckling. Iris creamy-yellow; bill horn-brown on culmen, otherwise pale whitish-horn tinged with yellowish; tarsi and toes ivory.

Adult female. Anterior crown sepia with large creamy tips shading to sandy buff posteriorly, which is finely vermiculated with blackish and ochraceous tipped; lores, area round the eye and ear-coverts sandy buff, the latter slightly streaked; chin and throat whitish with a creamy wash; upper hindneck sandy buff very finely vermiculated with blackish while the lower part, though similarly coloured, has in addition large creamy spots at the end of each feather, these spots outlined in blackish; foreneck with ground colour as back, but with elongate creamy central marks which become more rounded towards the base of the neck and widening out on the upper breast and outlined distally with blackish thus forming narrow angular marks; breast, abdomen, flanks and under tail-coverts creamy, the long feathers on the side of the body with black shaft-marks and slight blackish vermiculations; mantle, scapulars and long inner secondaries with small arrowhead blackish marks accentuated distally with creamy, the remainder of each feather being creamy with blackish vermiculations (the back is thus more pale and blackish and the central marks less conspicuous than in *melanogaster*). Wing-coverts creamy with sparse sepia vermiculations and blackish shaft streaks; outer primary black, the remainder black with dentate white marks on the inner webs; the secondaries black with white tips and whitish freckling as indistinct bars on the inner webs. Rectrices buffy with blackish vermiculations and irregular blackish bars (more ill-defined than in *melanogaster*); under tail-coverts buffy with distinct black bars and widely-spaced black vermiculations.

Immature. Undescribed but presumably female-like.

Identification

In the hand. This small to medium (wing 299–359 mm) bustard may be identified from the preceding species by its considerably darker (black in males, dark grey in females) rump pattern, and its generally darker upperparts. Both sexes have wings that are mostly white above, the primaries tipped with black, and with some brown on the larger upper wing-coverts.

In the field. Except in the limited area of possible sympatry with *melanogaster* (see preceding account) in north-eastern Africa (Kenya, Ethiopia and Somalia), the black underparts of the male, contrasting with mostly white wings (in flight) and a neck that is black in front and mostly grey behind (in males), serve to identify the species. In areas of sympatry, this species is found in drier habitats than the black-bellied bustard, and generally at lower elevations of less than 1000 or exceptionally to 1600 m. Females have white underparts and similarly black-tipped but otherwise mostly white wings. The species' calls include a three-part advertisement vocalization consisting of a weak click, a pop, and finally a deep booming sound.

General biology and ecology

This is a species adapted to semi-desert and relatively open areas, such as short-grass acacia savannahs, and very thin brush in open grassy plains or semi-deserts. Like some other bustards, the birds are also attracted to recently burnt ground. They are generally restricted to lower elevations of less than 1000 m in Kenya, but extend up to about 1600 m in Somalia. The birds occur singly or in pairs, and in general behaviour are much like the black-bellied bustard, with which they are easily confused. They are believed to be quite sedentary, but at least in the Serengeti of Tanzania they occur only during January–February and again in September–October (Someren 1933, Urban *et al.* 1986).

Foods have not been studied in any detail, but are known to include insects such as orthopterans, and some vegetable matter.

Social behaviour

The mating system is still unknown but probably similar to that of the black-bellied bustard. However, Someren (1933) noted that he had never observed the birds doing 'aerial stunts'. A photograph by D. Richards in P. Osborne *et al.* (1984) shows a male with a strongly inflated upper neck as compared with its normal posture (Figure 44), suggesting that the oesophagus is inflated during display, but no additional details are available on this point.

Fig. 44. Social behaviour of male Hartlaub's bustard: (a) normal alert posture; and (b) neck-expansion posture. After photos by J. F. Reynolds and D. Richards, in P. Osborne et al. (1984).

Reproductive biology

The breeding season in Ethiopia occurs during April, whilst in central Kenya and eastern Tanzania breeding records are for January and June, or during both the periods of rainfall when the grasses are at their highest. Nothing else seems to be known of their breeding biology, not even a description of their eggs or information on clutch size.

Evolutionary relationships

This is certainly a very close relative of *melanogaster* but is sufficiently sympatric with it to indicate that the forms must be regarded as constituting a superspecies (Snow 1968). The plumage of *hartlaubi* is considerably paler and more ashen, which is appropriate in view of its more generally desert-like environment of short-grass plains and acacia scrub. It is rather curious that this species should be of such restricted distribution and comparative rarity in Africa, whereas *melanogaster* should be so widespread and relatively abundant.

Status and conservation outlook

Although reported as fairly common in southern Sudan, this species is otherwise uncommon, and indeed has been reported as rare in Somalia (Urban *et al.* 1986). Its range extends slightly into eastern Uganda, but Pitman (1957) noted that in a quarter-century of experience there he never knowingly observed the species.

WHITE-BELLIED BUSTARD (Plate 27)

Eupodotis senegalensis (Vieillot) 1820

Other vernacular names: blue-necked bustard, Senegal bustard, Somali blue-necked bustard, Somali white-bellied florican, white-bellied korhaan; Natalse Korhaan (Afrikaans); petite outarde du Sénégal (French); Senegaltrappe (German).

Distribution of species (Map 23)

From Senegal (and formerly also Gambia) east to Ethiopia and Somalia, and south discontinuously to eastern South Africa. Absent from much of central and south-western Africa. The southern forms *mackenzei* and *barrowii*, which occur from the lower Congo Basin and Lake Tanganyika southward almost to the Cape, are sometimes separated specifically.

Distribution of subspecies

E. s. senegalensis (Vieillot): Senegal and Mauritania to north-western Ethiopia. Extirpated from Gambia.

E. s. canicollis (Reichenow) 1881: Ethiopia to Kenya and eastern Tanzania. Includes *parva* Moltoni 1935 and *somaliensis* Erlanger 1905 (Mackworth-Praed and Grant 1952).

E. s. erlangeri (Reichenow) 1905: southern Kenya and western Tanzania.

E. s. mackenziei White: 1945: southern Zaïre, eastern Angola and Zambia (where rare).

E. s. barrowii (J.E. Gray) 1829: Botswana, Transvaal and eastern Cape Province. Previously generally referred to as *E. c. cafra* (Lichtenstein) 1793.

Measurements (of *senegalensis*, mm)

Wing, males 263–287 (av. of 14, 276), females 264–276 (av. of 4, 269); tail, males 121–145 (av. of 14, 129), females 109–132 (av. of 5, 120); tarsus, males 87–99 (av. of 15, 93), females 85–93 (av. of 5, 89); culmen, males 30–35.5 (av. of 15, 33), females 30–33 (av. of 5, 31.6) (Urban *et al.* 1986). Egg, av. 49–55 × 39–47 (Schönwetter 1967). The average of *barrowii* is 51.1 × 41.1 (Urban *et al.* 1986). Johst (1972) provided measurements of three eggs of *erlangeri* that average 47.3 × 41, but these were laid in captivity and thus may not be representative of wild birds.

Weights

Males c. 1.1–1.6 kg (Someren 1933); 1.4 kg (Urban *et al.* 1986). A hand-raised male weighed 1880 g, and a hand-raised female 1190 g (Johst 1972). Estimated egg weight 44–67 g; Johst (1972) noted that three fresh eggs of *erlangeri* weighed 37–42 g, the average being 39.6 g, or slightly less than the estimated range just given.

Description (of *canicollis*)

Adult male. Forehead black, shading on the crown to blue-grey, which extends to the nape but is surrounded marginally by black that runs into a point of elongate feathers at the nape, forming a crest. Lores, supercilium, ear-coverts, chin and cheeks pure white, the central ear-coverts black-streaked, while a line of similar colour extends from above the gape to below the posterior angle of the eye (this line lacking in other races). A large V-shaped black throat patch, the apex towards the chin, and the angle of the V white. Entire neck bluish-grey (hindneck tawny buff in *barrowii*), this colour extending to the upper breast where it becomes slightly vermiculated with fine alternate lines of black and sandy. Mantle, scapulars, back, rump and upper tail-coverts all sandy buff, minutely freckled and vermiculated with blackish, the centre of the mantle feathers darker; the long scapulars and long inner secondaries with more marked sandy-buff areas and wider black speckling and wavy black lines. Lesser

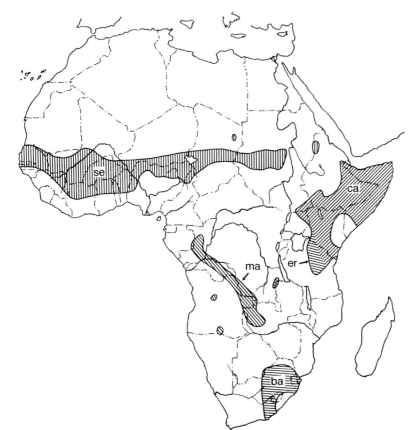

Map. 23. Distribution of the white-bellied bustard, including the locations of its races *barrowii* (ba), *canicollis* (ca), *erlangeri* (er), *mackensiei* (ma) and *senegalensis* (se).

coverts, bend of wrist and primary coverts slaty grey with slight sandy vermiculations; median coverts rather more tawny and less vermiculated at tip and outer web, giving this area a more uniform tawny appearance; greater coverts greyish distally shading to tawny buff proximally, mostly black-tipped. Secondaries black at tip and along outer web, with increasing whitish on inner web; primaries brownish-black, increasingly whitish on inner webs. Rectrices ashy grey at base, shading to buffy, vermiculated with blackish and with two blackish crossbars. Sides of breast washed with tawny buff and slightly vermiculated; rest of underparts to vent white except for dusky axillaries (these whitish in some races); under tail-coverts greyish at base, becoming white and then buffy at tip, crossed by greyish bars and freckling. Iris grey-brown; bill horn-coloured at tip, pinkish to yellow at base, lower mandible more whitish; tarsi and toes ivory to golden yellow.

Adult female. General colouration as the male, but with less black on the forehead and round the crown, also very much less black on the throat (absent in *senegalensis*), these feathers being largely buffy or white-tipped. The foreneck and the upper breast tipped with sandy-ochraceous finely vermiculated; the mantle and back more crossed with sandy-buff markings, the wings with less grey and the blackish freckling and vermiculations more distinct though more widely spaced.

Juvenile (1–3 months). Forehead rufescent shading to black towards nape, all feathers buff or white tipped; streak below eye rufescent and black streaked; throat patch as in the female; hindneck greyish, foreneck and upper breast tawny washed, all white or buffy tipped and a black line proximal to pale tip; mantle, scapulars, lesser and median coverts, back and rump with sandy and black vermiculations and freckles, the feathers with rust-tipped to whitish ends; the whole of the lower surface white; secondaries blackish on outer webs, whitish internally and with buffy and black vermiculations; primaries blackish with paler inner webs and buffy freckling at ends. Rectrices buffy at proximal half and sparsely freckled with black, more heavily freckled distally and with two black bars. Iris greyish; bill horn brown; tarsi and toes dirty greyish-pink. The change from this juvenal stage to an immature stage closely resembling the adult female requires about 4 months in captive birds (Someren 1933).

Identification

In the hand. This small (wing 264–287 mm) bustard occurs broadly in western, eastern and southern Africa (being second only to the black-bellied bustard in the extent of its overall African range), and it varies considerably in appearance through this region. However, it always has white underparts and a mostly bluish (males) or tawny (females) neck, with white surrounding the eyes. Black is mostly absent from the face of females except for scattered spotting, and is present in males only on the throat and as a superciliary stripe reaching the nape.

In the field. This fairly common, widely distributed and white-bellied bustard is fairly small, and blue is present only on the necks of adult males. The head is mostly whitish except for a black throat and superciliary striping in males. Generally associated with grasslands or thin acacia savannah, the species' three-note guttural advertisement or crowing call is variously said to sound like *rue-a-akh*, *kuk-pa-wow*, or *kuk-kaatuk*. Yet another description of the call is that it begins with a series of loud, frog-like *aaa* notes, followed by harsh and guttural three-syllable *tak-warat* notes. Such calls, which may be performed in an antiphonal manner, may possibly serve to advertise a group territory, represent interactions of adjacent territorial holders, or perhaps represent duetting among paired birds. Advertisement calls are uttered mainly during mornings and evenings, or during overcast periods associated with rainfall. A snorting alarm call is sometimes also uttered, and when feeding, the birds may utter soft, low whistling notes.

General biology and ecology

This very broadly distributed bustard occurs primarily in grassy plains, often with some acacias present, or in taller grasslands bordering acacia savannahs. However, in some areas it also moves temporarily into desert regions during the wet season, such as in Chad, and in southern Africa it moves to lower elevations during winter. At least in Somalia, it occurs most commonly at elevations above 1500 m, whereas in Zambia and Angola it is typical of dry plains (Clancey 1967, Urban *et al.* 1986). According to Pitman (1957), in Uganda it prefers drier areas than do either kori or Denham's bustards, and is typically found in thorn-scrub and low acacia bush.

The birds visit waterholes, and often occur in the general vicinity of rivers or streams, suggesting that the species is not highly desert-adapted. Its foods consist of the usual diversity of terrestrial invertebrates such as termites, ants, locusts, beetles, caterpillars, spiders, scorpions, and snails, but also includes lizards, grass seeds, bulbs, berries, and flowers. Most feeding is done during morning and afternoon hours; during the hottest part of the day the birds often roost in long grass under the shade of an acacia (Someren 1933).

Social behaviour

At least under captive conditions, a monogamous pair-bond seems to be present in this species. Johst (1972) noted that a captive pair maintained their bond throughout the year, and that when they were separated they called continuously and urgently. The female also sometimes accepted food from the bill of the male. The structure of the male's oesophagus has not yet been described, but there is no indication of neck-inflation behaviour during display. One brief account of male courtship indicates that the male approaches the female with his neck stretched forward, and his crown and throat feathers erected so as to form a ruff, in the same manner as also occurs in blue and black-bellied bustards (Kemp and Tarboton 1976).

It has been suggested that a kind of 'group territoriality' might exist, inasmuch as single birds or groups have been observed to utter loud, deep and frog-like advertisement calls, which are sometimes answered by another group (Kemp and Tarboton 1976). Mwanghi's (1988) recent observations shed some additional light on, and perhaps reinforce, this concept of 'group territoriality'. He heard a good deal of vocalization during early dawn and late pre-dusk periods, without apparent provocation. He considered these calls to serve a non-alarm communication function, and they were typically responded to by nearly all white-bellied bustards within earshot. He believed that such group vocalizations might serve to maintain cohesion within groups, and social integrity between members of different groups. A somewhat similar vocal behaviour has been observed in the blue bustard, but both species need additional study in this regard. It is quite possible that other social functions exist for such antiphonal calling, such as responses to territorial challenges and responses by individual birds (or pairs), as is known for example to occur in lesser prairie-chickens (*Tympanuchus pallidicinctus*).

Reproductive biology

The breeding season of this species is quite varied throughout its broad range. In South Africa it breeds during the summer months, from November to February, and similar summer breeding in December is reported for Zambia. To the west in Angola the birds probably breed in spring, during September and October. In most of Tanzania the breeding records are from December to February, and from April to May, during the early part of the wet season, whilst in north-eastern Tanzania and central Kenya the records are from October, November and May. Ethiopia records are for December and from March to June; June breeding has also been reported for Somalia. In sub-Saharan Chad and Sudan, breeding probably occurs between June and September, during the wetter season. In West Africa (Mauritania, Senegambia, Mali, Nigeria) the records are likewise mostly from June to October, but with a February record for Senegambia (Urban *et al.* 1986).

The eggs are laid on bare ground in a shallow scrape, and two eggs are the usual clutch, although exceptional clutches of one and three eggs have also been reported. Apparently only the female incubates, although she may be visited by the male. The incubation period was found to be 23 days under conditions of artificial incubation, and three newly hatched chicks weighed 28–33 g, averaging 30.3 g. When about 5–6 months old the young birds were calling in the same way as the adults, and were beginning to moult their juvenile feathers and develop adult-like plumages. At that time they weighed 1050–1250 g, or about the same as an adult female (Johst 1972).

Evolutionary relationships

It seems likely that the white-bellied, Rüppell's and black-throated bustards are all fairly closely related to one another, as there are strong similarities in adult plumages among these forms. Snow (1978) added the blue and little brown bustards to this list of apparently quite closely related species, but noted that *senegalensis* is not apparently extremely closely related to any of these. I have tentatively placed it in a rather central position (Figure 2) within the genus *Eupodotis*; it is the most variable geographically of any species in this genus, and this variability points up the ease with which an adult plumage pattern typical of one of these species can be transformed into one that is very similar to another's.

Status and conservation outlook

Because of its very broad range, this species is perhaps one of the most numerous of the African bustards, although it has been eliminated in some areas such as Gambia, and is rare in others. It is still generaly common to relatively abundant over much of its traditional range although, as is also true of other bustards, uncontrolled human development is probably the single most serious threat to its future. Perhaps because of its relative abundance, the birds have been extensively trapped to provide targets for falconers of the Middle East (Osborne *et al.* 1984), and such activities could easily change the status of the species.

LITTLE BROWN BUSTARD (Plate 28)

Eupodotis humilis (Blyth) 1856

Other vernacular names: Somalia little florican; outarde somalienne (French); Somalitrappe (German).

Distribution of species (Map 24)

Desert areas of northern and west-central Somalia and eastern Ethiopia. No subspecies recognized.

Measurements (mm)

Wing, males 242–260 (av. of 11, 253), females 228–253 (av. of 9, 246); tail, males 106–141 (av. of 13, 115), females 92–134 (av. of 11, 109); tarsus, males 56–71 (av. of 13, 65), females 59–65 (av. of 11, 63); culmen, males 28–31.5 (av. of 13, 29.6), females 27.5–33 (av. of 11, 29.5) (Urban *et al.* 1986). Egg, av. 47 × 36 (Schönwetter 1967).

Weights (g)

Unsexed adult, 698 (US National Museum). Egg, 33 (estimated).

Description

Adult male. General colour above pale sandy, minutely vermiculated with blackish lines, and blotched with black spots and bars on the mantle and back; wing-coverts like the back, but more coarsely and irregularly vermiculated with black, with wide spaces of pinkish isabelline, on which very few black lines appear; inner secondaries like the coverts; outer median and greater coverts lighter and more ashy and creamy white on the outer web, the greater series with a black band at the tip; alula and primary coverts creamy white, with the terminal half black; remiges blackish, ashy on the inner web; the outer primaries brown and the secondaries white at the base of both webs; rectrices pinkish-sandy colour, crossed with wavy lines of black, and having two or three narrow bars of black; head and neck bluish-grey, sandy-coloured on the crown, which is finely vermiculated with black; edges of crown and sides of face bluish grey; feathers round the eye and

Map. 24. Distribution of the little brown (stippled), black-throated (cross-hatched) and Rüppell's bustards (hatched), including the locations of the black-throated bustard's races *namaqua* (na) and *vigorsii* (vi), and of Rüppell's bustard's races *fitzsimmonsi* (fi) and *ruepellii* (ru).

cheek-stripe white; chin white; throat black, concealed by white tips; neck ashy but somewhat shaded with sandy, and minutely vermiculated with blackish lines; breast and remainder of undersides white, as are the under wing-coverts; axillaries black. Iris brownish-yellow; bill brown, with blackish culmen and becoming yellow basally, the lower mandible darker towards the tip; tarsi and toes yellow.

Adult female. Differs from the male in being much more coarsely marked above, without fine vermiculations; the back and wing-coverts mottled with heart-shaped patches of pinkish isabelline, in which are coarse black bars and heart-shaped blotches of black; head and neck like the back and streaked with isabelline; throat white; breast and under wing-coverts creamy white; axillaries black, tipped with white.

Immature. Like the female, but with light tips to the primaries (Mackworth-Praed and Grant 1952).

Identification

In the hand. This is the smallest (wing 228–260 mm) African bustard and the least patterned, with white underparts and very little bluish or black patterning present on the body and head, save for a small nuchal crest and mottled black throat in males.

In the field. This is the only very small bustard in north-eastern Africa, and although it is sympatric with the somewhat larger white-bellied bustard the two tend to be ecologically separated, the more generally brownish little brown bustard being more associated with drier and sandier areas of thornbush, and the more contrastingly patterned white-bellied with mixed acacia bush and grassland. The little brown bustard has distinctly sandy brown upperparts and white underparts, with black mostly limited to the wings and axillaries (visible in flight), and with little or no bluish or blackish patterning on the neck and head. Its calls include a series of rattling notes, *ka-ki-rak-ka-ki-rak*.

General biology and ecology

This is one of the least-studied of all bustards (its downy young and immature plumages are still undescribed in detail, and it has apparently not been maintained in captivity or even photographed in the wild). Its remote, desert-like habitats apparently generally consist of red sandy soils where there are patches of light to dense thorn scrub alternated with open spaces, and sometimes extending to open tussock grasslands. It occupies desert edges, and its need for surface water is still undetermined (Urban *et al.* 1986, Snow 1968).

Foods are said to consist of insects, small molluscs and seeds.

Social behaviour

These birds are reported to live singly or in pairs, but not in larger groups. As a result, the species is believed to have monogamous pair-bonds and to lack elaborate displays, but nothing definite is known on either of these points. The call is said to be a series of rattling notes, often uttered in the evening. Whether this or a similar vocalization actually consists of a duet by a mated pair (as is true of the closely related black-throated and Rüppell's bustards), as might well may prove to be the case, is still not known.

Reproductive biology

The breeding season in Somalia extends from April to August, the majority of the breeding occurring in May and June, during the relatively wetter period in this very arid region. The eggs reportedly are laid on sandy soil, without an associated scrape, and their streaked clay-colour appearance probably blends extremely well with such substrates. The clutch is typically two eggs, but there is at least one record of a three-egg clutch. Nothing further of significance is known of the breeding biology (Urban *et al.* 1986).

Evolutionary relationships

This small and rather nondescript bustard has plumage similarities somewhat like those of the sympatric white-bellied bustard, but it is even more similar to the black-throated and Rüppell's bustards, which are geographically rather far removed from it, although all of these adapted to similar semidesert or desert-edge habitats. Snow (1978) suggested that it may be most closely related to the *vigorsii–rueppellii* superspecies, but that too little evidence exists to try to group it in a formal manner with these or any other species of *Eupodotis*.

Status and conservation outlook

This bustard has probably the smallest geographic range of any African bustard; Snow (1978) showed only twenty locality records for it, the fewest of any of the bustard species. It occurs in an area that has been seriously disrupted by both drought and warfare, and its current status is unknown but possibly quite depleted.

BLACK-THROATED BUSTARD (Plate 29)

Eupodotis vigorsii (A. Smith) 1830–1831 (1831)

Other vernacular names: black-throated korhaan, Karoo (or Karroo) korhaan, Vigor's bustard; Vaal korhaan (Afrikaans); outarde de Vigors (French); Knarrtrappe, Namatrappe (German).

Distribution of species (Map 24)

Central Namibia south and east to Cape Province and Orange Free State.

Distribution of subspecies

E. v. vigorsii (A. Smith): Transvaal and Orange Free State west to central Cape Province. Includes *karooensis* Vincent 1949 (Mackworth-Praed and Grant 1962).

E. v. namaqua (Roberts) 1932: north-western Cape Province and southern Namibia. Includes *harei* Roberts 1937, *orangensis* Roberts 1932 (Mackworth-Praed and Grant 1962) and *barlowi* Roberts 1937 (Urban et al. 1986).

Measurements (mm)

Wing, males 318–375 (av. of 7, 351), females 316–356 (av. of 5, 329); tail, males 141–202 (av. of 7, 164), females 139–163 (av. of 5, 151); tarsus, males 77–96 (av. of 7, 88.1), females 78–92 (av. of 5, 84.8); culmen, males 34–41 (av. of 7, 37.7), females 31.5–37 (av. of 5, 34.8). Egg, av. 61.3 × 44 (Urban et al. 1986).

Weights

No specific information. Estimated egg weight 64 g, or 4.3% of adult female weight, estimated at 1.5 kg (Schönwetter 1967).

Description (of *vigorsii*)

Adult male (after Sharpe 1894 and Clancey 1967). General colour above dull ashy with an admixture of sandy buff (suffused with pink in *namaqua*), the feathers finely stippled with blackish; scapulars and inner secondaries with large blotches or bands of blackish; lower back, rump and upper tail-coverts more uniform, with very tiny stipplings of black, the upper tail-coverts with a few indistinct blackish bars; lesser wing-coverts lighter than the back, very finely stippled with blackish vermiculations, the greater series rather more sandy-coloured, the markings coarser, and the feathers tipped with black; alula ashy like the wing-coverts; primary coverts and remiges blackish at the ends and tawny at the bases, the primaries being mostly of the latter colour, the outer secondaries tawny at the bases and crossed with black bars, the inner ones having the tawny bases mottled with black; innermost secondaries like the back; rectrices ashy, tinged with sandy buff like the upper tail-coverts, somewhat freckled with blackish, and crossed by a narrow bar of black near the tip; crown, occiput and nape ashy, freckled with tiny blackish vermiculations, as are the sides of the face and ear-coverts; a concealed spot of black feathers on the nape; chin greyish-white, this colour skirting the throat; centre of throat black, forming a triangular patch that extends in a narrow line down the lower throat; sides of neck, foreneck and most underparts (including the flanks and axillaries) ashy brown (more greyish in *namaqua*), very finely vermiculated with blackish lines; centre of breast and abdomen uniform creamy white; under tail-coverts like the flanks and similarly vermiculated; under wing-coverts mostly creamy buff to white, the small coverts round the edge of the wing ashy with blackish vermiculations. Iris greyish brown; bill dark horn colour, the base of the lower mandible pale flesh; tarsi and toes yellowish.

Adult female. Similar to the male, but mantle and scapulars more heavily vermiculated and barred, and with the barring somewhat extended on to the wing-coverts.

Juvenile. Female-like but spotted with whitish on the crown and back; below paler, and somewhat barred with dusky.

Identification

In the hand. This medium-sized (wing 316–375 mm) bustard may be identified by the brown to greyish-brown (not bluish) breast, with bluish colour limited to the foreneck, black limited to a throat patch and rudimentary crest (browner in females), and with little or no white present on the head or neck except as a narrow border around the black throat.

In the field. This bustard is limited to the drier portions of Namibia and South Africa, where it occurs parapatrically and immediately to the south of the very similar Rüppell's bustard (the replacement zone occurring in Namaqualand, in the vicinity of Maltahöhe and Mariental). Thus except within this limited area of contact it is perhaps likely to be confused only with the similarly dingy and comparably sized Ludwig's bustard, which lacks both black and blue tones on the neck and head. It is said to utter a croak-like duetting, sounding something like *squark-kok*, the first note uttered by the male and the second by the female. What is perhaps the same call has also been described as a deep, harsh barking, *waa-u-u* (Macdonald 1957), as a frog-like *graag-uurg-og-og* (Newman 1983), and as a *kirr-reck, arack, arack*. Calling is sometimes done during night-time hours as well as during the day, and especially during morning hours.

General biology and ecology

According to Clancey (1967) and Snow (1978), this species inhabits open, typical semi-desert steppe and desert-edge habitats of the Karoo. It usually occurs in pairs or small groups of up to about eight individuals, in relatively flat areas that are usually completely treeless but having thin grasses, low, well-spaced shrubs and a stony or gravelly substrate. Although associated with very dry habitats, it is mainly active during the cooler part of the day, and sometimes visits river pools to drink (Macdonald 1957). It is probably somewhat less desert-adapted than Rüppell's bustard, which replaces it geographically in the progressively drier areas to the north in Namibia.

In the Kaokoland area of northern Namibia an aerial survey of 445 birds indicated that the average group size was 2.35 birds; 41 per cent of the groups consisted of two birds, and three-bird groups formed the second most frequent category, 30 per cent. Only 2 per cent of the groups were of more than six birds. The birds occurred in a region having an annual rainfall ranging from 30 to 120 mm, and although gravel plains formed only 34 per cent of the area of the total region, they accounted for 65 per cent of the sightings. Sandy plains represented 20 per cent of the region, and accounted for 17.5 per cent of the sightings. Mountainous areas represented 43 per cent of the region and 17 per cent of the sightings, although only valleys and plateau areas were used. River courses were apparently of little ecological significance to the species. The optimum habitat probably consists of flat, dark basaltic gravel plains intersected by shallow watercourses and a low density of small, desert-adapted shrubs (Viljoen 1983).

In a more recent study (Boobyer 1989), it was found that the sizes of groups of this species do not vary seasonally, although densities do. The species exhibits relatively plastic territories, which in areas of low resource availability increase in size (and thus population densities decrease) as resources (mostly annual plants and insects such as orthopterans and curculionids) become spatially and temporally concentrated. Birds apparently favour monocultures of growing annual plants, as are typical of disturbed areas. The species evidently relies on a wide variety of plant and annual resources that are adapted to a pulse-driven environmental system. Its diet is highly variable, which suggests that the species is perhaps more strongly influenced by relative resource abundance than by resource quality.

Little else is known of the ecology, but probably some seasonal movements do occur. Foods include such diverse items as small insects and their larvae, reptiles, and also seeds and other plant materials (Clancey 1967).

Social behaviour

The mating system of this species is still uncertain; monogamous and polygamous matings have been mentioned as equally possible, but the presence of duetting calls would suggest that monogamy may be the prevailing system, a possibility strengthened by the fact that males lack an inflatable crop or oesophagus (Niethammer 1940). The birds are territorial and strongly site-faithful, and pairs or breeding groups defend an area of up to a few hundred metres in radius for an extended period of time. However, the overall home range may be appreciably larger, judging from the sighting of a distinctively plumaged bird at points up to 8 km apart (Quinton 1948). It may be this site-tenacity that is responsible for the rather marked plumage variation occurring in different local populations of this species (Osborne *et al.* 1984).

It seems very likely that these residential groups represent a pair and possibly their immature offspring. The latter are likely to be excluded from the breeding site during breeding, although a female has been observed sitting on a nest in the company of three other birds (Kemp and Tarboton 1976).

Reproductive biology

Nesting occurs in a shallow scrape, the birds mainly breeding during the summer months from November to March, but with some South African records as early as August. Generally two eggs constitute the usual clutch, but single-egg clutches have also been reported, according to Clancey (1967). Viljoen (1983) stated that a single egg represents the normal clutch. Other aspects of the reproductive cycle remain unreported, although Mark G. Boobyer of the Percy Fitzpatrick Institute (University of Cape Town) has been doing research on the species since 1986.

Evolutionary relationships

This form is clearly very closely related to Rüppell's bustard, and at times has been considered conspecific with it. However, Snow (1978) regarded the two as constituting a superspecies, and the same treatment has been followed by Urban *et al.* (1986). It is probably still a moot question, although the fact that the two forms have apparently virtually identical vocalizations (Urban *et al.* 1986) would cause one to wonder if reproductive isolation actually exists between them.

Status and conservation outlook

According to Clancey (1972–3), the status of the nominate race of this species in Cape Province and the western half of Orange Free State is not currently known. However, the race *namaqua* was reported to be locally common in Bushmanland, Brandvlei and Van Wyks Vlei district of Cape Province. Clancey judged that although the (currently unaccepted) form *barlowi* was known only from its type locality (Aus, in Namibia), it was probably not uncommon along the edge of Namib in this general region. In general he believed the status of the species to be 'quite satisfactory' in the early 1970s, although marginal populations such as those of northern Zululand and eastern Swaziland might be expected to decline.

RÜPPELL'S BUSTARD (Plate 30)

Eupodotis rueppellii (Wahlberg) 1856

Other vernacular names: Damara Vaal Korhaan (Afrikaans); outarde de Rüppell (French); Rüppelltrappe (German).

Distribution of species (Map 24)

Endemic to desert and desert-edge areas western Namibia and coastal areas of southern Angola, north to Benguela and Iona National Park. Sometimes considered conspecific with *vigorsii*.

Distribution of subspecies

E. v. rueppellii (Wahlberg): south-western Angola and north-eastern Namibia south to Windhoek.

E. v. fitzsimmonsi (Roberts) 1937: southern Namibia from Windhoek south to Maltahöhe.

Measurements (mm)

Wing, males 312–341 (av. of 18, 329), females 293–325 (av. of 13, 313); tarsus, males 78–89 (av. of 10, 85.3), females 74–85.1 (av. of 7, 80.4); tail, males 132–159 (av. of 20, 145), females 135–161 (av. of 7, 148); culmen, males 35–43.2 (av. of 9, 37.8), females 33.7–37.5 (av. of 7, 35.6). Egg, av. 57.6 × 40.9 (Urban *et al.* 1986).

Weights

No information. Egg, 53 g (estimated).

Description

Adult male. Forehead and crown pinkish buff with grey tinge and dark vermiculations; vestigial black nuchal crest; face greyish-white with black moustachial flecks and along supercilium; chin and malar region whitish, extending back under earcoverts and sides of upper neck, the lateral areas joining midway down hindneck, where the white is bordered with blackish; throat with a large black patch narrowing to a black line down the foreneck and broadening again on upper breast; rest of neck buffish grey; upperparts and tail pinkish-buff lightly suffused with salmon pink; pale grey on both sides of black upper breast; rest of underparts whitish; remiges, alula and greater primary coverts creamy buff with brownish-black tips, the lesser primary coverts white with black tips; scapulars and mantle pinkish-buff, the tertials and rest of wing paler; undersides of primaries brownish-black, with creamy white on basal half of outer feathers, this becoming less extensive toward inner wing, forming a large whitish patch; rest of wing mottled pale and darker brown. Iris pale brown; bill dark grey, becoming yellowish-grey at the base of the lower mandible; legs and toes pale yellow (Urban *et al.* 1986).

Adult female. Like the male, but the cheeks more mottled, and the tail often faintly barred (Urban *et al.* 1986).

Immature. Like the adult, but the head more mottled, more dark markings on the back, and the tail barred (Urban *et al.* 1986.)

Identification

In the hand. This is a small to medium-sized (wing 293–341 mm) bustard, with a bluish neck and head patterned distinctively with white on the lower cheeks and hindneck, and with black stripes down the foreneck, sides of the neck, and from the eyes backward to the nape.

In the field. Limited to south-western Angola and Namibia (occurring immediately north of the range of the preceding species), and recognizable by its patterned blue, white and black head and neck, with whitish underparts. It occurs sympatrically with Ludwig's bustard, which is appreciably more brownish on the head and neck, with no hint of bluish present on these areas. One of Rüppell's bustard's calls is said to sound like *waa-wu-wu*, and it often calls in duet, a deep and resonant *waaa-a-re-e* begun by the male, followed by three responses from the female, and repeated in long sequences. Recorded calls of this species are reportedly nearly identical to those of the black-throated bustard (Urban *et al.* 1986). The flight call is a rapid *quark-quark-quark* (Newman 1983).

General biology and ecology

The ecology of this species is quite similar to that of the preceding one, namely being associated with desert-edge habitats, particularly along the Namib desert, where thin grasses and low shrubs grow on flat, dark basaltic gravel substrates. It occurs in pairs and small groups in such habitats, especially in areas having low shrubs and scattered grassy cover (Urban *et al.* 1986).

Like Vigor's bustard it is mainly active in the cooler hours of the day, spending the hot midday period in the shade. Possible needs for surface water are unreported. It forages on the usual bustard diet of insects and their larvae, some vegetable matter, and probably also small reptiles (Clancey 1967).

Social behaviour

The mating strategy of this species is still uncertain, although it is reported to apparently form monogamous pair-bonds, and Niethammer (1940) reported that the sexes remain together throughout the year. In common with the black-throated bustard, duetting occurs, and males will warn incubating females of possible approaching danger, according to Niethammer. He stated that during courtship both partners utter their far-carrying *quak* calls alternately, but lack distinctive posturing. During this calling the neck is not inflated, and there are apparently no specializations of the crop or oesophagus permitting such inflation.

Reproductive biology

According to Clancey (1967), this species apparently breeds more or less throughout the year, but with most nesting occurring during the summer months, from September to February. The nest is placed in a scrape on bare ground, typically on a substrate of small stones. There are one or two eggs (Urban *et al.* 1986), although Clancey stated that a single egg is the typical clutch size. Nothing else has been reported on the species' reproductive biology.

Evolutionary relationships

This is certainly a very close relative of *vigorsii*, and one might well argue that the two forms are scarcely specifically separable. Behavioural and ecological studies in the area of their apparent contact would be very useful.

Status and conservation outlook

Urban *et al.* (1986) stated that this species, which occupies a very small range in Africa, is generally frequent to common in Namibia and Angola.

LITTLE BLACK BUSTARD (Plate 31)

Eupodotis afra (Linnaeus) 1766

Other vernacular names: black korhaan, white-quilled korhaan (*afraoides*); Swart Korhaan (Afrikaans); outarde korhaan, outarde noire (French); Gackeltrappe (German).

Distribution of species (Map 25)

Namibia and Botswana south to the Cape, and east to Lesotho.

Distribution of subspecies

E. a. etoschae (Grote) 1922: north-western Namibia (Ovambaland) and northern Botswana (Makgadikgadi). Includes *boemeri* Hoesch and Niethammer 1940 (Mackworth-Praed and Grant 1952).

E. a. damarensis Roberts 1926: from northern and central Botswana west into southern and central Namibia. Includes *mababiensis* Roberts 1932 (Mackworth-Praed and Grant 1952). Urban *et al.* (1986) chose the epithet *kalaharica* Roberts 1932 for this taxon without explanation, but Clancey (1989) used the nomenclature followed here.

E. a. afra (L.): Cape Province north to Little Namaqualand and east to Grahamstown.

E. a. afraoides (A. Smith) 1830–1831: southeastern Botswana, western Transvaal, Lesotho lowlands and northern Cape Province. Includes *centralis* Roberts 1932 and *chiversi* Roberts 1933 (Mackworth-Praed and Grant 1962). Clancey (1986, 1989) recognized this form as a separate parapatric species, suggesting that morphological criteria, range dispositions, and the apparent lack of intergradation in the zone of near contact with *afra* supports this position.

Measurements (of *afraoides*, mm)

Wing, males 262–308 (av. of 47, 281), females 251–298 (av. of 23, 270); tail, males 118–162 (av. of

46, 131), females 110–137 (av. of 23, 125); tarsus, males 78–100 (av. of 46, 91), females 79–95.1 (av. of 23, 87.7); culmen, males 27.5–38 (av. of 47, 31.3), females 27.3–36 (av. of 23, 29.5) (Urban *et al.* 1986). Egg, av. 53–56 × 43–46.

Weights (g)

Males (of *afraoides*) 536–851 (av. of 26, 716), females 500–878 (av. of 76, 669) (Urban *et al.* 1986). Estimated egg weight 52–64, or 7.7% of adult (Schönwetter 1967).

Description (of *afroides*)

Adult male. General colour above dark sandy buff, regularly barred across with black and finely stippled or vermiculated; scapulars like the back; lower back, rump and upper tail-coverts rather more dusky and shaded with ashy, the black bars not so pronounced and the sandy-buff ones paler, being replaced on the upper tail-coverts by white bars somewhat irregular in shape; lesser wing-coverts like the back, the external ones with fewer markings and becoming pure white like the rest of the coverts, the greater series white with a few concealed black markings or bars; alular feathers black with white tips; primary coverts and remiges black; primaries mostly white (mostly or entirely black in other races), the secondaries with white at the base of the outer web (this lacking in other races) increasing in extent towards the inner secondaries, but the innermost secondaries barred with sandy colour and black like the back; rectrices ashy grey tipped with white, mottled with fine black sandy markings, mostly on the outer web, and crossed by two black bands, the second subterminal, these black bands less pronounced on the centre feathers, which are like the upper tail-coverts; crown ashy, some of the feathers crossed with narrow blackish bars, with a circle of white feathers round the nape; rest of the head except the white ear-coverts black, as is the entire under surface of body; a large white patch also extends backwards across the lower hindneck and forms a collar that separates the black of the hindneck from the back; under wing-coverts white; axillaries black. Iris light tawny brown, darkest towards the centre, orbital skin yellow; bill light greyish-brown apically, with the basal half pale rose-pink to coral red; tarsi and toes dark orange-yellow.

Adult female. Differs from the male in the absence of solid black on the neck, throat and breast, the black areas being confined to the abdomen, under

Map. 25. Distribution of the little black bustard, including the locations of its races *afra* (af), *afraoides* (afr), *damarensis* (da) and *etoschae* (et). The form *afraoides* is considered a separate species by some authorities.

tail-coverts and axillaries. The crown is dark brown, variegated and spotted with tawny buff; sides of face and hindneck buff mottled with dusky, rest of upperparts less heavily barred than male, and buff shaft-streaks on mantle and scapulars. Upper breast neck and throat buff, finely barred with blackish-brown, the lower breast buffy white; otherwise black below, with buff barring on the under tail-coverts. The tail is buffy grey, vermiculated with black and crossed by two black bars (Clancy 1967).

Immature male. With arrowhead markings of sandy rufous all over the upper surface; the head blackish, spotted with sandy rufous and narrowly barred on the occiput, and with a concealed black patch on the nape; ear-coverts sandy buff, as are the anterior cheeks and throat, the latter mottled with black bases; posterior cheeks white, extending to the sides of the nape; lower throat, foreneck, and sides of the neck sandy buff, narrowly barred with white, the chest a little more broadly barred.

Immature female. Probably very similar to the adult, but with pale feather tips on the wings, back and crown (Urban *et al.* 1986).

Identification

In the hand. This small (wing 251–308 mm) bustard has underparts that are entirely black, and either an all-black neck and white cheeks surrounded by a black face and throat (males), or a face, throat and neck that are conspicuously streaked or mottled with black (females).

In the field. In its limited southern African range this species can be identified by the black underparts of both sexes, and the all-black neck of males. In flight both sexes exhibit flight feathers that are black both at their tips and basally, but with a variably large white wing-patch on the primaries and a separate white wing-bar on the secondaries. Males utter raucous and loud grating *kr-aaak-a* notes from small elevations or during their conspicuous flight displays.

General biology and ecology

Within its rather limited range this species occupies a fairly broad range of habitats, which generally include open-country areas having stands of low shrubs and thorny scrub, but it also occurs on open grassy flats (especially those with scattered termite mounds that are used as lookout or display sites), old cultivated lands, or even sand dune areas having succulent or other kinds of vegetation. Kemp and Tarboton (1976) consider it to be one of several savannah-adapted bustard species, but it perhaps might better be described as one favouring rather sparse grasslands having scattered brush cover, sometimes also occurring on grassy areas bordering savannahs.

Because of its conspicuous male plumage, this species is quite easily visible in low grassy cover, especially when the males stand on small rises calling or launching into aerial displays. On the other hand, females tend to be extremely shy and elusive. The birds are sedentary, and males probably defend their territories for much of the year.

The species forages on the usual array of insects, their larvae, and a substantial amount of vegetable matter, including seeds, roots, and shoots. The birds are reputed to be responsible for the dispersal of seeds of various invasive *Acacia* species.

Social behaviour

Although the little black bustard's mating system has recently been described as unknown (Urban *et al.* 1986), Kemp and Tarboton (1976) describe the species as being one of those for which there is no evidence for the establishment of permanent pair-bonds, and in which females apparently wander from one male's territory to another until they are fertilized. Similarly, Osborne *et al.* (1984) describe the species as being almost certainly promiscuous, inasmuch as males display throughout the breeding season, females nest far away from the males, and they alone incubate the clutch.

Territorial males are extremely noisy during the breeding season, repeatedly uttering a raucous *krracker, krracker . . .* from elevated sites such as termite mounts, during both daytime and night hours. This same call is uttered during flight display, which is initiated from the calling site. The birds take flight and circle with deep and exaggerated wingbeats while calling loudly about 15 m from the ground. They finally descend in a fluttery glide, with the yellowish-orange legs dangling conspicuously below. The territories must be fairly small or are possibly somewhat grouped in a lek-like manner, as Kemp and Tarboton (1976) mentioned that as many as five may call and circle with a female if she is flushed. The contrasting black-and-white male plumage pattern is especially apparent during flight display, and different subspecies of the black bustard exhibit differing degrees of white patterning on the otherwise black flight feathers, the more northerly populations having considerably more white present and sometimes being regarded as a separate species.

Aggressive males will sometimes chase one

188 *Bustards, Hemipodes, and Sandgrouse*

another, with outstretched necks and fanned, flattened tails (Figure 45b). At other times they retract the neck back almost into the shoulders, a posture of unknown significance (Figure 45a). Clancey (1989) states that the call is uttered with the neck much inflated, a point not noticed by other authors.

Reproductive biology

The breeding season of this species in South Africa is quite prolonged, from August to March; in Namibia the records are from October to January. The eggs are placed in a shallow scrape, and the clutch size reportedly consists of a single egg more often than two eggs. All the evidence indicates that only the female tends the nest and looks after the young (Urban *et al.* 1986). Observations from captivity indicate an incubation period lasting 23 days, and the young feeding themselves after about 4 weeks (Klös 1977). One chick weighed 21 g shortly after it was hatched (Bell 1970).

Other information on breeding little black bustards in captivity has recently been provided by Gregson (1986), who reported that in each of three instances the incubation period was 21 days. The wing quills were growing fast by the sixth day after hatching, and by the fifteenth day the chick was quite well covered with feathers. At 6 months old it

Fig. 45. Social behaviour of male little black bustard: (a) including male neck-retraction posture (after photo by P. Johnsgard); and (b) male running with raised crest (after photo by P. Goriup).

was still in immature plumage, but was 'barking' like a male. Immature males apparently begin to achieve their male colours at about 7 months, and a two-year-old male was in adult plumage but still had the eye colouration of a female.

Evolutionary relationships

Snow (1978) judged that the little black bustard has no close allies, but that its nearest relative may be *ruficrista*, which has a similar flight display. This may not in itself be a significant trait, but there do seem to be some behavioural similarities between these two species and, more inclusively, among these and the two Indian species of 'floricans', both of which also have mostly black male plumages and apparently quite similar male territorial and courtship displays. As noted earlier, Clancey (1989) regarded the black bustard as comprising two separate allospecies, with unclear phyletic relationships to the other African bustards.

Status and conservation outlook

This is one of the more common species of bustards in South Africa, and in many areas is relatively abundant. It is generally common over much of Cape Province, widespread in Orange Free State, and common in the Transvaal, especially the drier west and south-west regions. It is also very common in grassland habitats of Botswana, in the lowlands of western and north-western Lesotho, and over substantial parts of Namibia, as for example in the highlands of Damaraland, and in Swatzrand, where it almost replaces the larger black-throated bustard (Clancey 1972–3).

RUFOUS-CRESTED BUSTARD (Plate 32)

Eupodotis ruficrista (A. Smith) 1836

Other vernacular names: buff-crested little bustard, buff-crested florican, bush korhaan, crested bustard, Lynes' bustard (*savilei*), pygmy bustard, red-crested korhaan, Savile's bustard (*savilei*); boskorhaan (Afrikaans); outarde à huppe rouge, outarde houppette (French); Rotschopftrappe (German).

Distribution of species (Map 26)

Occurs as three discontinuously distributed populations, which are sometimes regarded as forming two or even three species: Senegambia east across the sub-Saharan Sahel zone to southern Sudan; Ethiopia south to eastern Tanzania, and from Angola and Mozambique south to northern South Africa.

Distribution of subspecies

E. r. savilei Lynes 1920: from Senegal discontinuously east to southern Sudan (sometimes separated with *gindiana* as a full species).

E. r. gindiana (Oustalet) 1881: Ethiopia and Somalia, south through north-eastern Kenya to north-eastern Uganda and central Tanzania. Includes *hilgerti* Neumann 1907 (Mackworth-Praed and Grant 1952).

E. r. ruficrista (A. Smith): southern Angola, Namibia, and southern Zambia south-east to central Mozambique, Swaziland and northern South Africa.

Measurements (mm, of *ruficrista*; *gindiana* and *savilei* average progressively smaller)

Wing, males 252–279 (av. of 12, 264), females 227–279 (av. of 12, 255); tail, males 125–144 (av. of 12, 136), females 111–139 (av. of 12, 130); tarsus, males 77–86 (av. of 12, 81), females 58–83 (av. of 12, 81); culmen, males 32–37 (av. of 12, 33.8), females 28.5–35.5 (av. of 12, 32.7) (Urban *et al.* 1986). Egg, av. 49–50 × 37–42 (Schönwetter 1967).

Weights (g)

Males of *gindiana* c. 680–900 (Someren 1933), three males of *ruficrista*, 550–770 (av. 680) (Urban *et al.* 1986). Egg, 37–49 (estimated).

Description (of *ruficrista*)

Adult male. General colour above black to ashy grey in the centre of the feathers, which are margined with sandy rufous and mottled with grey to black, the feathers often centred with lanceolate markings of pale sandy buff; lower back and rump sandy buff, crossed with blackish vermiculations and some black arrowhead markings (these reduced in *savilei*) and tinged with ashy in the rump; the upper tail-coverts and centre rectrices decidedly more ashy than the back and more coarsely vermiculated with blackish, and distinctly crossed by several black bars or arrowhead markings; the outer rectrices black, freckled with sandy buff and vermiculated towards the base, the outermost pairs almost entirely black; lesser and median wing-coverts like the back, with the same lanceolate markings of sandy buff, the outer median coverts white with black bases and black shaft-lines, the greater coverts black or margined with black, inner ones white, all the inner coverts and inner secondaries coloured and mottled like the back; alula black; primary coverts also black, with the greater part of the outer web sandy buff, this sandy-buff portion sometimes mottled and

Map. 26. Distribution of the rufous-crested bustard, including the locations of its races *gindiana* (gi), *ruficrista* (ru) and *savilei* (sa). The form *ruficrista* is considered a separate species by some authorities.

barred with black; remiges black chequered for two-thirds of their length with fulvescent to yellowish markings, these generally squarish but sometimes taking the form of bars; crown dark slaty blue; lores, eyelid, and a broad eyebrow sandy buff (more orange in *gindiana*), streaked with blackish feather edgings, these superciliary bands meeting on the nape; ear-coverts light brown; below the eye a band of slaty blue; on the nape a tuft of porphyrin-pigmented vinous-red to isabelline (more buffy in *gindiana* and suffused with olive in *savilei*) feathers, forming a crest; hindneck ashy brown tinged with grey; cheeks and throat isabelline, with a broad band of black in the centre of the latter (extending along underside of neck to abdomen in *gindiana*); lower throat and sides of neck ashy grey, spotted with pale sandy buff in the centre of the feathers; the foreneck and chest slaty blue; remainder of undersides black; sides of chest for the most part mottled with sandy-buff spots like the back, followed by a conspicuous patch of white, these white feathers shaded with bluish-grey; axillaries brownish-black. Iris dull tawny brown; bill ashy yellowish basally, the ends of the mandibles pale ashy an the culmen dusky; tarsi and toes dull yellowish-green.

Adult female. Similar to the male in the colouring of the upperparts and the black abdomen, but differing in the colour of the head and chest; wings and tail also similar to those of the male. Crown dark brown, with large spots of sandy rufous; occipital region ashy, freckled with lines of black; lores and a broad eyebrow isabelline buff, the feathers edged with black; below the eye a band of sandy rufous, the feathers margined with black; ear-coverts sandy buff; cheeks and throat white, but with no mesial band of black on the latter; neck, foreneck and chest brown, mottled with sandy-buff spots and markings and varied with a few cross-lines of black; across the chest a broad band of white; axillaries black; under wing-coverts white barred with black or black tipped with white. Iris light tawny hazel; bill yellowish basally; lower mandible pale ashy at tip, upper mandible dark brown; tarsi and toes yellowish-white.

Immature. Juveniles generally resemble adult females (Clancey 1967). Friedmann (1930) provided a detailed description of juvenile females, which have grey tarsi and olive-black (above) to yellowish-green bills. In subadult males the back is more rufescent, the black markings are larger than in older birds, and the amount of black on the abdomen is variable.

Identification

In the hand. The small to very small size (wing under 280 mm), black underparts, and a short crest that in males is pinkish to buff-coloured (rudimentary in females) provide a distinctive combination.

In the field. In this species' African range it is the only bustard with black underparts that also lacks white at the base of the flight feathers (apparent only in flight). When not in flight, the male's pinkish crest may be visible; the female is most similar to the female of the immediately following species, but has a white rather than blackish throat and the upperparts are mottled and streaked rather than barred. Although there are regional variations in vocalizations (see Social Behaviour below), the male's territorial call typically begins with a series of clicking notes that increase and change to a series of shrill whistles, repeated five to ten or more times. These calls typically precede an aerial display, during which the male rises in the air 20–30 m and then falls quickly, somewhat resembling a black feather-covered football.

General biology and ecology

This is a species of dry habitats having some trees or bushes, such as thornveld or bushveld or thin woodlands with a generally well-developed grassy understorey that provides screening cover for the birds. They are never found in completely open habitats, and often occur along the edges of heavier cover. At least in Kenya and adjoining parts of East Africa they are found at elevations of less than 1550 m. All three of the rather widely disjunctive races occur in more or less the same kind of habitats as just described (Chappuis *et al.* 1979).

The birds are mostly to be found as singles or in pairs, and are extremely adept at remaining undetected behind the shelter of grass or thorny cover, or will quickly disappear into it when disturbed. They are apparently quite sedentary, remaining on the same site throughout the entire year.

The species feeds on a wide variety of insects and also some vegetable materials. These include beetles (such as tenebrionids and scarabs), beetle larvae, grasshoppers, ants, centipedes, seeds of *Acacia* and *Brachystegia*, berries, fruits, and gum (Urban *et al.* 1986).

Social behaviour

According to Kemp and Tarboton (1976), this is one of the several savannah-inhabiting bustards of Africa that live relatively solitary lives, and in which males call from traditional display sites that at least in this species are used from year to year. There is no indication of permanent pair-bonding, and females may wander from one displaying male to the next. All of the three species considered by Kemp and Tarboton to belong to this woodland- or savannah-adapted group (the others being the black and black-bellied bustards) have black underparts in males and aerial displays. To this list of black-bellied species, Hartlaub's bustard might also be added, although its mating system is still undescribed, and males are also not yet known to perform aerial displays.

The male's display period is apparently concentrated in the breeding season, which varies considerably across the species' range, although in South Africa some display occurs all year. There is also good evidence that the vocalizations of the males exhibit geographic variation to such a degree that Chappuis *et al.* (1979) suggested that *savilei* should be regarded as a distinct species, and that it was 'probable that *gindiana* and *ruficrista* are not mere geographic races'. These authors noted that the male call of *savilei* is distinctly different from that of the other two forms in its frequency, rhythmicity, and monotonous structuring of phrases. The advertisement call of this West African form consists of a whistled note followed by a series of slightly lower, short and clear whistles. In the East African *gindiana* this same call begins with some frog-like notes but ends with notes much like those of nominate *ruficrista*. In this South African form the male begins its display with a series of loud tongue clicks (not bill-claps, as stated by Astley-Maberly 1967), followed by a series of strong whistles that increase in volume, each interspersed with a tongue-clicking sound, sometimes terminating in a scream-like *keeweep*. These calls may then fade away, only to be soon repeated.

At least in the nominate form, the male often terminates his vocal display by taking a few steps forward, tongue-clicking as he goes. He then flies vertically up as high as 30 m while still clicking, throws himself onto his back with his feet up and breast feathers fluffed, rocks forward, and falls vertically down. When he nears the ground he breaks his fall with short winnowing wingbeats (Kemp and Tarboton 1976). Although some illustrations of this display show the entire flight sequence, the birds are almost never seen taking off or alighting; rather, the flying male appears suddenly above the canopy of vegetation and just as quickly disappears into it on the way down. In both going up and coming down the bird looks like a round, fluffy ball, with the wings and legs not projecting as much as in the accompanying sketch (Figure 46a), these apparently being tucked in during the final part of the ascent and also during the descent (Tarboton personal communication).

The aerial display of the East Africn form is evidently fairly similar to that of the southern race (Lynn-Allen 1951, Pitman 1957), although it has not been so fully described. Goriup *et al.* (1989) observed this display on five occasions in Kenya, and provided some photographs of it, on which Figure 46a is based. A song immediately preceded the display flight, although such flights did not always follow vocalizations. Two types of flight were observed: nearly vertical and more parabolic. During the vertical flights the male rose some 10–15 m, with legs dangling, throat and neck feathers erected, and the nuchal crest fanned so as to form an inverted U shape. At the apex of the flight the wings were

Fig. 46. Social behaviour of rufous-crested bustard: (a) male flight display (after photos in Goriup *et al.* 1989); (b) male running with raised crest, compared with normal male head appearance (after sketch in Astley-Maberly 1967); and (c) male displaying to female (after a description by Cassels and Elliot 1975).

folded, the head was drawn back, and the bird parachuted down with legs dangling and the head somewhat raised. In the more parabolic flights the bird rose only 8–10 m and then glided down, covering some 20 m or so, but apparently not fanning his crest or retracting his head noticeably. It might be noted that postural displays of *savilei* are still undescribed, including any possible aerial displays.

When a female appears, the male utters a special call, described by Kemp and Tarboton (1976) as a croaking *wak wak wak* that rises in volume and speed to break into a deeper and double-note *wuka wuka wuka*, and finally with a thin piping note interspersed between each phase. In the presence of a female, the crown feathers are erected into a crest, the throat is puffed out, and the neck feathers are also erected, but otherwise the reddish to buff crown feathers are kept hidden. When encountering another male, the birds march back and forth in an upright posture, or may come together with their heads down and feathers fluffed, but rearing up at times to kick the opponent.

Two other displays of this species have been described, both for the South African race. In one instance Astley-Maberly (1967) observed a male standing stiffly with his wings loosely held, almost trailed, and his tail partly spread (Figure 46b). His nape crest was erected into a shaggy 'bonnet' resembling that of a crowned crane (*Balearica*), and he made shuffling hops with his head bobbing and wings trailing as he approached a female. As he neared her he paused, trembled, and uttered plaintive peeping notes while standing and facing her (Figure 46c). He then alternated stiff dancing movements with this standing, trembling posture for a considerable time, throughout which he continued to look in the direction of the female.

The other, apparently different, ground display was described by Cassels and Elliott (1975). They observed a male with his crest erected who began a prolonged series of clicking sounds, but without interspersed whistles. He then circled a bush with the nearside wing drooping and the farside wing raised, and the body twisted so that the back was orientated toward the centre of the circle. As he walked, he made a limping gait in synchrony with the clicking sounds. With each circuit around the bush the male would stop at the same point, stare at the centre of the clump, then take about four slow steps backward, raise his crest to the maximum, and then set off again on another circuit. This bush-circling activity went on for about 3–4 minutes. No female or other animal was observed in the bush, but the authors believed the display to have some kind of territorial significance.

Reproductive biology

In South Africa this species breeds during the summer months, from October to February, whilst in Zambia and Mozambique the records extend from September to February. Some Kenya records are from November and January (during or after the short rainy season), and more generally extend from March to August (during or after the major wet season). Breeding in Uganda is probably mainly between mid-December and March. Records from Ethiopia and Somalia are also from the wetter period of March to June. West African records (Senegambia and Mali) are from September and October, and those of Chad are from June to August, during the wet season (Urban *et al.* 1986).

The eggs are laid in shallow scrapes, often in the shade of a large plant or bush. The usual clutch is two eggs (Someren 1933, Pitman 1957), although single-egg clutches are also known. Only the female tends the eggs and young, so far as is known presently. Some large raptors such as tawny eagles (*Aquila rapax*) and pale chanting goshawks (*Melierax canorus*) are probably significant predators on these relatively small bustards (Urban *et al.* 1986).

This species has recently been bred in captivity at the San Diego Zoo, where a wild-caught female laid a total of nine eggs (average weight 48.25 g) over a period of 6 months. Five chicks hatched after artificial incubation periods of 19–21 days, and all were subsequently raised successfully.

Evolutionary relationships

Apart from the possible specific status of one or more of the three forms here considered to be part of a single species (see P. Osborne 1989 for recent comments on this question), the broader relationships of *ruficrista* are somewhat uncertain. Its buffy-coloured crest is unique, but in its male calling behaviour and aerial display it is seemingly similar to the little black bustard. I have tentatively placed it in a relatively isolated position within the genus *Eupodotis* (see Figure 2).

Status and conservation outlook

This species is still generally common in southern Africa; Kemp and Tarboton (1976) stated that in some areas as many as six birds may be seen displaying simultaneously at the peak of the display season in early summer. In East Africa it is likewise still generally common in suitable dry scrub and thornbush habitats, but in West Africa the species' distribution is more patchy; for example it is common in some areas of Mali but absent in others (Urban et al. 1986).

BLUE BUSTARD (Plate 33)

Eupodotis caerulescens (Vieillot) 1820

Other vernacular names: blue korhaan; Blou Korhaan (Afrikaans); outarde plombée (French); Blautrappe (German).

Distribution of species (Map 21)

South Africa, from the eastern Cape to Orange Free State, interior Natal, and southern Transvaal highveld; possibly also Lesotho. No subspecies recognized.

Measurements (mm)

Wing, males 315–356 (av. of 10, 336), females 325–337 (av. of 4, 331); tail, males 135–172 (av. of 10, 158), females 145–161 (av. of 4, 151); tarsus, males 90–106 (av. of 10, 99), females 87–97 (av. of 4, 94); culmen, males 24–30.5 (av. of 10, 28.2), females 26–29.5 (av. of 4, 28.3) (Urban et al. 1986). Egg, av. 56 × 44 (Schönwetter 1967).

Weights

No information. Egg, 60 g (estimated).

Description

Adult male. General colour above sandy buff, minutely freckled with blackish wavy lines or stipplings; the lower back, rump, and upper tail-coverts rather more tawny and more finely vermiculated; scapulars and innermost secondaries like the back, the latter being a little more coarsely vermiculated; wing-coverts mostly tawny rufous, the lesser series finely vermiculated with blackish lines, the median series uniform externally, and the small coverts round the bend of the wing blue-grey; alula and primary coverts slaty blue, with dusky or blackish ends; the outer greater coverts also slaty blue, black at the tips, the inner greater coverts tawny rufous like the median coverts; remiges black with slaty-blue bases, the latter colour increasing in extent on the inner primaries and secondaries; rectrices tawny rufous at base, black at tip, the central pair sandy buff, freckled with blackish lines; the outer feathers shaded with slaty blue subterminally and having the inner web mostly slaty blue; on some of the feathers the tawny base is freckled with a few black vermiculations; crown slaty blue; forehead black; hindneck, sides of neck and underparts from the lower throat downwards slaty blue; lores and a broad eyebrow white; below the eye a streak of white and a black streak behind the eye; ear-coverts greyish-white with fine black shaft-streaks; a large black patch extends from the front of the eye to the ear-coverts and the front of the cheeks; posterior cheeks, chin and sides of throat white; a large black central throat patch consisting of elongated and somewhat laterally dilated plumes; sides of upper breast sandy buff with blackish vermiculations; under tail-coverts tawny buff, with minute blackish vermiculations; under wing-coverts mostly slaty blue, but the greater series white with a bluish tinge; axillaries slaty blue; under wing-coverts bluish-slate and white. Iris dusky; eye-ring tawny; bill mostly dusky, paler at the base; tarsi and toes yellow.

Adult female. Much as in the adult male, but crown duller and mottled posteriorly with buff, and rusty buff on face, lores, supercilium and ear-coverts. Blackish throat patch somewhat scaled with buff (Clancey 1967).

Immature. Juveniles have a blackish crown mottled with buff. The wing-coverts are more heavily marked with blackish-brown, and the remiges have buffy tips (Clancey 1967).

Identification

In the hand. This small to medium (wing 315–356 mm) bustard is easily identified by the predominantly bluish colour of its body, neck and upper head (paler in females), with black on the throat and some white on the sides of the head.

In the field. The distinctly overall bluish cast of this species identifies it readily in the field; no other bustard is so bluish on the underparts. Males utter a frog-like *kakow* or *kuk-pa-wow* note independently or in response to the calls of other males while on the ground, and another (or possibly the same) call uttered in flight sounds something like 'knock me down, knock me down'.

General biology and ecology

This species occupies a very small overall range, being limited entirely to South Africa and possibly also Lesotho, where it occupies upland grassveld, usually above 1500 m elevation, particularly where the land is dotted with termite mounds. It probably prefers well-grazed lands, and also moves on to cultivated fields during winter (Urban et al. 1986).

It is generally found in small groups of up to about five birds, but as many as twenty may gather in highly attractive areas, such as recently burnt grasslands. The mean group size of 572 groups was 3.39, but 59 per cent of these were groups of two or three birds, and 83.6 per cent were groups of two to five birds. Groups averaged somewhat larger during the non-breeding period of February to August (3.53 birds) than during the breeding season (2.96). Judging from these data, it is likely that young birds remain with their parents until the spring after hatching, and some may remain with their parents for up to 2 years. The density of birds in the Transvaal was found to be very high, 1–1.3 birds per km^2 (Maclean et al. 1983).

Food includes a diverse array of insects, small lizards, and some vegetable matter. Generally the birds forage early and late in the day, spending the hottest part of the day resting in the shade (Clancey 1967). At night the birds may roost together, forming a tight huddle on open ground (Urban et al. 1986).

Social behaviour

The mating system of this species is still not certain, but monogamy is possibly the commonest pattern, judging from average group sizes noted above. However, some breeding groups may diverge from this pattern; Kemp and Tarboton (1976) mentioned seeing a male and two females attending a single nest, and Maclean et al. (1983) observed that as many as five apparently adult birds might consort together near an incubating female. It is possible however that the latter situation involved sexually immature birds of the prior breeding season or two, rather than indicating a more complicated mating situation. Kemp and Tarboton judged that this species, like other open-country forms including the black-throated Rüppell's and white-bellied bustards, tend to live in pairs or small groups that remain intact during the breeding period, whereas the more savannah-adapted species (black, red-crested and perhaps black-bellied bustards) are more solitary, with no indications of pair-bonding.

In common with the other open-country species mentioned above, this species utters deep, frog-like advertisement notes, which are presumably produced by the male. In the blue bustard these consist of a double- or triple-note sequence repeated several times. Other birds in the group may join in, and this chorus is sometimes answered by more distant groups, the vocalizations carrying about 2–3 km. Most such calling occurs during dawn and dusk periods, and although some calling occurs throughout the year it is mainly done during early summer, probably corresponding to the start of the breeding season. It is not known whether duetting between the sexes occurs in this species, as has been reported for the black-throated and Rüppell's bustards.

During courtship, the male approaches the female with his neck stretched to full length, his throat puffed out, and the neck and throat feathers erected to form a ruff. Holding the feathers in this position, the male bobs his head rapidly up and down, and runs after the female (Kemp and Tarboton 1976, Urban et al. 1986).

Reproductive biology

The breeding season usually occurs between November and February, but there are some egg records as early as August. The nest is a ground scrape, and two eggs form the normal clutch, with extremes of one and three also recorded. Incubation is by the female alone, although other adults including males may visit the nest. The young join a group of (possibly related) adults after hatching, and probably remain with the group for at least one and perhaps two years. It is possible, however, that females gradually disperse from the group while males remain in it longer (Vernon 1983). The adult sex ratio of the species closely approaches equality; of 277 birds, 51.3 per cent were males (Maclean et al. 1983), which might favour the maintenance of a monogamous mating system.

Evolutionary relationships

This is seemingly a very distinctive species, although Snow (1978) judged that it might be closely related to the superspecies *vigorsii*, and less closely related to *senegalensis*. I have tentatively placed it within a somewhat isolated part of the *Eupodotis* group (Figure 2), although it must be admitted that it shows some strong behavioural and plumage similarities to these forms just mentioned, and perhaps is much more closely related to them than I have suggested.

Status and conservation outlook

Although of very restricted overall range (almost exclusively limited to South Africa), this species is still locally quite common, especially in eastern parts of its range, such as in south-eastern Transvaal. It is also fairly common in south-central and southern Transvaal and Orange Free State. However, it is marginal in the north-eastern and eastern Cape, and has generally declined in the western parts of its range (Clancey 1967, Urban et al. 1986).

BENGAL FLORICAN (Plate 34)

Eupodotis bengalensis (Gmelin) 1789

Other vernacular names: chards, charg, charat (Hindi); ulu moira (Assamese); outarde Bengal, le florican du Bengale (French); Barttrappe (German).

Distribution of species (Map 27)

Extremely rare and endangered species originally occurring in grassland habitats from northern Bangladesh, Bihar and West Bengal north-east to Assam and perhaps to Bhutan, and west along the terai lowlands of Nepal and northern India to about the Yamuna (Jumna) River of north-western Uttar Pradesh; rare south of the Brahmaputra River. Now fully protected in India and Nepal, and with an estimated world population of only about 400, mainly occurring in Assam plus some in Nepal. Also historically known from Cambodia, probably as migrants, but no recent information on its status there. Recently reported from south-western Vietnam, where its status is uncertain.

Distribution of subspecies

E. b. bengalensis (Gmelin): Originally eastern and northern India, from Bihar and West Bengal northwest to Uttar Pradesh and the terai of Nepal and Arunachal Pradesh along the base of the Himalayas and east to north-eastern Assam; also originally in northern Bangladesh but doubtful if ever present in Bhutan. Now very local in south-western Nepal (especially the western end of Rapti Dun) and northern Assam, possibly also along the Nepalese border of Uttar Pradesh.

E. b. blandini Delacour 1928: Described and still known only from a few apparent migrants obtained at Soai Ring (Svay Rieng), southern Cambodia; perhaps also occurring in adjacent Cochinchina. Breeding area unknown but possibly western Cambodia (Delacour 1929); a potential breeding area has recently been found in south-western Vietnam.

Measurements (of *bengalensis*, **mm)**

Wing, males 338–348, females 338–368; tail, males 165–185, females 163–184; tarsus, males 127,

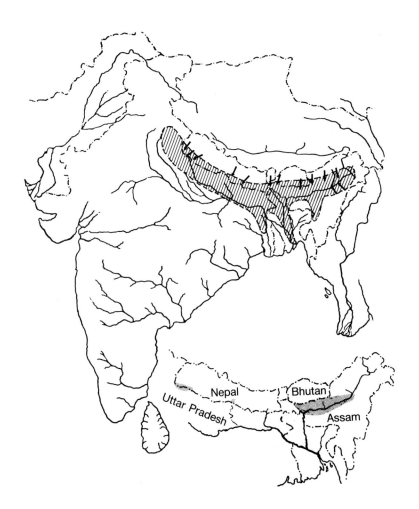

Map. 27. Distribution of the Bengal florican, including its historic range (hatched), and locations of recent sightings (arrows). The inset shows the areas (stippled) in Assam and Nepal where the population remnants now appear to be concentrated.

females 142; exposed culmen, males 30–31, females 38–39 (Baker 1921). Egg, av. 64 × 46 (Schönwetter 1967).

Weights

Males c. 1.1–1.7 kg, females 1.6–2.3 kg (Baker 1921); males c. 1.8 kg, females c. 2.25 kg (Inskipp and Inskipp 1983). Egg, 72 g (estimated).

Description (of *bengalensis*)

Blandini reportedly has somewhat richer overall plumage tones and the male has shorter ornamental feathers on the head and neck.

Adult male. General colour above black with sandy mottling, the latter coarse in character and the black taking the form of arrowhead centres to the feathers; the upper scapulars black, forming a patch on each side of the mantle; remainder of scapulars like the back; lower back and rump nearly uniform dark brown, with very slight indications of sandy mottlings on a few of the feathers; upper tail-coverts also dark brown, but a little more coarsely freckled with sandy buff and occasional cross-bands; rectrices black tipped with white, the outer feathers with indications of bluish-grey bands, which become more distinct towards the centre of the tail, where they are mottled with black and crossed by three black bars; entire series of wing-coverts white, as well as the alula, primary coverts and the greater portion of the remiges, the major coverts and remiges white, with black shafts for the greater part of their length, the outer primaries black, with white inner webs, the white extending to the inner primaries, which have the shafts and tips black; the inner secondaries white in the outer web, black on the inner one; the inner greater coverts bluish-grey mottled with black, and having blackish bases; the innermost secondaries like the back; head and neck all round glossy blue-black, the head crested and the sides of the neck ornamented with a very large frill, which also extends over the foreneck; remainder of underparts from the foreneck downwards dull black, the long under tail-coverts mottled with white; under wing-coverts mostly black, but the greater coverts and the edge of the wing white, with a conspicuous white base to the lower major coverts. Iris yellow; bill dusky, bluish above, bluish-grey to yellowish below and somewhat fleshy brown towards gape; tarsi and toes dirty straw colour. Unlike the lesser florican, the male does not have a female-like non-breeding plumage, but reportedly does undergo two moults per year, including a complete spring moult and an incomplete fall moult.

Adult female. General colour above black, coarsely freckled with sandy rufous and varied with deep sandy-coloured spear-shaped markings in the centre of the feathers; scapulars very coarsely marked, but otherwise like the back; lower back and rump nearly uniform brown, freckled with sandy buff; upper tail-coverts and centre rectrices sandy rufous, coarsely freckled with black and crossed by distinct black bars, the outer rectrices brown, with black frecklings, and crossed by three broad bands of black; lesser wing-coverts like the back but lighter and the mesial sandy markings rather paler; median and greater coverts light sandy rufous, paler at the ends and with very few cross-lines or mottlings of black; alula sandy buff and more thickly mottled with black; primary coverts and remiges black, the former mottled with white at their ends, the primaries with white bands on the inner webs more or less obscured by black mottlings, the secondaries freckled and mottled with sandy buff towards their ends, and with two or even three broad bands of black, the innermost secondaries like the back; crown sandy-coloured, with two broad black bands on each side, enclosing a mesial band of sandy buff, which like the nape is streaked with black; lores, eyebrows, sides of face and ear-coverts all sandy buff, with a few black spots below the eye; cheeks sandy rufous; chin and throat white; sides of neck sandy buff, with a line of black-tipped feathers on each side of the white throat; foreneck and hindneck also sandy buff, slightly freckled with blackish and with mesial streaks of pale sandy buff; breast and abdomen dull white, with a tinge of sandy colour, especially on the sides of the upper breast, which is freckled with black; under tail-coverts pale sandy buff, with a few blackish mottlings; under wing-coverts blackish, with white tips and sandy buff on the outer webs; axillaries black; undersides of remiges blackish crossed by white bars.

Immature male. Initially resembling the female, but gradually acquiring by moult the black on the head, neck, and under surface, as well as the white on the wings, which are among the first adult feathers to develop. Young males thus have much paler wings than adult females. An adult or subadult male plumage is attained with the first spring moult, but a female-like plumage may be acquired again the second autumn. Once the full adult male plumage is attained, presumably during the second spring, it is permanently retained (Ali and Ripley 1983).

Identification

In the hand. This medium-sized (wing 338–368 mm) bustard may be identified by the male's wholly black head, neck and underpart colouration, the wings being mostly white above except for black tips. Females have a buffy abdomen, an unspotted sandy whitish breast, and a sandy mid-crown striped on either side with black. Their upper wing-coverts are mostly buff-coloured rather than white.

In the field. Within its range this is the only bustard in which the males are almost entirely black on the head, neck and underparts, with a strongly contrasting almost entirely white upper wing surface (only the tips of the flight feathers black). Females are similar in size and colour to the lesser florican, but lack heavy breast streaking, have buff rather than white upper wing-coverts, and exhibit buffy and black banding rather than rufous and black colouration on the flight feathers. Normally silent, but a shrill and metallic repeated *chik* call is uttered by males during disturbance, during flight display, and during threat. The female produces a flatter and nasal *chock* alarm note when flushed or surprised.

General biology and ecology

It has long been recognized that this is a species dependent upon grasslands for breeding; Baker (1935) stated that it is found only in the enormous grass 'churs' on the banks of the Ganges, Brahmaputra, and other great rivers of northern India, and in the vast terai grasslands extending along the foothills of the Himalayas, and that it enters croplands only for the purpose of foraging. According to Inskipp and Inskipp (1983), floricans were typically seen by them in flat grassland areas (1–9 ha in area, sometimes with scattered trees present. During their observations, males were usually seen in fairly short-grass areas averaging about 25 cm tall, and females in taller grasses averaging 110 cm. In one area both sexes were found on a central ridge in which the grass averaged 31 cm. Such areas of fairly short grass may enable the birds to meet and associate more readily. Except for one site, they were found almost entirely in pure grassland habitats, which in the short-grass areas was frequently mainly of thatching grass (*Imperata*). The long-grass areas comprised a variety of genera of so-called elephant grasses. Periodic burning, which opens the grasslands, releases nutrients, stimulates grass growth, and probably improves foraging, may favour the birds, at least if it is done before the breeding season (Inskipp and Inskipp 1983).

Recent detailed studies in Manas Wildlife Sanctuary (part of Manas Tiger Reserve) of northern Assam correspond to the terai habitats just mentioned, in which the annual rainfall is about 4 m, mostly falling between June and September, and the general climate is moist tropical. Florican habitat there consists of large open grasslands having shorter types of grasses and shrubs growing in great density. These grasses rarely exceed 20 cm in height, but in other areas may reach 0.5 m or more, and are maintained by annual burning and to some extent by periodic inundation during the monsoon season. Scattered trees also occur, but their abundance is controlled by burning. The flat grasslands used by floricans are dominated by grasses such as *Narenga* and *Imperata*, but several other genera of grasses and sedges are also present. Tall grasses growing in and around nullahs that cross the grassland include genera such as *Saccharum*, *Phraqmites*, *Erianthus*, and *Arundo*. In typical florican habitat the average maximum height of the grasses during the approximate time of greatest male display in late April is about 50 cm (or the height of an adult male florican), and increases to nearly 1 m by the middle of June. Near the end of the monsoon the taller grasses may reach 2 m in male territories, and are much higher in surrounding areas (Narayan and Rosalind 1988).

Foods of this species are still unstudied in detail, but it is known to eat the shoots, seeds and runners of grasses, berries, stony fruits, and stems of small plants. It also consumes mustard shoots and flowers, wild cardamom, and various oilseed plants. It is apparently attracted to fires to eat insects and roasted seeds, and when locusts are abundant they may be almost exclusively consumed. Other animal materials that are eaten include grasshoppers, ants, slugs, beetles (especially cantharid beetles), lizards, frogs, and small snakes (Inskipp and Inskipp 1983). At Manas Wildlife Sanctuary the birds were often seen consuming grasshoppers, moths, beetles, and probably other insects obtained from the ground surface or vegetation. They also favoured the berries of an abundant shrub called 'falsa' (*Grewia*), and dug and scratched the ground for the roots and shoots of *Imperata* and other grasses and sedges. Territorial males foraged either in their own territories or in other patches of grassland, favouring areas of grass up to 50 cm. They were observed to drink rather rarely, as their foods contained considerable moisture, but during warm weather they sometimes flew to a river or other source of surface water. Foraging birds were often followed by black drongos (*Dicrurus adsimilis*) or small green bee-eaters (*Merops orientalis*) that hawked insects flushed by the floricans (Narayan and Rosalind 1988).

Little is known of possible migratory or local movements of this species, although the nominate race is generally believed to be residential. However, the birds are extremely difficult to see after the grass has grown taller than the birds themselves, and thus local or even long-distance movements would be difficult to verify at this time of year (Inskipp and Inskipp 1983).

Social behaviour

It is generally agreed that Bengal floricans do not form any pair-bonds, and that territorial males attempt to attract females to them during their month or two of active display. At Manas Wildlife Sanctuary the males established their territories during late January or early February, depending upon the timing of grass regeneration. They stopped display by the beginning of May but continued to occupy their territories until the latter part of May, or perhaps at times into early June, perhaps depending upon the height and density of the grasses. In one field a male maintained a territory of about 5 ha, while another male had a territory of about 3 ha, and used an open area of only about 1 ha for its displays. In one case the territorial boundaries of two males were only about 100 m apart, whilst in another case a male with a territory having exellent surrounding visibility had no other territory for at least 0.5 km around its 5 ha territory. A total population of about forty males was estimated (based on twenty-four actually seen) for the 391 km^2 Manas Wildlife Sanctuary in the 1988 season (Narayan and Rosalind 1988), making it perhaps the largest known single breeding concentration of this species.

In similar observations at Dudwa National Park, Uttar Pradesh, fourteen display sites were studied. In all of these the males used short-grass areas to display from, and eight of the sites were located on high ground. One display site had apparently been used since the mid-1940s, and another since the early 1950s, indicating an apparently high level of attractiveness to particular locations and probable site-fidelity, since at least three sites were in use for three successive years, and eight were in use for two successive years of study. Their display begins during the first half of March, and tapers to an end by the third week of June, with April and May the peak display season. Most of the territories there were not within visual or acoustic range of one another, with a minimum observed distance of about 500 m between sites (Sankaran and Rahmani 1988).

Observations by Inskipp and Inskipp (1983) were not so extensive as those just mentioned, but they observed display activities at three sites between April 13 and May 21. Displays were entirely confined to morning and evening hours, when the temperatures were relatively low. During morning periods, displays occurred between 30 min and 3.5 h after sunrise, and from 40 min before sunset until 14 min after sunset. At least one male seemed to display more often on cloudy days, although most displays were observed under clear skies with little or no wind. The average morning display frequency was 0.7 per hour, as compared with 2.4 per hour in the evenings, and no more than six displays were observed in a single evening period.

Similarly, at Manas Wildlife Sanctuary the flight displays started around sunrise, and sometimes continued for as much as 3 h, but might even be extended until noon if the weather was cloudy and cool. Afternoon displays began about 2 h before sunset and were terminated by 35 min after sunset. Afternoon displays were much more frequent than morning displays, with a maximum of six displays observed in a single hour and eleven in a 2 h period, but the average number of displays for two males was 1.34 per hour, for 156 total displays. No displays occurred during hot or rainy weather, and most were performed under calm to only mildly windy conditions (Narayan and Rosalind 1988).

Although earlier descriptions are somewhat different and probably less reliable in detail, those given by Inskipp and Inskipp (1983) and by Narayan and Rosalind (1988) provide an excellent indication of this species' advertisement display behaviour. Display flights (Figure 47e) are made from areas of relatively short grass, before which the male gradually erects its shaggy neck and breast feathers (Figure 47b). It then crouches, so that the raised neck plumes may touch the ground. It draws its head back, shakes its head, and springs up diagonally. During ascent the wings are flapped loudly, and the bird thus reaches a peak altitude of 3–4 m. At the apex, the male begins to utter a series of four to seven sharp and whistle-like *chip* notes, and glides for 1–2 m with its wings open, head drawn back egret-like, and neck pouch drooping. It then again flaps its wings to regain lost altitude, and after reaching its second apogee, floats more or less vertically down with neck-pouch drooping, legs dangling or paddling, and wings partly open. It lands with legs forward and stops after making up to five steps. It then looks about with feathers still fluffed, and finally shakes its body to assume its normal appearance (Figure 47a).

During the flight the bird may cover 20–40 m, and the flight lasts 6–8.5 s. After the display the bird may walk back to its launching point, but there are no favourite display sites, and successive jumps may be performed from various points within the more open parts of the male's territory. Should a female appear in the distance, the male will immediately take off, fly with its pouch drooping, and utter *chip* notes every few seconds. It comes close to the female, stalling upwards and then dropping vertically, as in the last phase of the flight display. When on the ground near a female, the male walks with its

Fig. 47. Social behaviour of male Bengal florican: (a) normal posture; (b) posture just before jumping flight display; (c) defensive posture; (d) walking and pumping display by male near female; (e) flight display; and (f) threatening posture. After sketches and photos in Rahmani *et al.* (1988b).

pouch drooping and breast feathers almost touching the ground, its head lowered on the back (Figure 47d, left), and moves with its anterior body undulating in a ludicrous manner. After every few such steps it stops and quickly pumps its head up and down (Figure 47d, middle), and then continues walking behind the female.

When threatening a territorial intruder (Figure 47f), the male partly fluffs its feathers, partly cocks and fans it tail, and charges at the other bird with open bill, uttering *chip* notes every few seconds. Actual fights, involving kicking, pecking or striking with the wings, may also occur when a male enters another's territory. Defensive responses, as per-formed towards a raptor, may include lowering of both wings and of the outstretched neck with the beak directed toward the opponent (Figure 47c).

Little information is available on copulatory behaviour (Finn 1915, Narayan and Rosalind 1988), but it probably differs but little from that described for the lesser florican and other *Eupodotis* species.

Reproductive biology

According to Baker (1935), the breeding season in Assám is mainly during March and April, when fifty-nine out of eighty-four eggs recorded by him were obtained. The earliest record he had was for

February 28, and the latest nests (in July) were found flooded and abandoned. According to his experience the clutch size is invariably two, but some single-egg clutches have also been reported. Based on such early accounts as these, the nest is typically situated in the centre of patches of *Imperata* or other grasses, and is frequently surrounded by dense jungle.

In their studies at Manas Wildlife Sanctuary, Narayan and Rosalind (1988) found only a single nest during more than 300 man-hours of searching. It was in a small opening in the midst of high grasses of about 1 m, with the grass heights diminishing towards the centre of the clearing in which the nest was located. This site was at least 1.5 km away from any known male territory, and had only a single egg present. It hatched in early May, and the chick was seen in the vicinity of the nest for only 2 days.

The incubation period has been estimated to last about a month, although this is probably little more than guesswork. Thus, the two eggs may be laid at intervals of 3–4 days, and incubated sets may exhibit differing degrees of embryonic development or show identical stages of development, suggesting that there may be some variation as to when incubation begins (Baker 1935).

Evolutionary relationships

The genus *Houbaropsis* is not easily separated from *Eupodotis*, and Ali and Ripley (1983) noted that the only distinction of the Bengal florican (which they included in *Eupodotis*) from *Sypheotides* is the presence in the latter of a female-like non-breeding plumage in the male. I believe these two species to be quite closely related (see Figure 2), and that both can readily be included in the genus *Eupodotis*.

Status and conservation outlook

A brief summary of this species' population status was presented in Chapter 5, but some additional points may be made here. Following a preliminary survey in Uttar Pradesh, West Bengal and Assam in 1985 and during which twenty-seven floricans were located (Lachungpa *et al.* 1985), a more detailed survey of these areas as well as Bihar was done in 1988 (Rahmani *et al.* 1988a). This latter survey included almost the entire terai zone of Uttar Pradesh and Bihar, the duars of West Bengal, and the Brahmaputra valley of Assam; between sixty and eighty floricans were located. Most recently, a 1989 survey of Assam and West Bengal was undertaken (Narayan *et al.* 1989).

In Assam, the largest florican population occurs in Manas Wildlife Sanctuary, an area of 391 km^2 in Barpeta, Kokrajhar and Nalbari districts. Between 1986 and 1989 at least thirty active territories were found, as well as some additional immature males. Assuming that an equal sex ratio exists, at least eighty floricans may exist there. The Kazinranga National Park (Golaghat and Nagaon districts) is managed to help support the endangered Indian rhinoceros, and contains many grassland areas. It is quite large (430 km^2), with much of it very difficult to survey, but has been judged to perhaps support as many as twenty-five to thirty floricans. The Orang Wildlife Sanctuary (75 km^2, in Darrang and Sonitpur districts) was believed to support thirty to forty birds in 1988, and thus probably represents the second or third most important area for floricans in Assam. Bornadi Wildlife Sanctuary (Darrang district) mainly protects pygmy hogs and hispid hares, but includes some seemingly excellent florican habitats. Likewise the Sonai Rupai Reserve Forest (Sonitpur district) includes some excellent grasslands and may support floricans. The Laokhowa Wildlife Sanctuary (Nagaon district) has some potential grassland habitats, and the even smaller Pabitara Wildlife Sanctuary (Nagaon district), which is also a rhino preserve, was observed to support a few territorial floricans.

In West Bengal the Jaldapara Wildlife Sanctuary (118 km^2, Jalpaiguri district) includes a few good grassland areas, and judged to perhaps support five to ten floricans. The Mahananda Wildlife Sanctuary (Darjeeling district) could not be surveyed, but was believed to perhaps have some suitable grassland habitats present. The Bahaluka Reserve Forest and nearby areas were judged unlikely to support floricans. Finally, some tea estates just south of Bagdogra in Darjeeling district were judged to be one of the few possible remaining habitats for the species in West Bengal, although only two males were seen in 1986, and none was found in 1987 or 1989.

In Bihar, the most likely remaining habitats in Valmikinagar Wildlife Sanctuary (West Champaran district) did not produce any evidence of floricans during a May survey, but a male was reported seen in Hariabara Protected Forest, in Purnea district, during this time.

In Uttar Pradesh two areas definitely supporting floricans were found. One of these is Dudwa National Park, which supports introduced Indian rhinos and other rare Indian wildlife. It had earlier been judged by Inskipp and Inskipp (1983) to support only two male floricans in 1983. In 1985 it had been found to have at least five males (Lachungpa *et al.* 1985). However, the 1988 survey revealed fourteen display sites, and thus a probable total park population of at least thirty-six to thirty-eight birds were judged present. In 1989 a total of twelve territorial males plus three immature males was observed

(Narayan et al. 1989). This area probably represents the largest single remaining florican population in Uttar Pradesh (Sankaran and Rahmani 1988).

Another probable location for floricans is a very small protected area (11.6 km²) at Lagga Bagga, which is located on the Indo-Nepal border adjoining Sukla Phanta Wildlife Sanctuary of Nepal, and where floricans are known to be present. There suitable habitat in two of three grasslands is almost certain to support floricans. Additionally, potential florican habitat occurs at Sohagi Barwa Wildlife Sanctuary (Gorakhpur and Deoria districts), Katerniaghat Wildlife Sanctuary (Bahraich district), and Kishanpur Wildlife Sanctuary (Lakhimpur Kheri district) (Rahmani et al. 1988a).

On the basis of these recent data, it may be said that as of 1989 about 150–200 floricans were known to exist in the state of Assam, and perhaps forty more were present in Uttar Pradesh. A very few birds also occur in West Bengal, mostly in Jaldapara Wildlife Sanctuary, which is situated in alluvial grasslands closely adjacent to western Assam. In addition to these areas, there is a small and apparently scattered population in Nepal (Inskipp and Collar 1984, Inskipp and Inskipp 1983, 1985a, b). Based on 1982 surveys, thirty-five to fifty birds (mostly males) were recorded in all of Nepal, and assuming an equal sexual ratio the adult population was judged by Inskipp and Inskipp to be fifty-six to eighty-two birds. A few already protected areas (Royal Chitwan and Royal Sukla Phanta Wildlife Reserves and Royal Bardia National Parks) held the majority of known birds, and although none were found in Kosi Tappu Wildlife Reserve it is possible that a few occur there too. It is possible, however, that no more than 100 birds still exist in all of Nepal, based on the data provided by the Inskipps.

Most recently, Dr George Archibald (personal communication) informed me that he and his associates observed three breeding-plumaged males and a female Bengal florican during March 1990 in the vicinity of Dong Thap reserve, a local sanctuary in south-western Vietnam that is not far south of the type locality of the little-known race *blandini*, whose breeding grounds have remained completely unknown. This area of Indochina receives its monsoon rains from May to December, and local residents report that the birds are present and breed there during the dry season. Dr Archibald did not observe any display, but it is possible that this area may indeed represent the breeding grounds of this extremely elusive race of endangered floricans, and additional studies on its status would be most welcome.

LESSER FLORICAN (Plate 35)

Eupodotis indica (J.F. Miller) 1782

Other vernacular names: charaz, chota charat, khar nor, leekh (Hindi); outarde passarage (French); Flagentrappe (German).

Distribution of species (Map 28)

Originally widespread in grasslands of eastern Pakistan and western India, from Punjab, Rajasthan and Gujarat (including Saurashtra and Kutch) south to Karnataka (Mysore) and Tamil Nadu (Madras). Resident and irregular local migrant; also nomadic during the non-breeding season, mainly moving to the east (to Orissa and West Bengal), but also to Sind and southern Nepal; rare west of the Western Ghats. Now generally extremely rare and local, most common in the grasslands of Gujarat, eastern Rajasthan and western Madhya Pradesh, with remnant groups on the Deccan Plateau in Andhra Pradesh and perhaps Karnataka. No subspecies recognized.

Measurements (mm)

Wing, males 180–204, females 209–248; tail, males 82–114; tarsus, males c. 85–95 (Ali and Ripley 1983). Egg, av. 49 × 41 (Schönwetter 1967).

Weights (g)

Males c. 390–570, females c. 510–740 (Baker 1921). Egg, 46 (estimated).

Description

Breeding male. General colour above dark sandy buff or ashy, coarsely but definitely vermiculated and mottled with black, and with numerous large spear-shaped markings or bars of black; the dorsal feathers varied with narrow ashy or whitish margins; inner secondaries like the back; lower back, rump and upper tail-coverts more dingy than the upper back and more finely vermiculated; the upper tail-coverts with well-defined cross-bars of black; rectrices tawny buff, with four distinct black bands, the light interspaces mottled with black markings; the central rectrices like the upper tail-coverts but more ashy, freckled with black, and crossed with black bars; all the lesser wing-coverts pure white, the inner median coverts greyish, freckled with blackish lines; all the outer coverts black, the innermost of the median series white, freckled with black; greater coverts yellowish buff, while at the ends and freckled with black lines, occasionally forming cross-bars; the innermost greater coverts ashy freckled with black on the inner web, white on the outer web like the adjacent median coverts; primary coverts and remiges ochraceous, tipped with white, the feathers uniform towards the base, but freckled with black towards the ends, and crossed by two, or at the most three, black bands; primaries attenuated toward their tips; inner secondaries more like the back, but banded with black; head, neck and underparts all jet-black, including the decorative plumes of the neck; between the hindneck and the mantle a broad band of white descending on to the sides of the chest; chin and upper throat white; some of the longer under tail-coverts white or sandy buff, mottled with black; under wing-coverts and axillaries black. Iris variable from very pale yellow to brownish-yellow; bill pale yellow, somewhat fleshy towards gape, the culmen, tip, and some of the upper surface shaded with dusky horny brown; the tarsi pale and somewhat fleshy yellow, varying from hoary to more dusky.

Non-breeding male. Resembling the female, but with more white on plumage.

Adult female. Differs from the male in being much more rufous above, the spear-shaped markings wider than in the male and much more of a sandy buff; tail also not so ashy as in the male, more sandy-coloured, but similarly barred and vermiculated with black; wings as in the male, except that the coverts which are white in the male are pale sandy buff in the female, with a few black bars and scanty vermiculations; no white collar across the mantle; the hindneck entirely sandy buff with very minute frecklings; crown tawny buff in the centre with a broad black band on each side; the nape tawny, streaked with black; lores, eyebrow and sides of face sandy buff, with a few black shaft-streaks and a black streak under the eye; throat white, the lower throat and remainder of underparts sandy buff, isabelline on the breast and abdomen; on each side of the throat a long black line of feathers, the foreneck broadly streaked with black; the breast somewhat mottled with black vermiculations, these extending to the sides of the body; under tail-coverts tawny buff, with irregular bars of black; axillaries black, very slightly mottled with sandy buff; the lower major and primary coverts also blackish, the former with sandy-buff tips; rest of the under wing-coverts isabelline.

Immature male. At first resembling the female, but gradually assuming the plumage of the adult male.

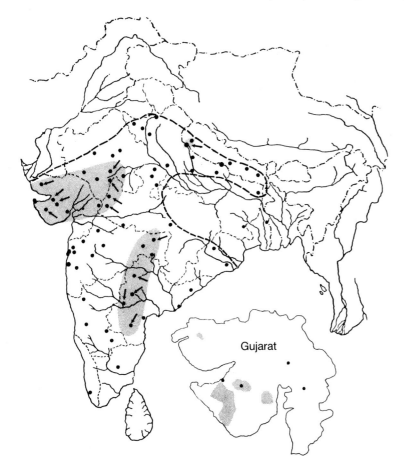

Map. 28. Distribution of the lesser florican, including its known historic breeding and wintering range (dashed line, with locality records indicated by dots) and probable maximum current range (stippled, with recent locality records indicated by arrows). The inset shows its known recent (since 1980) distribution (stippled) in Gujarat, with locality records indicated by dots. The hypothetical Cambodian range of *blandini* is not shown.

Identification

In the hand. Males of this very small (wing 188–248 mm) bustard are easily recognized by the long decorative black cheek plumes extending from the eye backwards, and otherwise are mostly black on the head except for a white chin. Females have strongly striped or streaked black and isabelline breast colouration; the remiges of both sexes are distinctively barred with ochre and black, and the upper wing-coverts are white (males) to sandy-ochre (females) in colour.

In the field. This is the only Indian bustard in which the adult males in breeding plumage are mostly black below, with white upper wing-coverts and ochre-and-black barring on the flight feathers (visible only in flight). Females, non-breeding males and young are mostly sandy and rufous-toned, with heavy breast streaking and with white lacking completely on the wings or limited to the upper wing-coverts (which are more tawny in immatures). Males utter a distinctive frog-like croaking during display, and a short whistle when frightened. Females produce croaking calls similar to those of the male, but shriller, and a whistle that seems to attract the male's attention. A clucking note may be uttered in flight or when a female has been flushed from her nest.

General biology and ecology

Early records of this species as summarized by Baker (1921) indicate that at one time it was widely distributed through the grasslands of India, although more recently it has become apparent that the species is distinctly migratory over much of India (Ali and Ripley 1983). Although the records are spotty, it would appear that breeding now, and perhaps in early times too, is concentrated in the relatively dry grasslands of north-central and north-western India, where the birds arrive with the monsoon rains. Thus Dharmakumarsinghji (1950) reported that they arrive on the Kathiawar peninsula from the end of May to August, probably having started their migration as soon as the heavy rains have begun in southern and central India. They apparently cross the Gulf of Cambay (Khambhat), and then spread out to all of Kathiawar, generally seeking cover of dry grasses, hedges and babul thickets. Some of the new male arrivals may still be in winter plumage on arrival, whilst others may be starting to grow their breeding body plumage and auricular plumes. The birds remain in Kathiawar through the breeding season, and following breeding they disperse in a southward and eastward direction, mainly during October and November. Some may remain through the winter months, and ringing data indicate that although birds may return to their same general area for breeding the following year, this is not invariable, at least in the case of males (almost no females having been ringed).

Ecological studies at Sailana, Ratlam district, in western Madhya Pradesh have provided some detailed information on vegetational and other ecological characteristics of florican breeding habitat. In this area the birds arrive with the start of the monsoon (late June or early July), and remain until the grasslands are ready for harvesting in October and November (Sankaran and Rahmani 1985a). Before the monsoon the area is nearly bare of vegetation, but with the start of the rains it is soon transformed into a grassland dominated by tall grasses of such genera as *Cymbopogon*, *Pseudanthesterea*, *Iseilema*, *Chrysopogon*, *Heteropogon*, and *Apluda*. At the peak of their growth in October these grasses average about 75 cm high, and at about this time the population of orthopteran insects also peaks. Reportedly, the species historically prefers breeding in areas having large expanses of grasslands about 1–1.5 m tall, especially where there are some thinly grown places in which the birds can readily move about (Baker 1935).

Feeding mainly occurs during early morning and evening hours, although birds that have recently arrived on their breeding areas typically feed throughout the day, and later during the breeding season males may feed and display alternately. They sometimes capture insects with an egret-like manner of creeping up on the prey and suddenly catching it in the beak, and at least in captivity have been observed moving slowly about with the neck arched and swaying with a jerky body motion as they stalked for hidden insect prey (Dharmakumarsinhji 1950). Large grasshoppers may be captured in flight with dumping movements, or chased on the ground. Their foods are quite variable, consisting of grasshoppers, cantharid beetles, ants, centipedes, worms, frogs, small lizards, and vegetational materials such as plant shoots, grasses, berries, and herbs. At least at times wide areas may be covered during foraging; in one 7 h period a foraging male was observed to walk more than 2 km (Ali *et al.* 1985).

Of the coexisting species occupying the same grassland areas for breeding, the floricans pay little attention to sarus cranes (*Grus antigone*), crows (*Corvus* spp.), or most of the common raptors. However, some of the larger falcons (*Falco* spp.) probably represent serious threats, and there is a record of a florican being killed in flight by a Bonelli's eagle (*Hieraaetus fasciatus*). Alarm calls by red-wattled lapwings (*Vanellus indicus*) are sometimes used by the birds as an alerting mechanism (Sankaran and Rahmani 1985b).

Social behaviour

Baker (1935) suggested that floricans are 'indiscriminate' in their sexual behaviour, with females wandering about during the breeding season, and males attempting to attract such birds to them by active display. Later, Dharmakumarsinhji (1950) suggested that male floricans may mate with one or more females during the breeding season, and similarly a female may probably mate with one or more males until it is fertilized, although further evidence was required. This general pattern of conspicuous male display from an established and traditional territory has now been well documented and certainly bears strong similarities to patterns found in the smaller African species of *Eupodotis* such as the little black and rufous-crested bustards.

Overall sizes of territories are still unknown, but in good habitat ten to fifteen males may be present on areas of about 100 ha (Sankaran and Rahmani 1985b). According to these authors (1986), distances between male territories (presumably measured from the centres of the territories) ranged from 275 to 450 m. This estimate was based on a population of eight or nine displaying males centering on a protected and predominantly grassland 'bheed' of about 350 ha near Sailana, Madhya Pradesh. According to Dharmakumarsinhji, the male's actual display ground or 'arena' may vary in size from as little as about 1 m square to one up to about 15 × 15 m in size. This is often an area of raised ground, where a view of the countryside is available and where the calls can carry longer distances. On the arena the grass is often trampled flat from the bird's activities, or it may choose to display on natural open spaces where vegetation is nearly lacking.

Ridley *et al.* (1985) analysed the topographic and vegetative characteristics of male display sites, and found that grass height was an important feature. Early in the season, when grass was short everywhere, the birds chose sites well covered with fairly long grass, but later males no longer preferred, and possibly even avoided, the areas of thickest grass. The birds were commonest on open grassland, but shrub density did not seem to be an important factor in the choice of display sites. Likewise the substrates varied from rocky ground to soft soil. Compared with surrounding areas within 500 m, eighteen sites were higher, eleven were lower, and five were on the same level. The display site itself was usually on level ground, with slopes seemingly avoided. Within a single territory one male spent about 73 per cent of his total display time at one of three sites, but six other sites were also used to some degree. All of these sites were located within an area of 1.2 ha.

Based on observations at Sailana sanctuary, it is probable that males begin to establish territories soon after their initial arrival, or at least as soon as the grass conditions have become suitable following the onset of rain. One male that probably arrived in late July of 1985 (first seen a day after the first summer shower) had established his territory by early August. He was joined by two males that apparently arrived a few days later, and who established their territories in prominent areas overlooking the first male's display site. In 1986 the first male was sighted only a week after the first shower of the south-west monsoon in June, and within a week of very heavy rain in late July at least four additional males arrived. Subsequent arrivals in July and August were also associated with periods of strong rainfall. One ridge was clearly the most preferred location for territorial establishment, being occupied first in both years, and with other later males taking up positions overlooking prime display locations. Thus it is likely that later arrivals attempt to establish themselves near already occupied territories. However, it is not yet known if males use the same territory from year to year, nor has it been proved that territorial establishment is triggered by a specific environmental factor such as minimal grass height (Sankaran and Rahmani 1985a, 1986). In 1985 and 1986, active male display began at Sailana a week or two after the initial arrival of birds, which in turn was seemingly related to the start of the summer (south-west) monsoon. Territorial behaviour apparently began about the time the newly sprouting grass had reached an average height of at least 10 cm.

Most of the agonistic behaviour between males is apparently associated with territorial establishment, and sometimes involves direct fighting, with interlocked beaks and mutual pushing until one of the birds is beaten and takes flight. Jabbing movements towards the other bird's head are common. Tail-cocking and raising of the mantle feathers are also associated with aggression, and in two cases the opponents were observed to turn sharply about when less than a metre apart, and exhibit their posteriors to one another (Sankaran and Rahmani 1985b).

Territorially active males were observed at Sailana during 42 of 71 days (59 per cent) of breeding-season observations in 1985, and in 1986 they were observed during 67 of 101 days (66 per cent) (Sankaran and Rahmani 1986). During the peak display period of late August males typically occupied and displayed from particular territorial sites for about half the daytime hours. During one display period, a male was observed by Ridley et al. (1985) to be engaged in display for 38 per cent of the daylight hours, and it averaged a display leap every 61 s. It displayed for an average of 17.3 min at each of its display sites, and perhaps jumped as many as 400 or more times per day. Similarly, during the peak display period at Sailana, males performed as many as ten jumps in 5 min, but averaged far fewer (Sankaran and Rahmani 1985b).

As was also found by Ridley et al. (1985), overall display activity at Sailana was generally greatest during periods of cloudy or cool weather, and on a daily basis was concentrated during morning and evening hours. Toward the end of the display season in September, displays were progressively restricted to early morning and late evening hours, as the daytime temperatures became increasingly hot. A larger number of males (fifteen vs. five) were observed displaying during a good monsoon year than during an unfavourable (dry) year, but during both years the display period had terminated by the first week of October, in spite of considerable cloudy weather and rains during one of these years.

The display leap of the male (Figure 48a–c) is preceded by an alert posture, with a stretching of the neck to full length. Then the legs are flexed, and the bird springs vertically in the air with wings flapping in shallow beats, and producing a loud rattling sound that can be heard for up to about half a kilometre under favourable conditions. When at the zenith, he arches his enlarged neck and nearly rests it on his back, and emits a frog-like croak, with his legs pulled upwards. He then falls back down, his feet pedalling and his wings held out at his sides in a parachute-like manner, landing in nearly the exact spot from which he took off. The height of the display varies from about 1 to 2 m, and the total duration of each display leap is about 4 s. If the bird is disturbed from its primary display site it may move off some distance and begin from that point, and the appearance of a female flying over the area will set off all the territorial males displaying simultaneously. (Just as birds sometimes perform the display leap without calling, they may also call without leaping, as may happen during the middle of the day.) Additionally, there is a considerable difference in the acoustic characteristics of the call in different birds, with some having deeper and more resonant calls and others (usually those not yet in full breeding plumage) having shriller notes (Dharmakumarsinhji 1950). It has been suggested that the male's loud rattling sound during the display leap may be produced by the tongue (Ridley et al. 1985), although it seems equally possible that the noise is produced as a result of the unusually attenuated primary feathers that are characteristic of this species.

Although the auricula plumes are normally directed backwards (Figure 48d), they can be erected antenna-like above the head and directed forward, as happens during close courtship display. At this time the throat appears slightly inflated, the neck is extended and slightly arched, and is swayed from side to side (Figure 48e). In a near-crouching posture (Figure 48f) the male approaches the female, with the posterior end of the body becoming distinctly depressed, so that the bird's body axis is held at a distinct upward angle relative to the ground (Figure 48g). Then the head is suddenly jerked backwards until it almost rests on the back (Figure 48h), and a low, wheezy croak is usually uttered. At the moment that the head hits the back, the wings are thrust outwards somewhat, thus exposing the contrasting white wing pattern (Sankaran and Rahmani 1985b).

In one case a male attempted copulation with a decoy immediately following this posture, and in each of several other observed cases the male always approached the decoy with the neck arched, and the crown and chin feathers erected. At times the head may be held on the back for a time (in a posture very similar to that of Figure 48c), while the tail feathers are erected and spread, and the body is lightly swayed to and fro. During attempted copulations with a decoy the male invariably pecked at its head before mounting, and while copulating rested on his tarsi, the wings drooping down on either side of the decoy (Dharmakumarsinhji 1950).

Reproductive biology

The egg-laying period of this species is mainly during the early monsoon period of August and September, which also corresponds to the period of maximum male display, although a considerable amount of egg-laying also occurs in July and October. There are occasional records for as late as January, probably reflecting year-to-year variations in rainfall pattern. The clutch size is typically four eggs, although three- or five-egg clutches have rarely been reported (Baker 1935). Six nests found at Sailana had three or four eggs, averaging 3.6, and the clutch size of the seventh (which had already hatched) could not be determined. Six of these were found within the limits of male territories 75–115 m from the display site), and the other was found within 200 m of a displaying male but outside its territory. Grass heights around the nests averaged about 25 cm at the time of laying and 65 cm at the time of hatching. The incubation period of one nest

Fig. 48. Social behaviour of male lesser florican: (a–c) stages in jumping display; (d) normal position of male head plumes; and (e–h) courtship display sequence. (a–d) after photos by A. Rahmani; (e–h) after sketches in Ali *et al.* (1985).

was found to last 21–22 days. No good information exists on nesting success, but six of seven nests found at Sailana during two seasons hatched successfully (Sankaran and Rahmani 1986).

Evolutionary relationships

Although the lesser florican is traditionally separated as a monotypic genus *Sypheotides*, this distinction is based on the combination of its relatively long tarsus, its unusually attenuated primaries, and the sex-based and nuptial plumage-limited trait of distinctive male auricular plumes, all of which hardly constitute an adequate basis for erecting a genus. I would suggest that it can readily be included in the genus *Eupodotis*, along with *bengalensis*, which probably represents its closest relative.

Status and conservation outlook

The general status of this species has been described in Chapter 5, but some additional points may be provided here. As has been summarized by Lachungpa and Rahmani (1985) and by Goriup and Karpowicz (1985), it is apparent that this species has suffered a massive decline in numbers during the present century and especially during the past 30 years, as native grasslands have been increasingly converted to cash crops. Additionally, it could still be shot under licence as small game as recently as 1980, when the schedules of the Indian Wildlife (Protection) Act of 1972 were last revised. Unfortunately, the small size and elusiveness of the species in its grassland habitat makes it much more difficult to census than is the endangered great Indian bustard. Complicating the problem of estimating the species' numbers are its migration and the associated uncertainties as to where birds that breed in the few important remaining nesting habitats might spend their non-breeding periods, as well as determining how to protect these areas.

Generally it would appear that the last major remaining stronghold of the species is the Kathiawar peninsula, in the state of Gujarat. Of secondary importance are the grasslands of western Madhya Pradesh and southern Rajasthan. Probably of only minor importance are the grasslands of northern Uttar Pradesh, along the Nepalese border, and the southern portions of Nepal.

During a 1982 survey of 80 grasslands ('veedis') throughout much of Kathiawar, Magrath *et al.* (1985) found a total of seventy-seven floricans located within the three districts of Jamnagar, Rajkot and Junagadh. An extrapolated collective population estimate of 362 birds for these three districts was obtained, based in part on the assumption of an equal sex ratio. Magrath *et al.* concluded that the species is rare there, and has declined considerably but is probably not in immediate danger of extinction. In 1985 an extensive ecological survey of Gujarat was made by Rahmani *et al.* (1985), who visited the three districts just mentioned as well as Amreli, Bhavnagar, and Kutch. Few specific records resulted from this survey, as it was carried out during the winter season, but it was found that especially in Bhavnagar district some possible florican habitats still exist, such as at Velavadar National Park, where a male was observed displaying in 1984.

In Madhya Pradesh, the western districts of Ratlam, Dhar, Jhabua and Neemuch, and in Rajasthan the southern districts of Kota, Bhilwara, Udaipur, Banswara and Ajmer evidently still support fair numbers of floricans. In Madhya Pradesh there may be about 150 floricans in Ratlam, Dhar and Jhabua districts. No birds were seen during a one-day survey of Neemuch district, but some possibly suitable habitat exists. In Rajasthan, short visits to Kota and Bhilwara districts failed to produce any sightings at Kota, although several birds were seen in Bhilwara district. It is possible that less than 200 floricans existed in these two states in 1984, based on an actual count of forty-seven birds (Haribal *et al.* 1985). A survey made during September 1986 revealed forty-six males in the Ratlam district of Madhya Pradesh, plus forty-five in Dhar district. Additionally, evidence of birds still existing in Rajasthan was obtained from Kota, Bhilwara, Tonk and Ajmer districts (Sankaran and Rahmani 1986).

Some surveys have been made of non-breeding or wintering habitats of floricans in Andhra Pradesh and Karnataka (Lachungpa and Lachungpa 1985), suggesting that the birds still exist in very small numbers in Cuddapah, Anantapur, Kurnool and Medak districts.

Recent sightings of lesser floricans along the southern valleys of Nepal and adjacent Uttar Pradesh have also been made in conjunction with a survey of Bengal floricans, suggesting that this area should perhaps be considered part of the species' breeding range (Inskipp and Inskipp 1983, Goriup and Karpowicz 1985). It is probably mainly a winter visitor to Nepal, but has also been reported present there between May and July (Inskipp and Inskipp 1985b).

In 1983 this species was placed on Schedule 1 of India's Wildlife (Protection) Act, to protect it more effectively. It is now especially important that the few remaining grassland areas of Gujarat be surveyed for floricans and that these be preserved so far as possible as critical areas of habitat for the species.

C · Sandgrouse (Family Pteroclidae)

KEY TO THE GENERA AND SPECIES OF SANDGROUSE

A. Hind toe absent; the rest feathered: *Syrrhaptes*, 2 spp.
  B. Abdomen black; outermost primaries acute: Pallas's sandgrouse
  BB. Abdomen white; outermost primaries normal: Tibetan sandgrouse
AA. Hind toe present, all toes unfeathered: *Pterocles*, 14 spp.
  B. Central tail feathers pin-like and much longer than the others
    C. Undeparts and under wing-coverts white: Pin-tailed sandgrouse
    CC. Underparts and under wing-coverts not white
      D. Lower abdomen lacking black; shaft of outermost primary white; under wing-coverts buffy: Namaqua sandgrouse
      DD. Lower abdomen black or black and rufous, shaft of outermost primary dark, under wing-coverts at least partly blackish

E. Under wing-coverts blackish: Chestnut-bellied sandgrouse
  EE. Under wing-coverts buffy except for blackish primary coverts: Spotted sandgrouse
BB. Tail wedge-shaped to rounded, the central feathers only slightly longer than the others
  C. Flanks and lower underparts all black; the under wing-coverts white or slightly tinted with khaki
    D. Underparts uniformly black; wing-coverts pure white: Black-bellied sandgrouse
    DD. Underparts with white to buffy band anteriorly; wing-coverts tinted with khaki: Black-faced sandgrouse
  CC. Flanks and lower underparts not black; underwing usually buffy to blackish
    D. Under wing-coverts white or pale buff; Palaearctic species
      E. Abdomen buff or spotted with buff, shaft of outermost primary white dorsally: Crowned sandgrouse
      EE. Abdomen narrowly barred with black, shaft of outermost primary black dorsally: Painted sandgrouse
    DD. Under wing-coverts buffy to blackish; mostly Afrotropical
      E. Tarsus barred with blackish or blackish brown; under wing-coverts grey to greyish brown
        F. Greater secondary coverts blackish brown, tipped with white; upperparts distinctly spotted or narrowly edged with white: Double-banded sandgrouse
        FF. Greater secondary coverts with a subterminal black bar; upperparts with few or no white spots or feather edges: Four-banded sandgrouse
      EE. Tarsus uniform rufous to whitish, underwing buff to dark brown
        F. Body finely barred above and below; tarsus white: Lichtenstein's sandgrouse
        FF. Body not finely barred; tarsus buff to rufous-buff
          G. Underwing buff brown to rufous; breast salmon with white spots: Burchell's sandgrouse
          GG. Underwing dark brown; breast not spotted
            H. Under tail-coverts deep chestnut: Yellow-throated sandgrouse
            HH. Under tail-coverts buff: Madagascan sandgrouse

PALLAS'S SANDGROUSE (Plate 37)

Syrrhaptes paradoxus (Pallas) 1773

Other vernacular names: syrrhapte paradoxal (French); Steppenhuhn (German).

Distribution of species (Map 29)

The USSR and China, from the Aral–Caspian region of Kazakhstan south to the lower Amu Darya and Syr Darya valleys of Uzbekistan and the adjoining Kirghiz steppes, and east to the foothills of the Kunlun range of northern Xizang Zizhiqu (Tibet), across north-western China through Zinjiang Zizhiqu (Sinkiang), northern Qinghai (Tsinghai), central Gansu (Kansu), Nei Monggol (Inner Mongolia) and southern and north-eastern Mongolia sporadically east to Heilongjiang (Heilungkiang). Mostly found between 1300 and 3200 m in summer, and often lower in winter. Mainly sedentary but irregularly migratory or irruptive. Sporadic breeding has occurred in Europe following irregular irruptions, including nesting in both Scotland and England (Vaurie 1965, Cramp 1985). No subspecies recognized.

Measurements (mm)

Wing, males 243–259 (av. of 14, 253), females 214–235 (av. of 12, 222); tail, males 165–228 (av. of 11, 190), females 131–160 (av. of 11, 144); tarsus, males 20–23 (av. of 7, 21.6), females 19–23 (av. of 8, 20.9); culmen, males 8.9–11.5 (av. of 15, 10), females 8.8–10.7 (av. of 15, 9.6). Egg, av. 43 × 30 (Cramp 1985).

Weights (g)

Summer, males 255–300 (av. of 8, 274), females 235–270 (av. of 5, 252); autumn, adult males 250–280 (av. of 6, 220), adult females 200–250 (av. of 6, 220) (Cramp 1985); season unstated, males 251–283 (av. of 3, 165.7), females 230–260 (av. of 8, 246) (Cheng 1963). Egg, 21 (estimated).

Description

Adult male. Forehead, superciliaries, cheeks and nape yellowish-grey, shading to dove grey on the nape and sides of the neck; chin whitish, throat and rest of upper face rust or yellowish rust; nape, chest and breast buffy grey, the two latter separated by a scalloped band, each feather white with a narrow black subterminal band; belly black; most of the scapulars with black tips, and some of the median coverts with round black spots near the tip of the outer web; primaries grey, becoming black inwardly, the outermost the longest, with a stiff attentuated tip, and with the outer web black, the rest margined with buff; secondaries buff on the inner webs, darker towards the tips, the outer webs black nearly to their buffy margins; greater primary coverts like the secondaries, but with the black on the outer web extending to the base and a wide buff margin; greater secondary coverts buff, with a dark chestnut patch near the end of the outer web, other coverts buff; tarsi, feathered toes and vent all white, the under

Map. 29. Distribution of Pallas's sand-grouse (pa), including its breeding (hatched), residential (solid) and wintering (stippled) ranges, and the residential range of the Tibetan sandgrouse (ti).

tail-coverts barred black and white; rectrices sixteen (sometimes eighteen), the central pair buff, marked with blackish-grey, and elongated into long black filaments; the rest blackish-grey, toothed on both webs with vinaceous buff and tipped with white; axillaries white, tipped with black; underwing lining buffy to white, the undersides of the remiges also pale silvery white excepting the tips of the secondaries and leading webs of outermost primaries, which are darker. Iris brown; eye-ring dark bluish-grey; bill greyish-blue; sole of foot brown.

Adult female. Like the male, but the feathers of the crown, nape, and ear-coverts with fine black shaft-streaks, the throat and extension of the eye-stripe buff-yellow or yellow, and the throat separated from the chest by a blackish-brown line. No band of black and white feathers divides the chest and breast, the sides of the chest, neck and the smaller wing-coverts are ornamented with small black spots, and the back barring is less regular than in males.

Immatures. Resembling the adult female, but with blackish-brown marks on the neck and chest, which lack the grey and ochre patterning of adults; the black bars on the interscapulars and the spots on the smaller wing-coverts all much less regular; the buff margins and brownish-black inner parts of the remiges and primary coverts indistinct, and the filiform tips of the outermost primaries and central rectrices not developed. The juvenal primaries are browner than in adults and have buffy tips with darker freckling, the young being recognizable as long as some juvenal outer primaries are retained (Cramp 1985).

Identification

In the hand. Both sexes lack hind toes, but have feathered anterior toes, highly attenuated tips on the outermost primaries, and black abdominal patches.

In the field. This pin-tailed species occurs well north of the Himalayas and the range of the Tibetan sandgrouse, and is the only Asian sandgrouse with black flanks and belly colouration that contrasts with the white under tail-coverts and underwing lining. Its calls are said to include rapidly repeated, low-pitched and trisyllabic notes, as well as various disyllabic and monosyllabic notes, often uttered in rapid series and sometimes resembling the trilled choruses of waders. The wings also produce a distinctive whistling sound when the birds are in flight, no doubt caused by the specialized outermost primaries.

General biology and ecology

In contrast to the Tibetan sandgrouse, this is a relatively low-altitude species that is adapted to open steppes and sandy deserts in the more typical manner of sandgrouse, and seems to be especially closely associated with wormwood shrubs (*Artemisia absinthium*) and the spiny xerophytic chenopod *Agriophyllum gobicum* of these steppes. Generally the birds favour flat or hilly grassy or shrub semideserts and steppes having clay, rocky, sandy or stony soil, and avoid drifting sands. Extensive areas of waterless desert are also avoided, and the birds must have sources of surface water available within their daily flying ranges. They also apparently enjoy having locations available for dust-bathing (Vaurie 1965, Cramp 1985).

The bird has been recorded at elevations as high as 2400 m in north-western Mongolia, and to 3250 m in the Soviet Tien Shan range, but more often occurs in mountain valleys or high plains at altitudes of 1300–1900 m, according to Dementiev and Gladkov (1951). On the other hand, Meyer de Schauensee (1984) reported that in the western parts of the species' range it sometimes occurs at elevations of up to 4720–5500 m (elevations as high as those typical of the Tibetan sandgrouse), which seems rather unlikely, especially since the mountains in this part of the species' range are of generally much lower maximum elevations.

Foods are fairly well documented, and consist primarily of seeds, but green shoots are also utilized. The seeds consumed include a very wide range of taxa summarized in Cramp (1985). They include representatives of numerous plant families, such as legumes (Leguminosae), docks (Polygonaceae), pinks (Caryophyllaceae), cranebills (Geraniaceae), dill (Umbelliferae), bugloss (Labiatae), eyebright (Scrupulariaceae), onions (Liliaceae), nettles (Urticaceae), myrtle (Ericaceae), hemp (Moraceae), wormwood (Compositae), stonecrops (Crassulaceae), goosefoots (Chenopodiaceae), grasses (Gramineae), sedges (Cyperaceae), crucifers (Cruciferae), buttercups (Ranunculaceae), and others. Of these many taxa, the nutritious seeds of legumes (which are high in proteins), and the frequently abundant seeds of chenopods and crucifers are perhaps particularly important, along with those of various widespread xerophytes such as *Artemisia* (Compositae). Of nineteen birds collected during summer in Kirghiz SSR, eleven had eaten leaves or seeds of legumes, eight had cruciferous plants represented, six had *Corispermum* (Chenopodiaceae) seeds, and five had seeds of *Lycopus* (Labiatae). Seeds of *Alium* (Liliaceae) and *Triticum* (Graminae) were present in two. A few animal remains, including various insect pupae, have been identified among food samples of this species, but perhaps at least in adults the consumption of these is relatively incidental.

In captivity it has been observed that chicks only 8 h old immediately picked up and consumed a variety of seeds (grass, millet, poppy, etc.). Ant larvae and other soft foods were also offered but were only rarely taken by the chicks (Grummt 1985). By contrast, Wilkinson and Manning (1986) observed that small mealworms were relished by newly hatched chicks, although they were otherwise fed on mash and other non-animal foods, apart from boiled eggs.

Social behaviour

This is a distinctly gregarious species for much of the year outside the breeding period, and apparent non-breeders also sometimes gather in flocks during the breeding season itself. It has been suggested these latter birds perhaps consist of immature yearlings, although at least in captivity, year-old birds have been known to breed (Grummt 1985, Wilkinson and Manning 1986). During migration, flocking tendencies are especially well developed, and migrant groups sometimes number hundreds. Sandgrouse flocks tend to exhibit well synchronized and coordinated aerial movements reminiscent of pigeon flocks. When alarmed, they may circle about, or the flock may split up into smaller units when threatened by raptors. Their daily flights to water are apparently not performed at any highly predictable times. Most often they have been reported as occurring during morning hours starting between 6.00 and 10.00 a.m., perhaps peaking at around 9.00–10.00 a.m., and generally terminating before noon. However, some watering activities may persist into afternoon hours, and an evening watering period may occur around sunset. Typically the birds

Plate 24. Ludwig's bustard, pair. Watercolour by H. Jones

Plate 25. Black-bellied bustard *(Eupodotis melanogaster notophila)*, pair. Watercolour by H. Jones

Plate 26. Hartlaub's bustard, pair. Watercolour by H. Jones

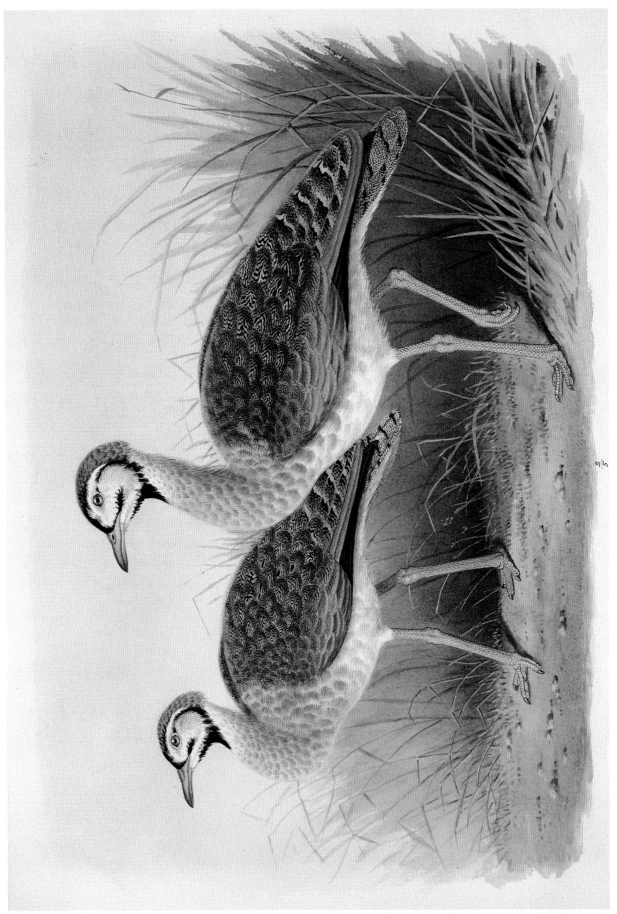

Plate 27. White-bellied bustard *(Eupodotis senegalensis canicollis)*, pair. Watercolour by H. Jones

Plate 28. Little brown bustard, pair. Watercolour by H. Jones

Plate 29. Black-throated bustard *(Eupodotis v. vigorsi)*, pair. Watercolour by H. Jones

Plate 30. Rüppell's bustard *(Eupodotis r. ruppelli)*, pair. Watercolour by H. Jones

Plate 31. Little black bustard *(Eupodotis a. afra)*, pair. Watercolour by H. Jones

Plate 32. Rufous-crested bustard *(Eupodotis r. ruficrista)*, pair. Watercolour by H. Jones

Plate 33. Blue bustard, pair. Watercolour by H. Jones

Plate 34. Bengal florican *(Eupodotis b. bengalensis)*, pair. Watercolour by H. Jones

Plate 35. Lesser florican, pair. Watercolour by H. Jones

Plate 36. Tibetan sandgrouse, pair. Watercolour by H. Jones

Plate 37. Pallas's sandgrouse, pair. Watercolour by H. Jones

Plate 38. Spotted sandgrouse, pair. Watercolour by H. Jones

Plate 39. Pin-tailed sandgrouse *(Pterocles a. alchata)*, pair. Watercolour by H. Jones

Plate 40. Chestnut-bellied sandgrouse *(Pterocles exustus floweri)*, pair. Watercolour by H. Jones

Plate 41. Namaqua sandgrouse, pair. Watercolour by H. Jones

Plate 42. Black-bellied sandgrouse *(Pterocles o. orientalis)*, pair. Watercolour by H. Jones

Plate 43. Crowned sandgrouse *(Pterocles coronatus atratus)*, pair. Watercolour by H. Jones

Plate 44. Madagascan sandgrouse, pair. Watercolour by H. Jones

Plate 45. Black-faced sandgrouse *(Pterocles d. decoratus)*, pair. Watercolour by H. Jones

Plate 46. Lichtenstein's sandgrouse *(Pterocles l. lichtensteinii)*, pair. Watercolour by H. Jones

Plate 47. Double-banded sandgrouse *(Pterocles b. bicinctus)*, pair. Watercolour by H. Jones

Plate 48. Four-banded sandgrouse, pair. Watercolour by H. Jones

Plate 49. Painted sandgrouse, pair. Watercolour by H. Jones

Plate 50. Yellow-throated sandgrouse *(Pterocles g. saturatior)*, pair. Watercolour by H. Jones

Plate 51. Burchell's sandgrouse, pair. Watercolour by H. Jones

cautiously circle the watering area before landing, and may loaf, drink and feed for extended periods at the site (Cramp 1985).

One of the most remarkable features of this species is its tendency to perform irruptive migratory movements over long distances, for reasons still uncertain. Various suggestions as to the causes of these invasions have been proposed, such as sunspots, abnormal displacements during the previous winter resulting from ecological effects such as heavy snowfall or snow-crusting, poor seed production during the prior year, and the like, but none is yet proven. It is most likely that relative food supplies are involved in some way, since local movements within part of the breeding range may precede irregular massive movements by a year or two. During the most spectacular of such movements, flocks have moved into Europe during April and have bred locally for a year or two before disappearing again. The most recent of these major incursions into Europe was in 1908; since then only a few birds have reached beyond the borders of the USSR (Cramp 1985). Similarly, massive movements also extended into northern China and adjacent Manchuria (now Heilongjiang province of northeastern China) in the late 1800s, in 1912–3, and again in 1922–3. Kuleshova *et al.* (1968) describe more recent eastern invasions. Generally these eastern versus western irruptions have occurred asynchronously, suggesting that local ecological phenomena rather than continent-wide conditions probably precipitate them (Dementiev and Gladkov 1951). Bannerman (1959) has provided a detailed discussion of the history of the species' historic occurrences in Britain, as well as some comments on its periodic incursions into Europe in general.

Very little has been written on the courtship behaviour of this species, and there are no descriptions of copulatory behaviour available, even though it has been bred in captivity several times. Loren Grueber (personal communication) has informed me that Pallas's sandgrouse is more gregarious and displays less aggressively around food than does the yellow-throated sandgrouse, and is also considerably more timid in captivity. It also much more conspicuously exhibits a vertical tail-fanning display (Figures 49b, 50b) during courtship and also when disturbed. According to Grueber, one social display used by this species, as well as by the yellow-throated sandgrouse, consists in erecting and fanning the tail while pivoting. He has observed the display more often in Pallas's than in the yellow-throated sandgrouse, and it most often occurs when the birds are in groups. He believes that the display may serve to help identify eligible males to females. The display begins abruptly, with a bird jumping and pivoting while flapping its wings and making a brief call. Sometimes the bird runs a few steps before leaping, and in addition there is some head-jerking as it is bouncing along the ground. A few more general accounts of apparent courtship have also been published, such as the statement that males run around the courted female like pigeons, though neither inflating the neck nor making bowing movements, but instead sometimes flying up and circling the female (Dementiev and Gladkov 1951).

Loren Grueber (personal communication) noted that during aggressive encounters the birds tend to avoid one another by moving sideways, but on occasion do exhibit a more typical sandgrouse frontal threat posture, with the head held low and the body crouched. However, they have also been described as having an aggressive posture in which the forepart of the body is raised, the neck, breast and upper back feathers are ruffled, and the wings are slightly raised (Cramp 1985). This posture may occur when the interacting birds are in close proximity, and raise their heads as high as possible to strike forwards and down with the bill (Loren Grueber, personal communication).

Grueber has also observed that this species seems to exhibit no pair-bonding during winter months (in California), but that when the birds were released in early March into an outside aviary, pair-boding began immediately and was completed in only 2 weeks, compared with more than a month in the yellow-throated sandgrouse. The female of one pair laid its first egg only 10 days after the breeding pairs were placed in separate enclosures. Grueber did not observe any elaborate display during this pair-forming period, other than seeing the male walking in front of the courted female with his head held arched downward. Pair-bonds also are apparently weaker in Pallas's sandgrouse than in the yellow-throated; for example, in Pallas's the paired birds do not sleep in the head-to-tail and side-by-side position as is often the case in the latter species. Furthermore, when an extra female was introduced into the cage where a pair had been established recently and the female had already laid one egg, the male immediately began to court the newly introduced female. However, the new female showed no interest in him, and soon began to chase him aggressively with hopping movements.

Reproductive biology

In the USSR the egg-laying season may begin as early as mid-April in Kazakhstan, and more generally in the Soviet Union and Mongolia extends into the latter half of June (Dementiev and Gladkov 1951, Johansen 1959). Down-covered young have

Fig. 49. Social behaviour of Pallas's sandgrouse: (a) low-intensity crouching by a male; and (b) tail-raising with fanning by a female toward an alert male. After photos by L. Grueber.

been observed as late as August 3. This fairly extended, nearly 3 month, egg-laying period may reflect the laying of supplementary clutches to replace lost ones, possible late laying by birds breeding for the first time, or perhaps even the production of second broods during the same breeding season (Dementiev and Gladkov 1951). However, this last possibility is unproven and seems unlikely, given the limited available time for both nesting and rearing the young in this region.

The nest is situated on the ground, either fully in the open or sometimes sheltered by some vegetation such as a clump of grass or wormwood bushes. Nests by different pairs are sometimes placed as close together as 4–6 m, producing at such times an apparent colonial pattern of nesting, which is unusual if not unique among sandgrouse. Three, or less commonly two, eggs are typically laid. Clutches of three eggs have on at least four occasions been laid within a period of 4 days, and once within a 5-day period by captive birds (Grummt 1985, Wilkinson and Manning 1986). Intervals between successively laid clutches within a single breeding season were reported by Grummt as 9 days on two different occasions, and 8 days on a third, whilst Wilkinson and Manning (1986) reported 4-day and 14-day laying intervals. The latter authors also reported a 25–26 day incubation period, using artificial incubation. Grummt observed a range of incubation periods from 22 to 26 days, with three of six eggs hatching on the twenty-third day. One of these clutches was incubated by the birds themselves. A range of 23–27 days of incubation has also been observed by Etienne Dierick (personal communication), who further observed that (in Belgium) as many as six clutches of two or three eggs may be laid in a single season by captive birds. Similarly, a female in Loren Grueber's aviary laid nine clutches, totalling 23 eggs, in a period of less than three months.

It is generally believed that incubation begins with the laying of the first egg, or shortly thereafter, and that both sexes participate, inasmuch as males with brood-patches have been collected. However,

Fig. 50. Behaviour of Pallas's sandgrouse: (a) normal resting posture of male; and (b) male's tail-fanning display posture toward a female. After photos by L. Grueber.

Grummt (1985) never observed the male of a captive pair to incubate, in spite of repeated checks during evening and nights, when males are believed to relieve the female. At times both during the daylight and night hours the males sat near the female, but only the female was actually observed incubating.

There is no information on development of the young under natural conditions, but Grummt (1985) has provided some information on growth rates as well as photographs of captive-raised young at various ages. Juvenal feathers began to appear on the back at 5 days, and the emerging primaries were then 12 mm long. Within 31 days after hatching, two chicks had increased in weight from 14 to 94 and 95 g, and their wing-lengths had reached 107 and 112 mm (or about half their definitive lengths). Attempts at flying were made by chicks as early as 25 days after hatching, and by 3 months they had fully moulted into their juvenal plumages. By that age the facial markings of a young male raised by Wilkinson and Manning (1986) were already distinct. A female raised by Etienne Dierick (personal communication) laid fertile eggs the following spring, and breeding by yearling birds has also been reported by Grummt as well as by Wilkinson and Manning.

Evolutionary relationships

This species is rather obviously a close and congeneric (Vaurie 1961) relative of *tibetanus*, but has the interesting distinction of having filamentous tips on its outermost primaries in both sexes (see Figure 50a). This feature no doubt accounts for the distinctive humming or whistling wing noise made in flight. This sound carries quite far, and is perhaps an important social signal.

Status and conservation outlook

This species has one of the widest ranges of all sandgrouse, and seems to be able to adapt to short-term ecological changes in breeding conditions by its irruptive migrations or dispersals. It also breeds in many submarginal semidesert areas that, at least at present, have not been seriously affected by development. As such it is probably fairly secure, based on the small amount of currently available information.

TIBETAN SANDGROUSE (Plate 36)

Syrrhaptes tibetanus Gould 1850

Other vernacular names: Mountain sandgrouse, Tibetan three-toed sandgrouse; syrrhapte du Tibet (French); Tibetische Steppenhuhn (German).

Distribution of species (Map 29)

From eastern Soviet Tadzhikistan eastward through western China from Xinjiang Zizhiqu (Sinkiang) to the eastern Kunlun and Altun (Astin Tagh) ranges, also in Qinghai (Tsinghai) to Qinghai Hu (Koko Nor), and north-east to western Nei Monggol (Inner Mongolia), also south through the Tibetan Plateau and Himalayas, west probably to the Karakoram ranges of eastern Afghanistan and Kashmir (Ladakh), and occasionally east to Sikkim. Sedentary, breeding mostly between 4700 and 5500 m, but moving to lower altitudes (up to about 3300–4000 m) in winter (Vaurie 1965, Meyer de Schauensee 1984). No subspecies recognized.

Measurements (mm)

Wing, males 245–265 (av. of 10, 256.9), females 245–255 (av. of 4, 250); tail, males 196–230 (av. of 9, 214.1), females 171–205 (av. of 5, 186.4). Eggs, av. 49.2 × 31.9 (Dementiev and Gladkov 1951).

Weights

No information. Egg, 24.1 g (estimated).

Description

Adult male. Forehead, lores, cheeks and chin white, the feathers with black shafts; sides of head, throat and a band around the nape buffy yellow. Crown black, barred with white; chest, sides and back of neck white, with narrow, wavy black bars, the ground colour becoming vinaceous on the upper back, and the black bars becoming vermiculations; wing-coverts, scapulars, and tertials vinaceous buff, vermiculated with black, the inner webs of the scapulars with large black blotches subterminally, sometimes forming a triangular interscapular patch; lower back, rump and upper tail-coverts whitish, vermiculated with black; primaries, secondaries and greater wing-coverts black, the inner primaries with an extensive terminal whitish buff patch; upper breast greyish white, belly, flanks and legs (including the feathered toes) all white; under tail-coverts chestnut, tipped with white and barred with black; rectrices sixteen, the central pair coloured basally like the upper tail-coverts and extended into long blackish filaments; the rest similar to the under tail-coverts; underwing lining dark greyish-brown, with grey marbling; axillaries black; the underside of the remiges also mostly black, but the trailing edge of the inner primaries silvery white. Iris brown; bill and claws bluish horn; sole of foot brownish.

Adult female. Differs from the male in having the upperparts narrowly and irregularly barred with black, especially on the tertials and axillaries, the upper breast and chest lightly barred with black, and the central pair of rectrices less developed. Bill greyish horn, claws blackish horn.

Immature. Resembling the female, but with only a trace of yellow on the sides of the head, and the barring on the upperparts coarser and more irregular. The secondaries are generally tipped or margined with buffy, and the juvenal outer primaries are reportedly retained, as in *paradoxus* (Dementiev and Gladkov 1951).

Identification

In the hand. Both sexes lack hind toes, and have white feathering extending to the claws, and white underparts.

In the field. This Himalayan species of sandgrouse has distinctive white underparts, a pin-like tail, and blackish underwing linings. Its calls include deep, double-syllable 'guk-guk' or 'caga-caga' notes, said to be more musical than most other sandgrouse.

General biology and ecology

This is a very little-studied species, but it generally is believed to occur on bleak, bare stony plateaux and rocky hillsides, as well as stony river valleys and lacustrine depressions. It also occurs in dune areas and sandy valleys in Tibet. During the summer months it occurs at elevations of around 3615 to 4897 m in the Pamirs (Tadzhikistan), between 4400 and 5700 m in Tibet, and in the Karakoram Range (northern Kashmir and Pakistan) at about 5660–5947 m (Dementiev and Gladkov 1951). It may descend to as low as about 3500–3700 m (Vaurie 1965) or 4000 m (Meyer de Schauensee 1984) during the winter. It often nests on plains at the edges of snow fields, and during winter typically occurs on southern slopes free of snow or only slightly snow-covered.

Outside the breeding season the birds are gregarious, and during early autumn gather in flocks of from as few as five to occasionally as many as 100 or more. Although the species is generally residential, during the latter part of winter small-scale altitudinal movements may be carried out in accordance with snowfall patterns, the birds occasionally appearing on pastures where argali sheep (*Ovis ammon*) have dug in the snow in search of food.

The birds forage mainly on vegetation, especially legumes, including green parts, buds, and seeds. Insect fragments (beetle elytrae) have also been observed among ingested foods (Meinertzhagen 1927). The crop of a bird from the Pamirs contained buds and seeds of *Oxytropis immersa* (Leguminosae), *Smelovskia calycina* (Cruciferae) and *Stellaria hemifusa* (Caryophyllaceae). Unlike most other sandgrouse, it drinks only irregularly, the birds perhaps getting their moisture from snow or green vegetation, and during the winter there are no regular flights in search of water. However, flights to water have been observed at noon and dusk during spring (Dementiev and Gladkov 1951).

Social behaviour

The species is little studied in the wild, and has apparently never been maintained in captivity. According to Dementiev and Gladkov (1951) the pair-bonds may be permanent, although no specific evidence for this seems to exist. The birds are said to call only while in flight, and they are sometimes heard at night.

Reproductive biology

The best descriptions of the nesting habitat of this species come from Baker (1935), based on his observations in the vicinity of Gyangze (Gyantse), southern Tibet. There the birds occur in the hills above the Gyangze Plains, which are at an elevation of about 3650–3800 m, on desert plateaux around high-elevation lakes at about 4275–4900 m. The birds sometimes nest directly on the hard, caked mud of lake shores, or often on the leeward side of a ridge or hill crest. The nest is a scrape among the small stones or earth, without any real lining or direct shelter. The breeding season there is primarily during May and June, but extreme dates for clutches are April 17 and July 24 (Baker 1935).

There are typically three eggs, but occasionally two-egg clutches have been reported. Both sexes are known to incubate, but beyond this very little is known of the reproductive biology. Dementiev and Gladkov (1951) suggested that second clutches are probably laid, or that two clutches may perhaps occur normally, based on the approximate 3-month length of the breeding season (nestlings found from mid-June to mid-August). Replacement laying is much more likely than the production of second broods, given the relatively short time available for breeding in this relatively hostile environment.

Evolutionary relationships

Obviously this species and Pallas's sandgrouse are closely related, although a few differences in proportional lengths of the remiges exist. Additionally, the outermost remix lacks the specialized pointed tips typical of the Pallas's, which perhaps may justify subgeneric recognition (Dementiev and Gladkov 1951) but hardly seems an adequate basis for recognizing a separate monotypic genus.

Status and conservation outlook

Of all the world's sandgrouse, this species is perhaps the most geographically and ecologically isolated and lives in some of the most inhospitable environments imaginable. As such, it is unlikely to be affected by human activities in the foreseeable future.

SPOTTED SANDGROUSE (Plate 38)

Pterocles senegallus (L.) 1771

Other vernacular names: Saharan sandgrouse; ganga tacheté (French); Wüstenflughuhn (German).

Distribution of species (Map 30)

Western Sahara from southern Morocco east across the northern Sahara to Egypt, and south to Somalia, thence across the Arabian Peninsula and through Iraq, southern Iran, southern Afghanistan, and Pakistan to Sind, Kutch and the Thar (Indian) Desert. Mostly sedentary, but with some winter or nomadic movements.

Distribution of subspecies

None recognized by Vaurie (1965). Birds from India are reportedly slightly darker and at times have been separated as *remotus* (Neumann 1934).

Measurements (mm)

Wing, males 203–212 (av. of 5, 206), females 189–203 (av. of 6, 196); tail, males 120–146 (av. of 5, 132), females 99–120 (av. of 9, 109); tarsus, males 21–24 (av. of 3, 22.7), females 21–22 (av. of 3, 21.3);

culmen, males 11.1–13.3 (av. of 6, 12), females 10.7–13.3 (av. of 9, 11.7). Egg, av. 41 × 28 (Cramp 1985).

Weights (g)

Two males 264 and 280, one female 255 (Cramp 1985). Both sexes 250–340 (Urban *et al.* 1986). Egg, 18 (estimated).

Description

Adult male. Crown, back and rump isabelline, shading into yellowish-buff on the upper tail-coverts; a band of pale grey extends from the lores around the back of the head; throat and rest of face ochre, base of throat pale grey; chest, breast, sides of belly and flanks like the back but paler; centre of belly blackish-brown; under tail-coverts white; primaries and primary coverts isabelline, the former shaded inwardly and toward the tips with dark brown, and the inner ones tipped and partly margined inwardly with buff, while the latter have incomplete dark brown shaft-stripes; secondaries brownish-black, edged inwardly with pale buff; scapulars and rest of wing-coverts dull isabelline brown basally and otherwise dull vinaceous grey and grey, the coverts tipped with buff and the scapulars with dull ochre; central pair of rectrices (sixteen) yellowish-buff like the upper tail-coverts and extended into long black filaments, the others isabelline brown, becoming dull black and tipped with white; underwing lining and axillaries whitish-buff to yellowish-buff; the undersides of the remiges medium grey. Iris brown, eye-ring yellow, bill bluish-white to bluish grey, toes bluish-white to pale lead-grey.

Adult female. Differs from the male in having the crown and rest of upperparts nearly uniform pale isabelline, and the head-stripe whitish buff rather than grey; these areas and the chest ornamented with round black spots and lacking the grey tones of the male; the throat is slightly paler yellow, and the central rectrices are isabelline barred with blackish-

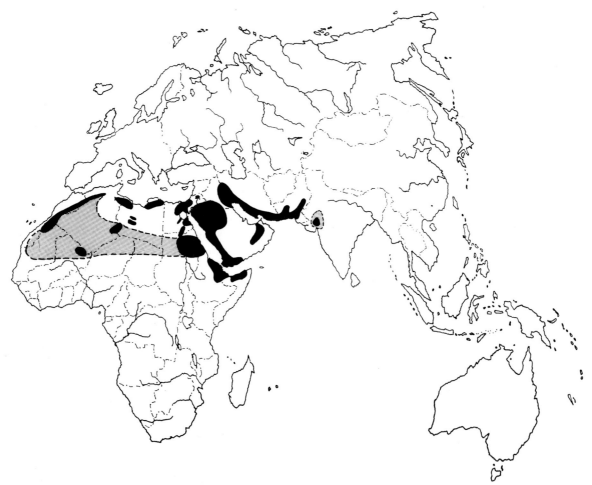

Map. 30. Distribution of the spotted sandgrouse, including its usual residential (solid) and nomadic or irregular (stippled) ranges.

grey and shorter than in males; the rest of the plumage is generally like the male, but the undersides of the remiges are greyish-buff rather than medium grey.

Immature (both sexes). Juveniles lack elongated central rectrices, and have primaries with pale buff outer webs and brown and buff mottling at tips. The definitive plumage, including elongated central rectrices, develops soon after the juvenal plumage has grown, but immatures may be recognized for as long as some of the (outer) juvenal primaries are retained (Cramp 1985).

Identification

In the hand. The combination of a pin-like tail, a belly that is black, surrounded by a fawn-coloured breast and flanks, and an unspotted ochre throat and face identifies this species.

In the field. This is the only pin-tailed species having fawn-coloured flank and breast areas, and similar pale ochre-tinted underwings, contrasting with a black belly. Its flight calls are an extended series of staccato notes, a flock sounding at some distance like the yapping of small dogs, but individual birds utter liquid double notes described as sounding like *cuddle-cuddle, waku-waku,* etc.

General biology and ecology

This is a species of semidesert and desert habitats, being most common in flat, stony and open semidesert habitats, but it also extends into true desert areas that have only isolated patches of vegetation. Stony plains having a light grey to brown background colour are preferred. In contrast to the locally sympatric crowned sandgrouse in Africa, it typically does not occupy completely barren desert, or areas of rocks and stones where wadis are scarce (George 1970, Urban *et al.* 1986). However, it is comparable to that species and to the Namaqua sandgrouse in terms of its ability to utilize extremely dry deserts having only rare and erratic rainfall (Thomas 1984a, b).

In Iraq the species occurs sympatrically with the pin-tailed sandgrouse. There both nest in areas of often totally barren desert without a single plant to be seen, the substrate consisting of hard, baked clay with small sandy or gravelly patches and low, irregular ridges that mark the remains of ancient irrigation canals. In that region winter rains are used locally for supporting submarginal cultivation, and during the time of the bird's breeding season the lower and once wetter areas have become covered either by barley, barley stubble, or low, thorny or prickly xerophytic plants (Marchant 1961). According to Meinertzhagen (1954) the birds never occur near or on arable lands and do not consume any grain, thus differing from both the pin-tailed sandgrouse and the chestnut-bellied sandgrouse. However, Marchant (1961) did find barley seeds in a pair of birds collecting during October in Iraq.

Foods consist of the usual assortment of small, hard seeds from desert plants, with perhaps some small insects and other plant parts also consumed. In Morocco the birds concentrate in wadis where their favourite food plant *Euphorbia quyoniana* grows (George 1969, 1970). They also are known to feed on the liliaceous forb *Asphodelus tenuifolius* in that area (Thomas and Robin 1977). The crucifer *Sisymbrium cinereum* is also apparently a locally favoured food plant in Algeria (Koenig 1896). Of forty-one stomachs of birds collected in the Spanish Sahara, 35 per cent contained food, and in these the seeds of six kinds of plants were the only foods present (Valverde 1957). Similarly, the food contents of more than fifty specimens obtained in Iraq, Sahara, and Arabia consisted entirely of small, hard seeds (Meinertzhagen 1954).

Drinking is done in early morning hours, the birds leaving at daybreak, typically flying in flocks at a considerable speed, estimated to be as much as 100 km h^{-1}. Given the rapid flight, the only major predators of flying birds are probably some of the larger falcons, such as the Barbary falcon (*Falco pelegrinoides* and the lanner (*F. biarmicus*) (George 1970, 1977). Upon arrival at the waterhole considerable circling is done before the birds land near the water's edge. Like the other sandgrouse, they drink by inserting the bill into water to eye level and lifting the head to swallow between draughts. Drinking is usually finished in 15 s or less, after which the birds run back to their original landing site, then take off in flocks to return to foraging areas. During the hottest part of the day, roosting may occur in the shade of shrubs, by both non-breeding birds and those brooding chicks. Very little drinking occurs during the evening hours, except perhaps during extremely hot, dry weather. Birds that forage together during the day may also roost together at night, in assemblages of up to about fifty (George 1969, Urban *et al.* 1986).

Social behaviour

This is certainly a monogamous species, but very little information is available on its social displays associated with pair formation. George (1970) noted that pre-breeding flocks broke up into pairs over a very short period, as brief as a few days, before the dispersal of pairs to breeding areas. In early May he observed an apparent aerial display flight involving six birds that he believed might be five males and a single female. This chase ranged in altitude from ground level to about 100 m and it was not always clear as to which birds were being chased and which were doing the chasing.

Nest-site selection is evidently done as a pair, the female searching out locations where larger pebbles and stones are present, and in one case observed by George (1969, 1970), making as many as five nest-scrapes in an hour, while the male stood guard nearby. George reported that each of the four nests that he found was placed near a conspicuous stone. In contrast, four of the five nests found by Marchant (1961) in Iraq were in natural hoof-marks, probably of camels, on a dried clay substrate. Both observers found evidence to suggest an approximate 24–48 h interval between the laying of successive eggs, and found clutches to number consistently three eggs. The average of twenty clutches from northern Africa was also 3.0 eggs (Urban et al. 1986), although a few two-egg clutches have been reported from that region (Etchécopar and Hüe 1967).

Reproductive biology

Egg records from northern Africa generally extend from March or early April to mid-July, and in Iraq, Marchant (1961) found nests of this species and the pin-tailed sandgrouse during June and July. He believed that the breeding season there probably starts in late April or early May, and lasts until late July or early August. It is possible that double-brooding may occur locally during years of favourable rainfall, in the opinion of D.H. Thomas. There is a general tendency to locate the nest fairly near a waterhole; George (1969, 1970) judged that nesting areas were influenced by the amount of vegetative cover. In one instance twelve nests were located within an area of no more than 1.5 km^2, and in another, five pairs occupied an area of about 2 km^2. Most nests were within 4 km of water, and none was more than 8 km away.

Like the other sandgrouse so far studied, the male normally incubates at night, sitting on the eggs until about 9.30–10.00 a.m. He remains on the nest until the female is alongside him, then steps off and waits for the female to settle down on the eggs before flying away, sometimes making pebble-throwing movements at the nest rim. The female then incubates (Figure 51a) until about 7.00 p.m., when she is again replaced by the male. However, she may take short breaks in the afternoon to feed, at which time she is immediately replaced on the nest by the male. Marchant (1961) believed that the eggs are tended only during daylight (namely, by shading them to keep them cool) until the final egg is laid, but George (1969) reported incubation to begin with the first egg and the total period from the laying of the first egg to the hatching of the first chick to vary from 29 to 31 days (for three nests), or about 27 days on average for all eggs. Midday temperatures of 48°C in the shade and soil temperatures of 68°C were recorded by George (1970) near incubating birds. Both sexes feign injury when disturbed from the nest (Meinertzhagen 1954), the female at least also utters a distraction call (George 1970), and an incubating female was once observed to raise and fan her conspicuously patterned tail when approached by camels, which caused the animals to make a detour around the nest (George 1969). There is also at least one observation of a female removing a hatched eggshell from the nest and dropping it about 20 m away (George 1970).

After hatching, the male alone is responsible for the watering of the chicks. When they are about 3 or 4 days old he takes over brooding from the female shortly after sunrise, and when she has returned from her drinking he immediately flies off to drink himself. On returning, the pair achieve vocal contact by repeated calls, and he lands near the brood. The chicks also react only to the calls of their own parents returning from waterholes, and thus provide clues to their parents as to their location. As with other sandgrouse, the male's abdominal feathers are soaked with water when he returns from drinking (Figure 51b). Then the male stands motionless and erect in a distinctive 'watering pose', his legs widely spread and the body held diagonally upward (Figure 51c). The abdominal feathers are held in such a way as to form a longitudinal groove in the mid-abdomen, into which the chicks insert their beaks to remove the water (George 1969).

The chicks are self-feeding from only a short time after hatching, and when 4–6 days old tend to try to hide among brown stones similar to their own down colour. Later, when they are mostly covered by sand-coloured down, they hide in sandy depressions. Along with their parents they feed mostly during morning and late afternoon hours, roosting in the shade during the hottest part of the day. The fledging period is probably less than a month, since birds about 5 or 6 weeks old are able to accompany their parents on flights to waterholes (George 1969).

There is no information on productivity or mortality rates in this species, although rarely are more than two chicks fledged. Nor is there any definite information on the time required to attain reproductive maturity, which presumably occurs in the first year.

C. Sandgrouse (Family Pteroclidae)

Evolutionary relationships

Urban *et al.* (1986) include this species in a species-group that also includes *coronatus*, *exustus* and probably also *namaqua*. Maclean (1984) included these same four species in a single, mainly Palaearctic, pin-tailed assemblage. I have tentatively (see Figure 3) placed it nearest *alchata* within this general group, although in adult plumage traits it might well be most closely associated with *exustus*.

Status and conservation outlook

This is one of the most arid-adapted of all sandgrouse, and at least in northern Africa it is generally frequent to abundant over much of its range, especially in Egypt, Sudan, Ethiopia, and northern Somalia (Urban *et al.* 1986).

PIN-TAILED SANDGROUSE (Plate 39)

Pterocles alchata (L.) 1766

Other vernacular names: Large pin-tailed sandgrouse, pintail sandgrouse; ganga cata (French); Spiessflughuhn (German).

Distribution of species (Map 31)

Spain, southern France (Rhone Valley), northwestern Africa (Western Sahara to Libya), and from Israel, Jordan, Syria and south-eastern Turkey east

Fig. 51. Parental behaviour of spotted sandgrouse: (a) female incubating near rocky cover (b) male with soaked feathers about to water a chick; (c) chicks drinking from male's soaked belly feathers; and (d) male standing in water, about to soak feathers. After photos by George (1977) and Maclean (1968).

disjunctively to Iran, Afghanistan, and Soviet Central Asia (from western Kazakhstan east to the Aral Sea and the vicinity of Lake Balkhash), locally or accidentally south to Afghanistan and Pakistan. Resident, nomadic and locally migratory, wintering south to Pakistan and north-western India.

Distribution of subspecies

P. a. alchata (L.): endemic to Spain and southern France (La Crau); formerly also Portugal. Winters on breeding range and irregularly outside it.

P. a. caudacutus (Gmelin) 1774: Africa (Western Sahara to Libya), and from Israel, Jordan, Syria and probably southern Turkey east locally to southern USSR (Turkmenia, Uzbekistan and Kazakhstan). Winters mostly in breeding range, but locally migratory, especially in breeding population of USSR, where wintering regularly occurs south to Pakistan and adjacent India.

Measurements (of *alchata***, mm)**

Wing, males 205–214 (av. of 10, 209), females 201–211 (av. of 7, 205); tail, males 147–173 (av. of 8, 157), females 129–146 (av. of 7, 138); tarsus, males 27–30 (av. of 9, 27.9), females 25–27 (av. of 7, 26); culmen, males 13.1–14.3 (av. of 9, 13.6), females 13–14.5 (av. of 7, 13.8). Egg (*alchata*), av. 47 × 31; (*caudacutus*), av. 45 × 31 (Cramp 1985).

Weights (g)

Males (*caudacutus*), 250–408 (av. of 48, 315), females 207–274 (av. of 23, 302) (Urban *et al.* 1986). Egg (*alchata*), 25; (*caudacutus*), 24 (both estimated).

Description

Adult male, breeding plumage. Crown greyish-ochre, chin and throat black or brownish-black, as is

Map. 31. Distribution of the pin-tailed sandgrouse, including its residential (solid) and non–breeding (stippled) ranges, and the locations of its races *alchata* (al) and *caudacutus* (ca).

a postocular line extending from the eye backwards; rest of head bright rufous buff, shading to ochre on the neck; upper back and scapulars dull olive-ochre, the feathers margined with greyish, and most of the scapulars with a subterminal bright ochre patch (in early post-nuptial moult these feathers are barred with black and yellowish-buff, and in late post-nuptial moult they are plain dark greenish-grey); lower back, rump and upper tail-coverts yellowish-buff barred with black, chest pale rufous, separated from both the neck above and the white breast and underparts below by two narrow black lines; primaries and greater primary coverts grey, the outer web of the tenth primary and the tips of the inner webs of the five inner primaries blackish; the latter are also margined with white terminally; secondaries white, the distal half of the outer web and the last third of the inner web black, margined with white; the inner greater, median and all lesser wing-coverts white, with most of the outer web and, in the smaller coverts, the tips of the inner web bright chestnut, with a narrow submarginal white or yellow and black band. Outer greater and median coverts with yellowish-buff to white tips, margined with black; rectrices (sixteen) blackish-grey, mixed with ochre and barred with pale rufous; the central pair extended into long blackish filaments, and the rest tipped with white. Axillaries white; underwing lining mostly white to buffy proximally, but the primary coverts medium grey, as is the undersurface of most remiges excepting the secondaries, which are white basally. Iris brown, eye-ring blue-grey, bill dull brown or slate, toes greyish-yellow to dusky green.

Adult male, non-breeding plumage. The crown, hindneck and sides of face becoming barred or spotted with dark brown; the centre of chin and upper throat becoming white, and the black eye-streak reduced or absent; back, tertials and scapulars distinctively barred with buff and black in a 'zebra' pattern. During the latter part of the post-nuptial moult in autumn an early version of the breeding plumage is gradually developed, with the crown, sides of head and chin, throat area becoming transitional toward the breeding plumage, and the barred mantle and scapular feathers are replaced by dark grey-green feathers without yellow spots. Lastly, the distinctive mantle and scapular feathers with large yellow-ochre spots and dark margins appear, presumably as the result of a third moult.

Adult female. Crown, nape, back, rump and upper tail-coverts yellowish-buff barred with black; chin and throat white to buffy; rest of the face as in the adult male but paler (the dark post-ocular stripe disappearing in the post-nuptial moult, the sides of the head becoming spotted, and the pale silvery-grey bars on the sides also disappearing); the front and sides of the neck covered by a wide black band edged with buff, followed by a more greyish band tipped with black (thus producing a double black upper breastband, rather than one narrow band as in males); chest and rest of the underparts similar to the male, but the chest paler; scapulars, outer secondary and median wing-coverts barred like the dorsal feathers, but the scapulars are tipped with bands of slate-grey followed by tawny yellow edged with black, and the subterminal band on the wing-coverts is white instead of slate; inner secondary and median coverts white (*caudacutus*) to yellow (*alchata*), margined with black along the outer web, the other median and smaller wing-coverts white to pale rufous with black tips. Soft-parts as in the male.

Immature male. Many feathers of the crown and some of the scapular and back feathers yellowish-buff, barred with black, the black throat-patch often with some white feathers, and the breast patterning often imperfect. Central tail feathers not as fully elongated as in adults, and some outer primaries not replaced. By the spring of the second calendar year the pre-nuptial moult occurs as in adults, and the primary moult begins again, proceeding distally, after having been suspended the previous autumn (Cramp 1985).

Immature female. Scapulars banded subterminally with black rather than grey, and tipped with pale buff; the black band on the upper part of the neck represented by a subterminal spot at the end of each feather, and the sides of the face and neck also spotted. Like non-breeding adults, the sexes of immatures can be distinguished by their upper greater wing-covert patterns, females having golden to bay-brown outer webs, often margined with black, rather than having deep brown to chestnut outer webs, with a yellow line along the margin. Juveniles of both sexes carry worn and pointed outermost juvenal primaries until the summer of their second calendar year (Cramp 1985).

Identification

In the hand. The combination of a pin-like tail, white underparts, and a chestnut breast terminated above and below by narrow black bands identifies both sexes at all seasons.

In the field. Over this species' range it is the only sandgrouse having the combination of pure white under wing-coverts and a white belly, contrasting in flight with a blackish undersurface of the flight feathers. The crowned sandgrouse is similar but lacks a pin-like tail, and is pale buffy rather than white below.

General biology and ecology

This is certainly the best-studied species of sandgrouse, owing mainly to the fact that it is one of only two species (the other being the black-bellied) that regularly breed in Europe, specifically in Spain and France. There these two species overlap in some habitats, such as on stony plateaux and plains as well as on relatively infertile soils used for dryland cultivation. In general the pin-tailed sandgrouse is somewhat more attracted to sand and less attracted to or dependent upon stony pavement or any vegetational cover such as grasses than is the black-bellied sandgrouse. Lowland plains, stony areas on the edges of deserts, expanses of alternating sand and bare clay, and arid flats or sand dunes are all commonly exploited for breeding by *alchata*. On the other hand, it sometimes also breeds near irrigation ditches or around marshes where dried mudflats occur, and where at times there is a good grassy or bushy vegetational cover (Dementiev and Gladkov 1951, Ferguson-Lees 1969, Cramp 1985).

Like the other sandgrouse, this species breeds within daily flying distance of a source of water, but in the absence of fresh water it will drink brackish water, or even alight on deeper waters such as rivers and swim gull-like on their surface. Although in south-western Europe and northern Africa it is not noticeably migratory but may be somewhat nomadic, the temperate-breeding Middle Eastern and USSR populations are distinctly migratory. Especially in the latter region, the post-breeding birds tend to move out of nesting areas during October and November, although some local overwintering may occur in mild winters. These birds return again in March and April, presumably after having wintered in the Arabian and Indian deserts, although in the Middle East the birds sometimes flock into river valleys during winter (Cramp 1985). The birds are quite resistant to cold temperatures, and whereas captive double-banded sandgrouse commonly huddle together during cold weather, this species can tolerate temperatures as low as $-18°C$ without signs of discomfort, nor do they seem to mind being exposed to snow or wind (Frisch 1970). Meade-Waldo (1897) also commented on the cold-tolerance of the species, having seen them 'dusting comfortably in 25 degrees of frost'.

Flock sizes numbering several thousands have been reported for this species, and during the winner, estimates of as many as 50 000 birds have been made at Turkish waterholes. However, migrating flocks are much smaller, often numbering dozens or perhaps at most 150. Within these flocks, paired birds are often evident, suggesting that pair-bonds may be permanent (Cheylan 1975). It is generally believed and probably true that sexual maturity occurs during the first year of life (Dementiev and Gladkov 1951), even though Frisch (1970) observed that wild-caught birds kept in captivity did not breed in their first summer and also did not develop their full breeding plumages until their second year.

Like the other sandgrouse, these birds are almost exclusively vegetarians, as least as adults, nearly all studies suggesting that insects or other animal materials are taken only incidentally. One possible exception to this generality comes from the work of Guichard (1961), who observed that about half of the materials he obtained from faecal pellets consisted of chitinous parts of beetles and other small insects. However, such pellets probably provide a highly biased estimate of food intake, as they would greatly exaggerate the percentage content of undigestible food components and correspondingly underestimate readily digestible items such as soft plant parts. Thus seeds, with smaller quantities of shoots and leafy materials, are probably the primary foods of wild birds (Ferguson-Lees 1969, Cheylan 1975).

In Spain a total of forty-one plant species have been identified as foods of the pin-tailed sandgrouse (Casado *et al.* 1983). More inclusively, Cramp (1985) has summarized the available literature on plants that have been reported as foods, which include the seeds of forbs, grasses, and other native plants, as well as grain crops such as wheat and barley. Seeds, shoots or leaves of plants that have been documented as foods include those of asphodel (*Asphodelus*), camel thorn (*Alhagi*), pulse (*Phaseolus*), clover (*Trifolium*) and various other legumes (Leguminosae), several taxa of Polygonaceae such as *Calligonum*, *Fagopyrum*, and *Polygonum*, and miscellaneous xerophytes such as *Salicornia* (Chenopodiacae), *Artemisia* (Compositae), and *Mesembryanthemum* (Aizoaceae). The high incidence of legume seeds among the food plants presumably reflects their high protein levels, whereas the leaves or other soft parts of some plants such as the succulent *Mesembryanthemum* and the xerophytic *Salsola* (Chenopodiaceae) may be eaten for their water or salt contents (Thomas 1984). Many of the other plants listed above are also highly xerophytic, often producing numerous but very small and hard seeds. Perhaps as a result, grit is commonly ingested and may constitute as much as a third of the stomach contents by weight.

Foods of the chicks are probably similar to those of adults, and include many small seeds (Ferguson-Lees 1969). Although chicks are able to drink water independently, they do so only when an insufficient amount of water is provided by the father (Frisch 1970). Frisch never observed the adults or young to pick up any animal foods; instead they consumed a mixture of small seeds and greens.

Social behaviour

As noted earlier, it has been suggested that pair-bonding may be permanent in this species, based on the observed presence of apparent pairs in non-breeding flocks. As in the other sandgrouse, pair-bonding is certainly monogamous, with both sexes tending the young until they fledge. The timing and mechanism of pair formation is still only very poorly understood, even in this relatively well-studied species of sandgrouse. Dementiev and Gladkov (1951) reported that in the Aral Sea region of Kazakhstan and Uzbekistan pair formation occurs during late April, as the migrating flocks are breaking up. Frisch (1970) observed that paired females would chase one another, with stretched-forward necks and 'gaggery' calls, when one approached another closer than about a metre. An extra female was also chased away, but she continued to maintain a limited degree of social contact with one of the pairs. Frisch also noted that the male followed the female for a few metres just before copulation, using a stiff-legged gait and a depressed and fanned-out tail, after which he jumped upon her and copulated. No special prior receptive display by the female was observed. Tail-fanning, but with the rectrices raised rather than lowered, has been observed in frightened or threatened birds.

Reproductive biology

The breeding season of this widely distributed species is probably somewhat variable from year to year and by geographic region. In southern France, the Middle East and northern Africa, clutches are usually found from mid-April onwards. The normal breeding peak is from the second week of May to late June, with some egg-laying continuing on into August (Ferguson-Lees 1969, Cramp 1985). Baker (1935) stated that in Mesopotamia (Iraq) breeding occurs from early May until mid-July, with most laying occurring during the last week of May and the first two weeks of June. In the USSR the season is apparently initiated somewhat later and may be more constricted than in Europe, with the first eggs appearing in late April, and complete clutches present by the first half of May. Clutches have also been reported as late as July 28, and unfledged birds seen as late as August 9 (Dementiev and Gladkov 1951). Most eggs in northern Africa are laid towards the end of the rainy season, between late April and June. It is possible that double-brooding sometimes occurs there (Urban et al. 1986).

Although many authors such as Baker (1935) indicate that the birds actively scrape out a nest-site, Marchant (1961) reported that fifteen out of twenty-three nests observed by him were in natural hoof-marks that had not been altered by the birds. In the rest of the nests the ground might have been slightly scratched, but in no instance was any vegetation added. Ferguson-Lees (1969) similarly found no lining materials in the nest that he observed. On the other hand, Frisch (1969a) observed scraping behaviour by the male, and later (1970) reported that the male had led the female to a specific site with a call that sounded to him like the distant calling of a black grouse (*Tetrao tetrix*). Both sexes then scraped the site with their feet, although in a manner quite unlike that of domestic fowls scratching for food. He further noted that the birds picked at the nest scrape with their beaks, and placed stones, feathers and droppings in the nest.

About 11 days after the birds studied by Frisch had begun their nest scrape, the female laid her first egg. Six days later, a clutch of three eggs had been laid, and the male began sitting on the eggs within half an hour of the laying of the last egg. Later clutches of this pair were also of three eggs, which were laid 2 days apart, and always in the late afternoon. Similarly, Marchant (1961) reported a probable average egg-laying interval of 48 h. He also observed incubation to begin with the laying of the last egg.

A clutch size of three eggs is evidently typical of the species; Marchant (1963) found this to be true in all but one of twenty-four nests, the exception being a two-egg clutch. Others such as Baker (1935) and Meinertzhagen (1954) have also described two-egg clutches, but quite possibly these represent incomplete sets.

As with most sandgrouse, incubation is performed by the male at night and by the female during the day (Meade-Waldo 1897, Marchant 1961, Frisch 1970). Frisch stated that his captive female typically incubated from 6 a.m. until 8 p.m., with day-to-day variations of half an hour. Marchant (1961) observed that among wild birds in Iraq incubation was virtually continuous, with the male arriving at about 6 p.m. (range in three cases 5.50 to 6.10), and settling on the nests from 3 to 15 min after the female had left it. Of four observed morning exchanges, the female settled on the eggs at times ranging from 8.08 to 8.37 a.m., usually immediately replacing the male at the nest. Except for these short changeover periods, the eggs were apparently never left uncovered, other than during disturbances caused by humans or animals. During incubation the birds sit facing into the wind; injury-feigning by disturbed incubating birds is also typical, at least of females (Meinertzhagen 1954, Ferguson-Lees 1969).

The incubation period among wild birds has been variably reported as from a minimum of 19–20 days (Marchant 1961) to 23 days (Guichard 1961). Similarly, in captivity a 21–23 day incubation period was reported by Meade-Waldo (1806), and 21–21½ day periods were reported by Frisch (1970). The eggs typically begin to pip on the same day, but hatching occurs over a prolonged period of about 48 h. Like shorebirds, the parents carry the empty eggshells away from the nest (Ferguson-Lees 1969). At the time of hatching, Frisch observed the male rubbing his belly on the ground with is abdominal feathers spread apart. Frisch believed this might have been a mistaken act of an inexperienced male, which had probably 'intended' to wet himself in the water.

There is no proof of second broods being produced in a single breeding season, but it is almost certain that replacement clutches must be fairly regularly produced, if not actually second broods. Each of two captive pairs kept by Frisch (1970) produced three clutches in a single season. One of these pairs laid the first egg of a second clutch 7 days after the death of their only chick, and later started a third clutch only one day after the single 3-week-old chick of their second brood died. This would suggest a remarkable ability of females to recycle back rapidly into laying condition.

Frisch found that as soon as the chicks hatched, the female kept them warm, including nights. The male performed the same brooding duties away from the nest. The young initially left the nest 24 h after hatching but soon returned to be brooded by the female. The pair guided the chicks away from the nest for increasingly longer periods thereafter, but returned to the site at night or during rains. When following their parents, the chicks followed closely, either remaining next to the parent's body or under the tail. The male always ran ahead and warned of possible danger, at which time the chicks would squeeze close to their mother. During times when brooding was unnecessary, the female picked up food items almost continuously, which the chicks then tried to pick up and swallow. However, young chicks would eat sufficiently only after such 'pecking demonstrations', and a single hand-raised chick might starve in spite of an abundant food supply if not taught to peck.

Almost as soon as the first chick has hatched, the male begins his watering behaviour. This typically consists of flying some distance to water and returning with it in his breast and belly feathers to allow the young to drink (Marchant 1961, 1962). During mornings and evenings the captive male observed by Frisch (1970) would go to water, sit for a few seconds in the water with his abdominal feathers widely spread, and then run to the chicks and stop beside them. Likewise as soon as the chicks saw their father in his water-carrying pose they would run to him and stick their beaks into a groove formed by his wet belly feathers. This occurred very regularly during both morning and evening periods, and when the chicks were thirsty they would seemingly 'ask' the male to drink by making searching movements in his belly region (Frisch 1970). However, Frisch observed that day-old chicks were able to drink from a dish when necessary (i.e. in the absence of parental watering), and he also observed independent drinking by 16-day-old chicks. He believed that precocious drinking behaviour by very young chicks occurs only when insufficient water is provided by the father.

The chicks become fairly independent rapidly; Marchant (1961) noted birds picking up food for themselves while certainly still less than a week old, and when only a third to a half grown they were able to fly quite well. In captivity they have been observed refusing to be brooded and even roosting separately when only 10 days old (Meade-Waldo 1897). The fledging period has been estimated at about 4 weeks, and by about 4 months the elongated central rectrices typical of adults have grown in (Urban *et al.* 1986).

Evolutionary relationships

In his review of the sandgrouse, Maclean (1984) included *alchata* in a group of mainly Palaearctic species all of which have long central rectries, and additionally include *exustus*, *senegallus* and the African species *namaqua*. These same species made up the genus *Pteroclurus* as recognized by Ogilvie-Grant (1893, 1896–7). Of these three, *alchata* is perhaps most closely related to *exustus*, but *alchata* is unusual among this entire group in that the female is almost as brightly patterned as the male, and is the only species of *Pterocles* in which both sexes have pure white underparts and the male has a well-developed and distinctive non-breeding plumage. Urban *et al.* (1986) did not include *alchata* in any larger species group, but instead considered it as being of 'independent' origin.

Status and conservation outlook

The somewhat nomadic (in Africa) to highly migratory (in Asia) nature of this species makes its status difficult to ascertain, a problem which is further complicated by the fact that it typically frequents areas of low accessibility to humans.

However, both of these traits are certainly to its long-term survival advantage.

The species' status in Europe, however, is better known and requires some comment. Hudson (1975) considered it to be 'local and decreasing', with its precise status in Iberia uncertain. He estimated from Cheylan's (1975) work that about 120 breeding pairs then existed in southern France's La Crau (Rhone Valley), and further judged that about 350–400 birds were present during winter, when the population is supplemented with juveniles. Although Cramp (1985) was unable to provide any more recent population data for France, he considered the species to be 'not scarce' in Spain but extirpated as a breeder from Portugal.

CHESTNUT-BELLIED SANDGROUSE
(Plate 40)

Pterocles exustus Temminck and Laugier 1825

Other vernacular names: common Indian sandgrouse, common sandgrouse, Indian sandgrouse (*hindustan*), Kenya pin-tailed sandgrouse (*olivascens*), lesser pin-tailed sandgrouse, singed sandgrouse (*erlangeri*), small pin-tailed sandgrouse, Somaliland pin-tailed sandgrouse (*olivascens*); ganga à ventre brun (French); Braunbauchflughuhn (German).

Distribution of species (Map 32)

Africa, in the sub-Saharan Sahel zone from Senegal and Mauritania east to Sudan, Ethiopia, and Somalia, and south to Kenya and northern Tanzania; also coastally along the southern and western Arabian peninsula and from southern Iran and southern Pakistan east probably to north-western India as a regular breeder, and to eastern and southern India (West Bengal and Tamil Nadu) irregularly or in winter. Previously also in the Nile Valley, Egypt. Mostly sedentary, but locally nomadic or migratory in parts of Africa, Pakistan and India. Introduced unsuccessfully into Australia and Nevada. Also introduced on Hawaii, where still only locally surviving in the South Kohala District near Waimea.

Distribution of subspecies

P. e. exustus Temminck and Laugier 1825: Mauritania east to Sudan.

P. e. olivascens Hartert 1909: south-eastern Ethiopia, Somalia, Kenya and Tanzania.

P. e. floweri Nicoll 1921: formerly endemic to Egypt (Faiyum to Luxor); now probably extinct.

P. e. ellioti Bodganow 1881: south-eastern Sudan east to northern Ethiopia and Somalia.

P. e. erlangeri (Neumann) 1909: western and southern Arabian peninsula.

P. e. hindustan Meinertzhagen 1923: south-eastern Iran, Pakistan and north-western India. Introduced locally into Hawaii during the 1960s (Paton *et al.* 1982). Considered part of *erlangeri* by Baker (1928) and by Ali and Ripley (1983).

Measurements (all races, mm)

Wing, males 175–190 (av. of 15, 182), females 160–184 (av. of 11, 175); tail, males 112–136 (av. of 15, 124), females 84–119 (av. of 13, 99.8); tarsus, males 22–24 (av. of 14, 23), females 20–23 (av. of 8, 21.8); culmen, males 10.8–13.4 (av. of 14, 11.9), females 10.3–13.8 (av. of 11, 11.5). Egg (*exustus*), av. 36 × 25 (Cramp 1985).

Weights (g)

Males of *hindustan*, 184–284, females 170–241 (Cramp 1985). Males of *ellioti*, 170–220 (av. of 18, 183), females 140–213 (av. of 15, 171); males of *exustus*, 174–295 (av. of 72, 202), females 170–225 (av. of 64, 191) (Urban *et al.* 1986). Egg, 13 (estimated).

Description

Adult male. Lores, cheeks, throat and nape dull ochre, shading into vinaceous buff on the chest; crown, upper back, rump and upper tail-coverts isabelline brown; scapulars and outer secondaries darker, shading to buff at their tips, and terminated by a pale brown bar; lesser and median wing-coverts yellowish buff, tipped with a rich brown band and sometimes also a subterminal white spot; primaries, primary coverts and inner secondaries black, the five inner primaries tipped with white; a narrow black band edged with white separates the chest from the yellowish-buff upper breast, which shades into deep chestnut-brown on the lower breast, flanks and belly; tarsi and under tail-coverts whitish-buff; centre pair of rectrices coloured like the scapulars but extended into long black filaments, the other fourteen dark brown tipped with buff; underwing lining and axillaries dark brown to greyish-brown or blackish, the undersurface of the remiges similarly greyish-brown except the tips of the secondaries and inner primaries, which are white. Iris dark brown, eye-ring yellow to greenish-yellow; bill pale blue to pale grey, toes pale yellow-grey to pale blue-grey.

Adult female. Differs from the male in having a broad blackish-brown bar down the middle of each feather on the top of the neck, neck, upper back and chest, and in the two latter areas the tip of this bar is swollen, producing a spotted appearance; back,

Map. 32. Distribution of the chestnut-bellied sandgrouse, including the locations of its races *ellioti* (el), *erlangeri* (er), *exustus* (ex), *floweri* (fl), *hindustan* (hi) and *olivascens* (ol). The distribution of *floweri* is the former range; current evidence indicates that it is now possibly extirpated. It is also questionable that *hindustan* breeds throughout peninsular India.

rump and upper tail-coverts vinaceous or yellowish-buff, thickly barred with black. Scapulars, lesser and median wing-coverts the same, but with yellowish-buff ends tipped with brown; upper breast whitish or yellowish-buff; belly and flanks blackish-brown, closely barred with rufous buff; rectrices coloured like the back feathers, the central pair extended into filaments, and the rest tipped with yellowish-white.

Immature male. Upperparts like adult, but ochre-grey feathers subterminally vermiculated with dark brown and pale buff marginally, breast lacking dark band, and upperparts with narrower barring. The central juvenal rectrices are not elongated. The outermost juvenal primaries may be retained longer than the rest of the juvenal plumage; these are different in colour and shape from the inner ones (Cramp 1985).

Immature female. Differs from the adult in having narrower barring on the upperparts, the outer primaries and the inner secondaries tipped with buff, the former vermiculated with black, the upper breast spotted with blackish-brown, and the central rectrices not elongated. The lower mantle and scapular feathers are tipped with pale ochre, and the outermost juvenal primaries are sometimes retained and recognizably different from adult feathers (Cramp 1985).

Identification

In the hand. The combination of a pin-like tail, a belly that is deep chestnut brown (males) or barred rufous and blackish-brown (females), and blackish-brown under wing-coverts identifies adults of both sexes.

In the field. The dark chestnut flanks, becoming blackish on the belly, and uniformly dark underside of the entire wing (excepting the trailing edge), identify this species. No other pin-tailed sandgrouse is so uniformly dark underneath. It often utters a series of double gurgling or chuckling *gutta* notes as a flight-call or contact-call.

General biology and ecology

This is an extremely arid-adapted species, adapted to desert and semi-desert habitats of the tropics and arid subtropics, and reaching elevations of about 1500 m or higher. It is primarily associated with level or sloping tracts of steppes and semi-deserts, especially those with silty or dusty soils, or with stony or rocky surfaces, but generally not sandy areas (Cramp 1985).

In north-western India it occurs in areas with annual precipitation ranging from about 12.5 to 35 cm, where the principal vegetation comprises scattered thorny shrubs (*Acacia*, *Ziziphus* and *Capparis*), with an understorey of grasses and weeds. The presence of leguminous forbs and their associated seed crops may be especially important, although in areas of dry-land farming (millet, pulse, wheat) the birds are often found in association with cultivated fields. In north-central India, where precipitation is considerably greater, the birds are more scattered, and they are largely associated with eroded areas of benchlands at the foot of low, rolling hills, where the vegetation is likewise mostly thorny brush with an understorey of grasses and forbs. The species commonly breeds in areas with summer temperatures reaching maxima of about 48°C, although in Baluchistan it occurs at higher elevations (usually about 1220 m) and associated cooler temperatures (Christensen *et al.* 1964).

In Africa this species also inhabits bare semi-deserts and arid scrublands, as well as cultivated fields and grasslands (Urban *et al.* 1986), but where it occurs sympatrically with *lichtensteinii* in Somalia it is found in open country habitats devoid of the trees and shrubs characteristic of *lichtensteinii* habitats (Archer and Godman 1937).

Apparently throughout its range it feeds almost exclusively on vegetable matter, primarily seeds. A study of forty-seven crops from Pakistan and north-western India included seeds of at least nineteen plant taxa, with those of various legumes (*Indigofera*, *Tephrosia*) especially well represented. Seeds from native leguminous plants were present in crops collected during all seasons, whilst those of cultivated plants such as pulse (*Phaseolus*) were taken only in small numbers. Grit was present in nearly half the crops, but no animal materials were found. Many of the legume seeds taken are extremely small; as many as 10 000 *Indigofera* seeds were present in one crop, and another contained 5600 seeds of this taxon (Faruqi *et al.* 1960).

Similarly, in Africa hard seeds of various forbs and grains are commonly eaten, including especially such legume genera as *Indigofera*, *Trianthema*, and *Heliotropium* (Kalchreuter 1980); young shoots and insects such as ants are also sometimes consumed (Urban *et al.* 1986). In a more general range-wide summary (Cramp 1985), the seeds of various other plants were mentioned as foods of this species, the identified taxa reportedly including a variety of genera of grasses (*Panicum*), sedges (*Cyperus*), xerophytic forbs or succulents (*Amaranthus*, *Crotalaria*, *Desmodium*, *Euphorbia*, etc.) and some woody plants (*Acacia*).

The birds forage both before and after drinking, namely very early in the morning and again, in a more desultory manner and usually at a different site, later on in the day. Foraging may also occur after midday roosting. The birds reportedly favour feeding in areas where antelopes have been lying, by scratching in dry dung. Much of their food is obtained by picking it up from the ground, sometimes scratching partridge-like, but they may also pluck parts off growing plants (Cramp 1985).

Drinking is done in early morning hours after sunrise, the birds sometimes flying as far as 16 km to sources of water. Watering is done every day, the birds usually starting to gather in the vicinity of waterholes about an hour after sunrise. About 2–3 h after sunrise the gathered birds rise and circle over the waterhole and, if undisturbed, will land near the shore and walk to the edge of the water. They apparently only very occasionally return at sunset for watering (Christensen *et al.* 1964). Similarly, in Africa the major watering period is 2–3 h after sunrise, averaging earlier on hot days and later on cloudy days, with some birds also drinking shortly before sunset during very hot weather (Urban *et al.* 1986).

The sizes of watering flocks are highly variable, probably depending on the relative availability of surface water, but extremes of from as few as five to as many as 50 000 birds have been reported. Such large flocks have been observed, for example, in the Sahel zone of north-central Mali between December and June. The ecological significance of such massed drinking flights is uncertain, but Ward (1972) has suggested that such flights may serve as 'information centres', being used by hungry birds to help them locate feeding areas by simply following others which, after drinking their fill, are heading back to favoured sources of food.

Roosting is done on the ground in compact groups; apparently the roosting sites are preferentially in open country with a minimum of cover.

Daytime roosting may also occur during the hottest part of the day, in shaded sites such as under shrubs (Cramp 1985).

Social behaviour

There is no direct information on the pair-bonding system or the duration of pair-bonds, but there is no reason to believe that monogamy is not the typical pattern. It has been reported that breeding occurs within a year after hatching, and that the adult male:female ratio approximates equality, and the probability of second broods in a single year is 'strongly indicated' (Christensen et al. 1964). Biparental care of the young extends at least to the time of the chicks' fledging, and perhaps somewhat beyond (Aldrich 1943). All these observations would suggest that prolonged monogamy is likely.

In spite of the fact that the species has bred in captivity on various occasions (e.g. St Quintin 1905; Meade-Waldo 1922, Meinertzhagen 1954), no descriptions of courtship behaviour have appeared.

Reproductive biology

The breeding season of this species in the Thar (Indian) Desert region is primarily from February to April, with a probable second period of nesting in November, December and possibly January (Christensen et al. 1964). More generally in the Indian region it probably breeds throughout the year, with the possible exception of August. In the southern half of its Indian range the largest number of eggs are laid between January and April, whilst in the northern regions most of the eggs are laid between March and May (Baker 1935).

In the desert regions of the Arabian peninsula and eastern Africa (Sudan, Ethiopia, and Somalia), the egg-laying period is during the wet season from the end of April to the middle of June (Archer and Godman 1937, Meinertzhagen 1954, Urban et al. 1986). Further south, the race *ellioti* breeds over an extended period lasting from February to perhaps as late as November, with a peak during May in central Kenya, and young chicks observed as late as September on the Serengeti plains of Tanzania. In a rain-shadow area west of Kilimanjaro, northern Tanzania (where the annual rainfall is only 24–36 cm and is centred in March), the breeding season occurs over a 6–7 month period, lasting from May to November. In this area the adults begin breeding in May and June at the end of the rainy season, when nest-flooding is no longer a problem and protein-rich leguminous plant seeds are becoming abundant. This nesting activity by adults produces an initial breeding peak, which is followed later by nesting on the part of renesters, older juveniles (possibly including those hatched only a few months earlier), and perhaps also second-breeders (Kalchreuter 1980). In western Africa (Mali, Senegambia) there is also an extended breeding season lasting from February to November, but mainly occurring between March and July, during the late dry season and early rainy season (Urban et al. 1986).

The nest consists of a simple ground scrape on barren, arid ground, and is usually unlined or at most scantily lined with a few bits of grass or grass stems. Only a very small proportion are ever placed under the shelter of grass tufts or roots according to Baker (1935), and nests have often been found in which the albumen of unprotected eggs has been semi-coagulated from the intense heat of such locations.

Nesting is done in a solitary manner, and eggs are usually laid at 2-day intervals (Böhm 1985). Three eggs certainly represent the commonest clutch size, although variants of two (infrequently) and four or even five eggs (very rarely) have been reported. However, clutches of more than three eggs quite possibly represent the efforts of more than one female. Incubation is by both sexes, the female normally incubating at day and the male at night, although the male may also cover the eggs during daytime hours when the female is away feeding. Incubation reportedly takes about 23 days (Meade-Waldo 1922), although Böhm (1985) reported an incubation period of slightly under 22 days. Only the male brings water to the young chicks. Although Meinertzhagen (1954) stated that in his experience water is brought to the young in the male's crop and then regurgitated, St Quintin (1905) had earlier documented water-carrying in the breast feathers in this as well as in several other sandgrouse species. Both parents brood the young, often in a sunny location (in England), and for the first few weeks after hatching they do not wander far, squatting closely when approached by humans (St Quintin 1905).

After fledging, the young birds probably begin engaging in drinking flights when about 2 months old, and probably also continue moulting into their definitive plumage. Moulting of wing and tail feathers by males was found by Kalchreuter (1980) to occur throughout the year, but with a peak between August and February. However, moulting of the primaries by females was largely confined to the period of low breeding activity between August and April.

Evolutionary relationships

It is possible that this species, along with *coronatus*, *senegallus*, and probably *namaqua*, form a species-group (Urban *et al.* 1986). I have likewise included it in this same assemblage, and have tentatively associated it most closely with *namaqua* (Figure 3). On the other hand, Snow (1978) questioned whether *exustus* could be linked to any single species-group or superspecies. He believed that *exustus* and *namaqua* were perhaps derived from the same ancestral sandgrouse stock, since both species have pintails and single dark breastbands, and both are characteristic birds of open desert or semidesert habitats.

Status and conservation outlook

This is probably one of the most common species of sandgrouse. In Africa it is generally frequent to locally abundant, and it is certainly also abundant in the drier regions of the Indian subcontinent. The introduced population on Hawaii, although initially believed to have failed, survived locally and increased sufficiently during about two decades as to allow hunting for the first time in 1981, at least in a very limited area (Paton *et al.* 1982).

NAMAQUA SANDGROUSE (Plate 41)

Pterocles namaqua (Gmelin) 1789

Other vernacular names: ganga Namaqua, ganga de Namaland (French); Nama-Flughuhn (German).

Distribution of species (Map 33)

From south-western Angola south and east to Namibia, Botswana, South Africa (Cape Province and western Transvaal; formerly also to Orange Free State) and Lesotho. Locally migratory or nomadic, especially the southern populations.

Distribution of subspecies

None recognized by Urban *et al.* (1986). Clancey (1967) recognized both *furva* Clancey 1959 and *ngami* Meyer de Schauensee 1931.

Measurements (mm)

Wing, males 167–179 (av. of 10, 173), females 163–173 (av. of 10, 167); tail, males 89–124 (av. of

Map. 33. Distribution of the Namaqua (wide-hatched), Madagascan (solid) and black-faced (narrow-hatched) sandgrouse, and the locations of the last-named species' races *decoratus* (de), *ellenbecki* (el) and *loverridgei* (lo).

10, 112.5), females 88–106 (av. of 10, 95.6); tarsus, males 21–24 (av. of 10, 22.6), females 20–23 (av. of 10, 21.8); culmen, males 11–13 (av. of 10, 11.9), females 11–13.5 (av. of 10, 12.2). Egg, av. 36.3 × 25.2 (Urban *et al.* 1986).

Weights (g)

Males 166–191 (av. of 13, 180), females 143–192 (av. of 15, 172) (Urban *et al.* 1986). Nineteen adults of both sexes averaged 186.2 (Dixon and Louw 1978). Egg, 12 (estimated).

Description

Adult male. Crown, nape and neck ochraceous buff, becoming vinaceous buff on the lower chest and orange-yellow on the throat and chin; upper back, rump and upper tail-coverts dull olive-brown, each feather with a buff-grey subterminal blotch; lesser, median and secondary wing-coverts and scapulars dark brown, with a subterminal buff or buff-and-white blotch; outer secondaries buffy brown; primaries, primary coverts and inner secondaries black, the outer two primaries with white shafts, and the five inner ones and the inner secondaries tipped and partially margined inwardly with white; a white and dark chestnut pectoral band separates the chest from the brown breast, which fades to buff on the belly, flanks, and under tail-coverts; feathers on front of tarsi white; the central pair of rectrices olive-brown and extended into long black filaments, the other fourteen dark brown tipped with buff; underwing lining pale buffy brown; axillaries light brown. Iris dark brown, eye-ring yellow, bill light greyish-horn, toes pinkish-grey to buff.

Adult female. Lores, crown and upper back pale chestnut, shading into pale yellow on the sides of the head, throat and nape, each feather with a black shaft-stripe, reduced to a subterminal spot on the throat, while on the upper back and chest it gives off lateral bars and terminates in a rounded blotch, producing a distinctly spotted effect; lower back, rump and upper tail-coverts similar to the back, but the ground colour mixed with buff. Secondary, median and lesser wing-coverts, scapulars, and outer secondaries pale chestnut-buff, each feather with a black shaft-stripe giving off lateral bars and tipped with buff or buff and reddish-brown. Breast, belly and flanks buff, barred with black. Rectrices black, barred with chestnut, and the central pair extended into black filaments, the rest tipped with buffy white.

Immature male. Similar to the adult female, but the top of the head and upper back more rufous and barred with narrow brownish-black bars, the feathers fringed with whitish; the chest buff, irregularly marked with dusky grey; the throat, breast and belly approaching the adult male, but the throat paler, the breast more rufous and the brown abdomen sometimes with sepia vermiculations. The juvenal primaries are tipped with buff and vermiculated with black, and the central rectrices are not lengthened and filament-like in younger birds. The juvenal plumage is apparently held only a very short time, and is replaced by an adult-like first basic plumage (Clancey 1967). Moulting patterns of the juvenal remiges have not yet been studied but are likely to be similar to those of other *Pterocles* species, so the presence of worn juvenal outer primaries may assist in ageing older immatures during their first year.

Immature female. Very similar to the adult female, but more cryptically marked, the tail paler and, at least initially, lacking elongated central rectrices; the wing-coverts also broadly tipped and fringed with buffy white (Urban *et al.* 1986).

Identification

In the hand. The combination of a pin-like tail, a belly that is brown, shading to buff (males) or barred brown and white (females), and a pale buffy brown underwing lining serves to identify this species.

In the field. In its limited southern African range, this species is the only sandgrouse with a pin-like tail. The birds generally appear rather pale brown (including the underwing lining in flight), without strong head or breast patterning in males, and have characteristic three-note *kwelkiewyn* flight calls. They also have at least four other adult calls, including a series of sharp *quip* notes on take-off, a strident *ki-kiii* alarm call, a high-pitched call uttered during nest-distraction display, and a soft, repeated *quip* note used by parents for summoning their hiding chicks (Maclean 1968).

General biology and ecology

This species frequents a broad range of habitats, but is generally most common in open steppe or rolling country having dry, sandy soil, and the ground variably covered with small stones, gravel, and thin grasses or well-spaced and low xeromorphic shrubs and succulents. In the Kalahari desert it also occurs on sandy savannahs having denser vegetation, but most typically is found where the annual precipitation is under 30 cm (Clancey 1967, Urban *et a.* 1986).

The birds are often found in small groups of up to

thirty during the non-breeding period, but much larger flocks have been observed at times, and when congregating at waterholes they may often be seen in the hundreds. The birds fly as far as 60 km or more each day to watering areas, and are joined by others on the way, so that at times quite large groups may arrive simultaneously. The usual time for watering is between 8 and 10 a.m., but watering is also done to a lesser degree in late afternoon hours. As the first birds arrive, they usually settle some distance away from the water and later make their way the remaining distance to the shoreline on foot after resting or crouching for a time where they first landed, or they may fly to within about 20 m and walk the remaining distance. After drinking and perhaps also wetting their breast feathers, which may require as little as 10–15 s or as long as 15 min in the case of prolonged feather-soaking, the birds break up into smaller groups or pairs as they return to feeding areas (Cade 1965, Maclean 1968, Urban *et al.* 1986). Rarely—namely in six times out of several hundred observed instances—females have been observed wetting their breast feathers, although this is normally only done by adult males and during the breeding period (Cade and Maclean 1967).

Typically, drinking is done by the bird sucking once, then raising the bill and swallowing. On average, about eight to ten sips of water are taken, equivalent to some 19–23 g of water, or about 11–14 per cent of body weight. Based on a few samples of birds collected just after drinking, up to about 23–25 ml of water, or 14–16 per cent of the bird's adult weight, can be temporarily stored in the crop. Evidently the birds can easily survive up to 3 days without water. However, adults that had been deprived of water for 25 days drank no more than 15 ml of water (Cade *et al.* 1966). Quite possibly they regularly go several days between drinks, perhaps to avoid the energy expenditure of daily flights to waterholes, with an associated reduction of exposure to aerial predators such as falcons. Reduced vulnerability to predators may also account for the birds forming close huddles at night, although probably the major advantage of such behaviour is in the conservation of metabolic heat that is provided by this activity (Thomas *et al.* 1981).

Foraging by this species is typically done by making a relatively cursory search over large areas, as opposed to searching a small area thoroughly, as appears to be the typical foraging technique of the double-banded sandgrouse of the same general region. These two different methods of foraging may help to reduce competition between these species. They also have somewhat different habitat preferences, the Namaqua sandgrouse occurring in more open habitats, and the double-banded occupying more rocky, bouldery habitats and extending into areas of heavier vegetation. The two species are otherwise apparently equally well adapted to hot desert-like conditions by water and energy conservation adaptations, and devices for the reduction of heat loads (Thomas and Maclean 1981, Thomas *et al.* 1981).

Foods consist of the usual array of small, dry seeds of grasses, desert herbs, and other low, xeromorphic plants. In the Kalahari a major food plant is the chenopod forb *Lophiocarpus burchelli*, which flowers and sets seed after the rains, and in the Namib the seeds of the legume *Tephrosia dregeana* and beeplants (*Cleome diandra* and *C. luederitziana*, Capparaceae) are important foods and are highly nutritious in both protein and energy content. In Namib National Park the very small seeds of seven plant species are reportedly most important, and in some cases these are as small as the sand grains that cover the desert surface. A few insect and mollusc fragments have also been reported, but these materials may have been only accidentally ingested (Maclean 1968, Dixon 1978, Dixon and Louw 1978, Urban *et al.* 1986).

Social behaviour

Like most other sandgrouse, this species is quite undemonstrative, showing a pacific nature. It also exhibits a seemingly egalitarian gregariousness, with no apparent hierarchical social structure apparent in mixed-sex flocks of non-breeding birds (Thomas *et al.* 1981). During aggressive threat the bird runs or hops toward the intruder with the head and tail lowered (Maclean 1968).

Courtship and other aspects of sexual behaviour are still only poorly described for this species. However, the species has a monogamous pair-bond, which Maclean (1968) believed quite strong. For example, pair members always waited for each other to finish drinking, and when one of the pair members was captured in a mist-net, the other pair member would sometimes land and wait for its mate to be released.

During sexual display the male is reported to follow the female on the ground, with his head hunched in and his tail raised and fanned. Maclean (1968) called this behaviour 'strutting', and he thought it might have something to do with pair formation. He also noted that a female would sometimes follow a male in this posture. The birds would typically respond to any sudden movement, or to other birds (of any species) flying overhead by adopting a ritualized flight-intention movement. This consists of crouching while facing away from the stimulus, raising the head, flicking the wings open, and raising and fanning the tail (Figure 52b). At higher density (as probably occurs during courtship) the head is lowered during this tail-raising display (Figure 52c) (Urban *et al.* 1986).

Reproductive biology

The breeding season of this species is highly variable, and probably depends entirely on rainfall, eggs having been found during every month of the year. However, most breeding activity throughout the species' entire range is concentrated during the period from July to November, during which 73 per cent of 175 clutches were reported (McLachlan 1985). According to Clancey (1967), in the centre of the species' range (i.e. within the limits of his nominate race *namaqua*) most breeding occurs between April and July or August, depending upon the incidence of seasonal rains, but McLachlan demonstrated a broader breeding season for this region, peaking in winter or spring. The breeding of the population recognized by Clancey as *furva* in the Karoo, Orange Free State and Transvaal occurs from August to November, and the population in the southern Cape region breeds during the spring and summer months. McLachlan further noted that in the Kalahari area breeding occurs after rainfall, mainly during winter and peaking in July, but with a second peak in November. Dixon and Louw (1978) stated that breeding records exist in southern Africa for all months except March and May, and breeding generally peaks between August and November.

Territoriality is only weakly developed in this species. Thus nests may be situated as close as only 20 m apart, although coloniality is absent (Urban *et al.* 1986). More closely intruding birds, including other sandgrouse as well as other harmless species, are driven away from the nest. The nest is typically located among stones, pebbles, thin grass, or scrub, or may simply be placed on bare soil. It is a simple scrape or natural depression in the ground only large enough to accommodate the eggs (Clancey 1967). Of thirty-six nests found in the Kalahari, all but six were located in areas of hard and rocky calcrete, with the remainder equally divided between sand dune and river bed substrates. The majority (two-thirds) of the nests were in completely exposed sites, and the rest were near shrubs but not so orientated as to be shaded. Bits of vegetation or small stones that sometimes accumulate at the nest are probably

Fig. 52. Social behaviour of Namaqua sandgrouse: (a) alert posture by a female and chick (after a photo in Maclean 1968); (b, c) agonistic tail-fanning with head erect and facing away from a fearful stimulus, as compared with (d) tail-fanning towards an aggressor while facing it (after sketches in Urban *et al.* 1986).

the result of side-throwing behaviour during nest-relief (Maclean 1968).

The clutch is usually three eggs, which are laid at approximately 24-h intervals. Two-egg clutches are only rarely found, and probably represent depleted clutches. The mean clutch size of sixty-two clutches was 2.9 eggs (Urban *et al.* 1986), and Maclean (1968) noted that all but one of thirty-two clutches had three eggs, the exception having two.

Incubation is done by both sexes, with the female starting to tend the nest with the laying of the first egg, but beginning actual incubation with the laying of the second or third egg. The male begins to attend the nest with the laying of the first egg, and begins his nocturnal incubation when the clutch is complete. The female rarely incubates for more than 10 hours per day, and the male does so the remainder of the time, which ranges from 14 h in summer to 18 h in winter. Incubation requires 21–23 days, typically only 21, but certainly is never as brief as the 16 days often indicated in the literature. During her daytime incubation periods the female sits so orientated that she is facing the sun, or into the wind if any is present. After being relieved by the male at sunset, the female flies to water to drink. She then returns and remains near the nest all night, to take over again after feeding and drinking the following morning (Maclean 1968, Urban *et al.* 1986).

In some breeding areas the daytime temperatures may exceed 45°C. However, even with a soil temperature of 45°C the female is able to keep her eggs at 30–35°C, by raising her body to shade the clutch while resting on her lowered wings, and cooling herself through gular fluttering and fluffing her mantle feathers. When soil temperatures exceed 50°C, the female has difficulty in keeping the temperature of her eggs below 45°C, yet the eggs are evidently able to tolerate such temperatures. When adults are disturbed from the nest, injury-feigning and anxiety-calling sometimes occur (Urban *et al.* 1986). Probably the major predators on eggs and young are the bat-eared fox (*Otocyon megalotis*) and the silver fox (*Vulpes chama*). Jackals have also been observed preying on eggs and chicks, and greater kestrels (*Falco rupicoloides*) and pied crows (*Corvus albus*) are known predators on chicks (Dixon and Louw 1978).

The chicks of a clutch sometimes all hatch on the same day, but usually one hatches a day earlier or later than the two others, or they may even hatch on up to three successive days. Thus if incubation does actually begin with the laying of the first egg, and if synchronous or nearly synchronous hatching is to occur, the rate of embryonic development must increase in the later eggs of the clutch. Various possible mechanisms for achieving this are known to exist, but which one might be operative in sandgrouse has yet to be established (Thomas 1984).

Of sixty-nine eggs in twenty-three clutches observed by Maclean (1968), slightly over two-thirds hatched, most of the losses being the result of predation. Although exact counts were impossible to obtain, Maclean judged that somewhat more than a third of the chicks leaving the nest failed to survive. Thus an estimated 23 per cent of the eggs that were laid produced fledged young, or an average of about 1.3 young per breeding pair. Assuming that a perfect 1:1 adult sex ratio exists, that all birds pair and nest, and furthermore that there is only one breeding attempt per year, this level of breeding success would permit a maximum recruitment rate of roughly 40 per cent. This hypothetical figure is reasonably close to but somewhat higher than my estimate of an approximately 30 per cent estimated recruitment rate for *decoratus* and *exustus* based on adult–juvenile ratios in flock samples collected by Kalchreuter (1980). Obviously, comparable recruitment rate estimates could be obtained using other reasonable assumptions, such as a smaller percentage of the adult population pairing and nesting, but some renesting or double-brooding by the pairs that do actually attempt to nest.

The young leave the nest as soon as they have all hatched, and within 24 h after hatching they are feeding themselves. The adults remove the hatched eggshells from the nest, and initially the female broods the young as the male flies off to drink and obtain water for them. Occasionally both parents go off to drink together, leaving the young hiding under shrub cover. As in the other sandgrouse species in which parental behaviour has been closely observed, only the male normally waters the young. Yet in a few cases females have been observed feather-soaking, so probably females sometimes take over watering responsibilities owing to the loss of their mates. Perhaps the normal pattern of male-only watering frees the female to brood the chicks on very cold mornings, while the male's breast feathers are still wet (Cade and Maclean 1967, Maclean 1968).

By the age of 3 weeks the young are half-grown and nearly fully feathered, but they are still then unfledged and flee on foot from danger, using their wings for balance. When about a month old they can fly a few metres, and become capable fliers when about 6 weeks old. They probably do not accompany adults on watering flights until they are about 2 months old, but within 4 or 5 months after the peak of breeding most of them are in full adult plumage. They are probably able to breed within a year of hatching (Maclean 1968).

Evolutionary relationships

This species is of somewhat uncertain relationships, but perhaps is part of a species-group that also includes *coronatus*, *senegallus* and *exustus* (Urban *et al.* 1986). Snow (1978) suggested that *exustus* may be the nearest living relative of this species, and Wolters (1974, 1975) placed these two forms in a separate subgenus (*Namapterocles*). I have tentatively regarded these two species as close relatives (Figure 3).

Status and conservation outlook

Although not widey distributed, the Namaqua sandgrouse is generally common to locally abundant in most areas of South Africa. However, at the eastern edges of its historic range in Orange Free State, as well as in Lesotho, it is now absent (Urban *et al.* 1986).

BLACK-BELLIED SANDGROUSE (Plate 42)

Pterocles orientalis (L.) 1758

Other vernacular names: imperial sandgrouse, large sandgrouse, oriental sandgrouse; ganga unibande (French); Sandflughuhn (German).

Distribution of species (Map 34)

From the Iberian peninsula south to north-western Africa (Morocco to Libya) and Fuerteventura Island (eastern Canary Is.), and east from Israel and Cyprus through Turkey to central Iran and Pakistan (Baluchistan and Sind), and across south-eastern USSR into adjoining Xinjiang Zizhiqu (Sinkiang), China. Partly migratory, wintering from Turkey east through Syria, Iraq and Iran to Pakistan and north-western India. Introduced unsuccessfully into the USA (Nevada).

Map. 34. Distribution of the black-bellied sandgrouse, including its breeding (hatched), residential (solid) and wintering (stippled) ranges, and the locations of its races *arenarius* (ar) and *orientalis* (or).

Distribution of subspecies

P. o. orientalis (L): Iberia, Fuerteventura Island, Morocco and north-eastern Africa east through Asia Minor to about the Turkey–Iran border area of the Caucasus. Includes *enigmaticus* Neumann 1934, *koslovae* Meinertzhagen 1934, and *bangsi* Koelz 1939 (Vaurie 1965).

P. o. arenarius (Pallas) 1775: From the lower Volga area (USSR) east to Xinjiang Zizhiqu (to Urumchi, the Turfan depression and Kashgaria), and south to Iran Afghanistan, and Pakistan.

Measurements (both races, mm)

Wing, males 227–244 (av. of 8, 237), females 221–242 (av. of 3, 228); tail, males 88–101 (av. of 8, 94.4), females 84–101 (av. of 3, 91.3); tarsus, males 26–34 (av. of 7, 30.3), females 26–35 (av. of 3, 31); culmen, males 12–15.6 (av. of 8, 13.4), females 12.4–14.4 (av. of 3, 13.4). Egg, av. 48 × 32 (Cramp 1985).

Weights (g)

Males (*arenarius*), in autumn, 400–460 (av. of 9, 428), females 300–420 (av. of 11, 383) (Cramp 1985). Males (*orientalis*), 480–550 (av. of 4, 514), females 410–465 (av. of 4, 434). Egg, 28 (estimated).

Description

Adult male. Chin and throat chestnut, becoming orange posteriorly on the nape; a triangular black patch at the base of the throat; rest of head and upper back grey, tinged with dull yellow; lower back, rump, and upper tail-coverts darker grey, pale vinaceous buff basally, and tipped with a rufous buff spot; primaries and primary coverts grey; the outer web of the outermost primary blackish-grey; secondaries and inner median wing-coverts whitish, with most of the outer web ochre; rest of wing-coverts and scapulars like the lower back, but the terminal spot mostly ochre tinged with rufous buff internally. Axillaries and under wing-coverts white; chest and upper breast isabelline grey, separated by a moderately wide black band; lower breast and rest of underparts black; tarsi and under tail-coverts white; central pair of rectrices (sixteen) rufous buff, barred with grey and tipped with greyish-ochre, the rest similar but tipped with white, and gradually becoming greyer outwardly, the outermost pair grey tinged with reddish-buff; underwing lining and axillaries white, the larger coverts spotted with grey; the underside of the remiges also grey. Iris brown, eyelids lemon yellow, bill pale to dark lead colour; toes brownish-grey to lead-coloured.

Adult female. Differs from the male in having the throat yellowish-white, the outer feathers with black shafts, and tipped by a narrow black and rather wide pale grey band, producing a narrow black lower neck stripe; the rest of the head, upper back and chest isabelline, a submarginal black spot at the tip of each feather; upper breast rufous to yellowish-white, separated from the spotted chest by a narrow black band; back and rest of upperparts pale rufous buff, closely barred and marked with black.

Immature (both sexes). Juvenile males initially resemble adult females. Later, immatures of both sexes differ from the adults mainly in having their primary tips and primary coverts edged with buff and vermiculated with black. Subsequent stages are like adults, but may be recognizable when some outer juvenal primaries are retained (Cramp 1985).

Identification

In the hand. This is the only species of sandgrouse lacking a pin-like tail that has uniformly black flank and belly colouration (in *decoratus* the blackish underparts posteriorly are bounded anteriorly by white or buffy and broken up by lighter feather margins).

In the field. The black flank and underpart colouration, combined with a wedge-shaped tail, identify this species, which is the largest of the *Pterocles* sandgrouse. In flight the white underwing linings contrast with the dark flight feathers and underparts. Calls include double chuckling notes, deep churring sounds, and other deep notes.

General biology and ecology

In general, this species occurs mainly on flat plains, saltflats, and soils ranging from clay-like to gravelly with dusty patches, or covered with stones or having hummocks, eroded slopes, or worn-down rocky outcrops. Saltflats with only scattered vegetation are sometimes used, as are airfields, but generally the species avoids both vegetationless areas and those that are extensively covered with trees or shrubs. Grassy steppe-like or semidesert vegetation is perhaps the ideal habitat, where scattered grasses and weeds are interspersed with dusty areas, such as occurs in moderately to heavily grazed arid rangelands or areas of marginal dryland agriculture. In the Mediterranean basin, this species is found both in desert-like habitats and in arid areas having some dryland farming, as for example in Turkey where semidesert areas have in part been converted to submarginal wheat or barley agriculture. In Iran, where dry steppes grazed by sheep and goats are interspersed with alkaline flats and where some dryland wheat farming is sometimes attempted, black-bellied bustards, houbaras, and great bustards have all been reported (Christensen *et al.* 1964, Cramp 1985).

Wintering areas in Spain, Turkey and Iran are similar to or identical with breeding habitats, often consisting of semidesert plains where there are sandy hillocks strewn with stones and dotted with grassy patches. The Indian or Thar desert is also an extremely important wintering area, and consists largely of scattered mesquite (*Prosopsis*) with an understory of large, spiny shrubs such as *Acacia*, *Zizyphus*, and *Euphorbia*. During years of favourable rainfall the desert floor is covered with a variety of forbs including the important sandgrouse leguminous food plant *Indigofera*, as well as representatives of various grass genera. Additionally, small areas of cultivation of such plants as pulse (*Phaseolus*) provide additional foods for the birds (Christensen *et al.* 1964).

Foods of this species have been studied in various areas, as summarized by Cramp (1985). They mostly consist of protein-rich seeds of a variety of legumes (*Astragalus*, *Indigofera*, *Melilotus*, *Phaseolus*, *Tephrosia*), seeds of grasses or cultivated grain plants (*Hordeum*, *Panicum*, *Triticum*), those of xerophytic forbs and shrubs (*Alhagi*, *Artemisia*, *Salicornia*, *Salsola*), and those of other miscellaneous native plant genera (*Ammodendron*, *Cyamposis*, *Gynandropsis*, *Heliotropium*, *Onobrychis*, *Polygonum*, *Sisymbrium*). Insects and their larvae have been reported in some samples, but perhaps represent materials taken in by chance, whereas grit is almost invariably present.

Seeds taken by this species typically are extremely small, with the result that as many as 30 000 seeds of *Melilotus* and *Astragalus* have been counted in the digestive tract of a single individual, whilst a bird obtained in India had an estimated 8700 seeds of *Indigofera*. However, this relatively large sandgrouse species also consumes somewhat larger seeds than are typical of most sandgrouse, including those of various cultivated grains. Of twenty birds collected in Turkey, thirteen had eaten only weed seeds, four of the crops had both weed and cultivated crop (wheat and legume) seeds, and three crops were empty. Most crops also contained grit, but insect and green plant remains were absent. Of thirteen birds collected in India, twelve contained the seeds of *Indigofera*, which accounted for 21 per cent of the volume. The larger seeds of the crop plant pulse (*Phaseolus*) accounted for 39 per cent of the total volume. The other two most notable plants present in the crop samples were those of *Panicum* and *Cyamopsis*, which volumetrically formed about 21 per cent (Christensen *et al.* 1964).

Like other sandgrouse, this species often flies long distances to water, perhaps as far away as 30 km. The birds apparently make the round trip once and perhaps occasionally even twice a day, although it has not been proved that individual birds must drink every day. They favour watering sites that are clear of vegetation, and like other sandgrouse, typically alight some distance away before walking down to the water to drink. Although Christensen *et al.* (1964) repeated the oft-quoted point that these sandgrouse can suck up water without raising their heads between drinks, this observation seems somewhat questionable in view of the observations of Maclean (1968) and others on various other sandgrouse species, and needs confirmation.

Flock sizes at such watering sites are of course highly variable, but Christensen (1963) observed an estimated 8000 wintering birds at one north-western Indian site. During the period April to October, pairs or flocks of up to a few dozen birds are apparently much more common. Typically the species comes to water just before sunrise, and before the chestnut-bellied sandgrouse, which follows it to water. The black-bellied is also reportedly more wary than the chestnut-bellied sandgrouse. After watering, the birds fly directly to feeding grounds where they tend to remain together and move about quite deliberately in search of food. They may also rest in dusty spots during the hot midday period. If not disturbed they may remain on their foraging areas until evening, when a proportion may again return to water. However, evening flights may be restricted or even absent, especially during colder weather (Meinertzhagen 1954).

Daily flights to water occur throughout the entire year, and some quite substantial seasonal migrations must also occur for the birds breeding in south-western Siberia. These birds probably cross the Hindu Kush mountains of Afghanistan to reach wintering areas in southern Iran, Pakistan and India, a migration route of at least 1500 km. They begin to arrive on wintering areas when the average monthly temperature has dropped to approximately 20°C, and begin to leave again in spring when the average monthly temperatures exceed this level (Christensen *et al.* 1964).

Social behaviour

Like other sandgrouse, this is an apparently monogamous species, with a one-year period to sexual maturity. Birds arriving on breeding areas in Turkey were usually paired off before or shortly after their arrival. Very little has been written on their courtship, but in one account a male displayed near a female with his tail raised and his wings drooping (Gavrin *et al.* 1962), and in another the male was said to fly after the female while calling, and then circle her on the ground (Dementiev and Gladkov 1951).

Also in common with other sandgrouse, this species is 'rather even-tempered' and non-aggressive, even when placed in confinement with various species of partridges and francolins, or with the smaller chestnut-bellied sandgrouse (Christensen *et al.* 1964).

Reproductive biology

The breeding season of this species rather consistently occurs during the spring and summer months throughout its range, generally between March and September. On the Canary Islands it breeds from mid-March to June, with most nests found in April (Bannerman 1963). In Spain, eggs are found from early May onward (Hudson 1975), and in northern Africa laying occurs from mid-April to July, beginning after the rains and continuing into the dry season (Urban *et al.* 1986). In Turkey and Afghanistan, eggs have been reported as early as mid-April, but probably most nesting occurs during May and June (Christensen *et al.* 1964). In the USSR the clutches are apparently sometimes completed as early as the first third of May, and late clutches, perhaps renests, have been found as late as mid-September, but probably June represents the maximum period of breeding (Dementiev and Gladkov 1951).

Both sexes help to build the nest, which is a simple scrape or a natural depression, in which a circle of small stones is sometimes placed. The nest is often placed in the shade of wormwood (*Artemisia*) or similar xerophytic vegetation. There are typiclly three eggs, but occasionally only two; the mean of seventeen clutches from North Africa was 2.6 eggs (Etchécopar and Hüe 1967), but all of twelve Turkish clutches had three eggs (Lehmann 1971). Replacement clutches are laid after the loss of the initial one, and it has been suggested (Christensen *et al.* 1964) that double brooding is probably 'not uncommon'. Double-brooding has also been suggested in Israel.

Tending of the eggs begins with the laying of the first one, probably because of the need to shelter the eggs from sunlight. In one Turkish nest the surface temperature in the sunshine beside the nest was over 60°C, when the surrounding air temperature was 39°C (Christensen *et al.* 1964).

Incubation is apparently done by the female during the day and by the male at night, as with other sandgrouse. It requires 23–28 days, based on observations in captivity, compared with estimates of 21–22 days in the wild. The young are cared for by both parents, the adult male providing water for the young with his soaked belly feathers, and the chicks feeding for themselves almost immediately (Meade-Waldo 1922). It has been suggested that fledging may occur at only 2 weeks of age, but judging from information on other *Pterocles* species, this seems unlikely, and a month-long fledging period is more believable. Nothing is known of productivity rates or mortality factors for the eggs and young.

Evolutionary relationships

This is seemingly a somewhat isolated species, which Maclean (1984) placed in a unique group of its own, although Wolters (1974, 1975) included it with *gutturalis* in a distinct subgenus *Eremialector*. Like the preceding species-group it has sixteen rectrices, but lacks a pin-like tail. I am inclined to believe that the number of tail feathers is more significant than the shape of the central feathers, and so have tentatively associated it with this group of species (see Figure 3).

Status and conservation outlook

Hudson (1974) included this species in his 'endangered' birds of Europe, although he admitted that the population size and precise status were then unknown. He judged that on the Iberian peninsula the species was becoming increasingly local and certainly reduced in numbers, for unknown reasons. More recently, it has been judged to be 'rather numerous' in Spain, but probably decreasing in Portugal. It is also present only on Fuerteventura in the Canaries, and is becoming very scarce there (Cramp 1985). Otherwise the species is seemingly still common over much of its historic range, although it seems possible that the extensive conversion of native steppes in the south-western USSR to agriculture, which has been so enormously destructive to bustards there, may eventually prove similarly detrimental to this sandgrouse.

CROWNED SANDGROUSE (Plate 43)

Pterocles coronatus Lichtenstein 1823

Other vernacular names: coroneted sandgrouse; ganga couronne (French); Kronenflughuhn (German).

Distribution of species (Map 35)

Africa, from Morocco across the northern and central Sahara to Israel and the Red Sea, thence east across the southern Arabia Peninsula and through the southern part of Iran to Afghanistan and Pakistan. Mostly sedentary but somewhat nomadic.

Distribution of subspecies

P. c. coronatus Lichtenstein: Africa from the central Sahara north to the Mediterranean, and from Morocco east to the Red Sea.

P. c. vastitas Meinertzhagen 1928: Sinai and deserts of southern Israel and Jordan.

P. c. atratus Hartert 1902: deserts of Arabia, Iraq and southern Iran, east to south-western Pakistan (Baluchistan) and Afghanistan.

P. c. saturatus Kinnear 1927: mountains of interior Oman.

P. c. ladas Koelz 1954: Pakistan (North West Frontier Province) south to Sind. Considered part of *atratus* by Ali and Ripley (1983), but recognized by Vaurie (1965).

Measurements (*coronatus*, mm)

Wing, five males 193–205, five females 183–195; tail, five males 79–86, six females 75–83; tarsus, five males 22–24, six females 23–24; culmen, both sexes 13–14. Egg, av. 39.4 × 27.4 (Cramp 1985).

Weights (g)

One male, 300 (Urban *et al.* 1986, Cramp 1985). Egg, 17 (estimated).

Description

Adult male. Very similar to *senegallus*, but with a black face, the central pair of rectrices are not extended into filaments, and the belly is not black; a black band on the chin and centre of the throat sur-

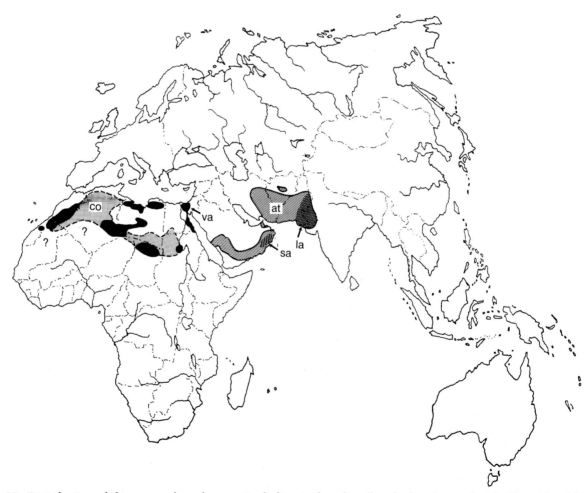

Map. 35. Distribution of the crowned sandgrouse, including its breeding (hatched and cross-hatched), residential (solid) and non-breeding (stippled) ranges, and the locations of its races *atratus* (at), *coronatus* (co), *ladas* (la), *saturatus* (sa) and *vastitas* (va).

rounds the gape and terminates on the lores in a black mask-like patch, the rest of the anterior face a contrasting white, becoming ochre-yellow on the throat, neck and sides of head; forehead and top of head dull vinaceous buff, this 'crown' separated from the ochre cheeks by a grey post-ocular stripe; rest of upperparts predominantly isabelline and the underparts mostly pale buff; under tail-coverts white; remiges and primary coverts blackish-brown; outer primaries with white shafts, the inner ones tipped with whitish-buff; secondary coverts buff, the rest of the wing-coverts and scapulars vinaceous buff, with a pale buff patch at each feather tip, edged internally with dusky grey; the central pair of rectrices isabelline like the upper tail-coverts; the other fourteen rufous, with a subterminal black bar and white tip; underwing lining and axillaries white, the undersurface of the remiges mostly light to medium grey, with paler tips to the outer secondaries and inner primaries. Iris dark brownish, eye-ring and bill pale bluish grey, toes pale grey to greenish-grey.

Adult female. Differs from the male in having the top of the head indistinctly patterned with narrow black shaft-stripes but no black facial 'mask' or definite 'crown', and all the upper surface and chest barred or spotted with blackish-brown, the ochre throat finely spotted with black, and the breast and belly spotted or otherwise extensively marked with dark brown. Very similar to *senegallus*, but the throat finely spotted and more barred than spotted dorsally.

Immature (both sexes). The tips of the primaries, central rectrices, and some of the secondary coverts and scapulars buff, vermiculated with black. Otherwise like adult female, but chin and upper throat creamy white, not golden ochre; older birds increasingly like adults but recognizable as long as some outer juvenal primaries are retained (Cramp 1985).

Identification

In the hand. This widespread species differs from the similar *senegallus* in lacking a pin-like tail, and has whiter underwing linings, and (in males) a black facial mask. Males differ from the two other 'masked' sandgrouse species in lacking any chest spotting or barring, and females have the outermost long primary (the tenth) with a white upper shaft (dark in *personatus* and *gutturalis*), a distinctly yellow throat with only slight spotting (unspotted in *senegallus*), a short tail (long in *exustus*), and a lightly barred upper breast (heavily spotted in *gutturalis*).

In the field. This species appears rather uniformly pale isabelline to buffy below (lacking the black belly of *senegallus*), has a wedge-shaped tail, has nearly white underwing linings (paler than those of *lichtensteinii*), and, in males, has a black throat. Its calls are a distinctly staccato series of soft triple-note *kla, kla, kla* or *cha-chagarra* notes.

General biology and ecology

This is one of the most arid-adapted of all the sandgrouse, inhabiting the hottest and driest portions of the Sahara desert, but avoiding sandy substrates and instead preferring to breed on stony desert, where dark reddish-brown sandstone closely matches its plumage colour (George 1969, 1970). It also occupies tracts of very sparse grasses, in both level and montane areas (Cramp 1985, Urban et al. 1986). In Oman it has been found in stony areas and ravines among lava fields, and perhaps occurs in all the Middle East deserts within daily flight distance of water (Meinertzhagen 1954). In Pakistan it occurs on the barest deserts, with scraggy grasses and stony wastes, and reportedly nests in barren, windswept sand-dune tracts, which is seemingly a quite different typical substrate than that reported for Africa (Baker 1935, Ali and Ripley 1983).

Drinking is done primarily during morning hours, the birds often arriving at waterholes somewhat after sunrise, and sometimes in company with the spotted sandgrouse where the two species occur together. However, unlike that species, the crowned sandgrouse typically made no preliminary reconnoitring before landing. Instead, they would land about 20 m from water and immediately crouch. The birds would then quickly run to water, drink, and then run back to stony areas. Falcons such as lanners (*Falco biarmicus*) would cause the birds to desert the area and fly considerable distances to the next waterhole (George 1970, 1977). Similar observations have been made in India and Pakistan (Ali and Ripley 1983), the birds arriving at waterholes from early morning until a couple of hours after sunrise, in groups ranging from pairs to groups of up to twenty. The birds thereafter left in groups, but some would return again in the evening to drink again when it was often almost too dark to see. The species is reportedly tolerant of saline water, and has the largest kidneys, relatively, of the three most desert-adapted sandgrouse species of North Africa (the other two being *senegallus* and *lichtensteinii*), indicating a fairly high salt turnover rate (Thomas and Robin 1977).

The foods of this species consist of the usual small, hard seeds of desert grasses and weeds (Meinertzhagen 1954, Ali and Ripley 1983). Little specific information on food plants is available, but at least in Morocco it and the other desert-adapted sandgrouse feed heavily on the seeds of the liliaceous forb asphodel (*Asphodelus tenuifolius*) that blooms and sets seeds abundantly after periods of heavy rainfall (Thomas and Robin 1977).

Social behaviour

Almost nothing is known of specific aspects of the social, pair-bonding or sexual behaviour of this species. Like other sandgrouse, it is generally gregarious outside the breeding season, but it is a relatively solitary nester. George (1970) found a slight tendency toward grouping of several pairs within a few kilometres of water (namely, three instances of two or three pairs all nesting in areas ranging from 1 to 3 km from water), but no obvious clustering of nests occurred as seemed to be characteristic of spotted sandgrouse breeding in the same general area.

Reproductive biology

Not many nest records are available for this species, but in north-western Africa the breeding records range from April 10 to June 20, although chicks have been found in Chad as early as April 12 (Heim de Balsac and Mayoud 1962). George (1970) observed a male near a nest site in late May, and found nests in Morocco during mid-June. May and June have also been reported to be the months of breeding in the Pakistan region (Baker 1935). However, it is likely that breeding is opportunistic, occurring after any significant rainfall has fallen in these desert areas.

The nest is located in fully exposed sites, perhaps most often on a pebbly or stony substrate, the smaller pebbles in the middle of the nest scrape being placed around the rim of the nest during the egg-laying period. Or the nest may be a simple depression in the sand. The clutch-size is typically two or three eggs, most commonly and probably normally three, but the egg-laying interval and incubation period are still unreported (Cramp 1985, Urban *et al.* 1986).

Incubation is by the female during the day and the male at night, the male usually taking over before sunset. One male crowned sandgrouse was found to have thoroughly wet breast feathers as it relieved a female on the nest, suggesting that wetting might be used to reduce or prevent desiccation of the eggs. Incubating females have been observed at air temperatures of 41–51°C, and even at these temperatures neither gular fluttering or dorsal feather erection was apparent, nor was any consistent alignment with respect to the sun observed. Indeed, gaping, gular fluttering and thermal panting were never observed in either this species or the spotted sandgrouse, but these cooling mechanisms were observed in pin-tailed sandgrouse under comparable conditions (Thomas and Robin 1977).

Although not studied in detail, the young are known to be provided with water by the male, who carries it back to the brood in his breast feathers, as described for several other sandgrouse. Like these species, the male stands with his back to the sun while watering the chicks, but in contrast to other species some watering occurs in the evening as well as during morning periods (George 1970, 1977).

There is no information on productivity or mortality.

Evolutionary relationships

Urban *et al.* (1986) include this species along with *senegallus*, *exustus* and probably also *namaqua* within a species-group, but Maclean (1984) included *coronatus* with *decoratus* and *personatus* in a separate group of species having black forehead colouration in males and non-elongated central rectrices. I have tentatively (see Figure 3) also included it with these latter two species.

Status and conservation outlook

This is an extremely desert-adapted species, and its local distribution and productivity are perhaps influenced by climatic conditions, especially rainfall, although breeding evidently occurs even in very dry years. It is generally less common than the other sandgrouse where it occurs with them, but it probably also ranges into more extreme desert types than these, and so is perhaps more widely if more sparsely distributed.

BLACK-FACED SANDGROUSE (Plate 45)

Pterocles decoratus Cabanis 1868

Other vernacular names: bridled sandgrouse, masked sandgrouse, pale black-faced sandgrouse (*ellenbecki*), Tanganyika black-faced sandgrouse (*loverridgei*); ganga à face noire, ganga masqué (French); Maskenflughuhn, Schmuckflughuhn (German).

Distribution of species (Map 33)

From north-eastern Ethiopia south across eastern Ethiopia to Somalia, Kenya, eastern Uganda, and Tanzania almost to the Mozambique border.

Distribution of subspecies (after Urban *et al.* 1986)

P. d. decoratus Cabanis: south-eastern Kenya and north-eastern Tanzania.

P. d. ellenbecki (Erlanger) 1905: southern Ethiopia south to northern Kenya, north-eastern Uganda and southern Somalia.

P. d. loverridgei Friedmann 1928: south-western Kenya and interior Tanzania.

Measurements (mm)

Wing, both sexes 154–175; tail, both sexes 60–75; tarsus, both sexes 25; culmen, both sexes 13–15. Egg, av. of two, 34.5 × 25.5.

Weights (g)

Males 160–216 (av. of 35, 187), females 149–210 (av. of 34, 178) (Urban *et al.* 1986). Egg, 12.4 (estimated).

Description

Adult male. A broad black band edged on either side with white extends up the middle of the throat and widens on the chin, surrounding the gape and forming a facial 'mask' both above and below the bill; a white-and-black superciliary stripe; crown and nape buff, with black shaft-stripes; sides of face pale vinaceous buff, becoming darker on the sides of the neck and chest; feathers of the back of the neck, back, rump and upper tail-coverts dull buff barred with black; scapulars darker, with the black bars becoming confluent and widely tipped with yellowish-buff; greater, median and lesser wing-coverts buff with one or two rather wide-set black bars; remiges and primary coverts black, the inner primaries tipped with white; upper breast dark isabelline, separated from the white lower breast and anterior abdomen by a narrow black band; lower breast, flanks and belly blackish, most of the feathers variably margined with white to pale rufous, the white increasing anteriorly, so that a distinct white lower breast-band is formed; lower thigh and tarsus whitish-buff; under tail-coverts bright buff, each feather with a subterminal black triangular mark; tail of fourteen rectrices, all barred with black and resembling the upper tail-coverts, but the outermost tipped with whitish-buff; underwing lining whitish, tinted with khaki, becoming very dark brown on axillaries; the undersides of the remiges also dark brown. Iris brown; eye-ring dull yellow; bill deep yellow; toes dull orange-yellow.

Adult female. Similar to *coronatus* but darker throughout, especially on the belly, which is blackish, the feathers having buffy edges. Differs from the male in having no black and white patterning on the head (the chin yellowish and the forehead spotted with blackish), the blackish markings on the back and wing-coverts following the shape of the feathers rather than crossing them, the dorsal parts more closely barred and the breast also barred with dark brown and lacking a distinct dark band separating the barred upper breast from the buff-coloured lower breast area, which becomes blackish posteriorly on the abdomen.

Immature (both sexes). Similar to the adult female, but the primaries with rusty tips and edged with whitish, and the rectrices more irregularly barred, with basal streaks. Immature males may have indications of the black facial patterning but lack a breast-band, and both sexes have most feathers that are edged with whitish (Urban *et al.* 1986). Kalchreuter (1979) described a second juvenal plumage, which is very similar to that of adults but is more downy and carried only for a very short time. He noted further that subadult birds have a uniform degree of wear and tear on their bleached outer secondaries, whereas older birds, because of varied moult sequences, exhibit a very irregular pattern of wear. This moult complexity includes an irregular tail moult and the moulting of the inner secondaries (and perhaps also the body feathers) twice per annual moulting cycle. Apparently the double moult of the inner secondaries relates to the fact that they are exposed to intensive solar radiation and hard mechanical wear and tear, whereas some outer secondaries may not be replaced every year.

Identification

In the hand. This wedge-tailed species of sandgrouse is the only one having blackish-brown feathers on the underparts that are narrowly margined with white posteriorly and on the flanks, and the abdomen grades rather abruptly from black to white or buffy, forming a broad whitish band behind the narrow black pectoral band in males, and a comparable buffy band in the same lower breast area of females. The underwing lining is likewise mostly whitish, contrasting with dark brown on the axillaries and remiges.

In the field. This is the only sandgrouse species in eastern Africa having a white (males) to buffy (females) band across the lower breast, behind which is a generally blackish belly. It is also the only sandgrouse having the black mask extending all the way down the base of the throat. It is wedge-tailed, and typically utters a series of three whistled *chukar* notes in flight.

General biology and ecology

This species occupies dry savannah, thornbush, semidesert scrub, and coastal dunes, as well as small bare areas in heavier vegetation, such as road verges or other open spaces in heavy bush (Urban et al. 1986). The species is bounded geographically on the north by the Lichtenstein's and four-banded sandgrouse, and on the south by the double-banded sandgrouse. It overlaps somewhat with the two former species, but it inhabits thicker bushvelt and wooded savannahs, which tends to separate it ecologically from them (Snow 1978). In one breeding area of northern Tanzania the topography consists of saline, poorly drained soils, with an associated semidesert vegetation of short grasses (*Sporobolus* and *Cynodon*) with scattered scrub and small trees, especially various *Acacia* species. The annual rainfall there averages 24–36 cm, and is especially heavy during March, when large areas are flooded, causing rapid vegetational growth (Kalchreuter 1980).

Drinking flights are made during morning hours, the birds flying in groups ranging in size from pairs to as many as twelve. Approaching birds typically fly quite high, and circle the waterhole a few times before landing (Urban et al. 1986). Little else is known of drinking behaviour, and water-transport behaviour by males is still undocumented.

Nearly all the seeds identified in the few crops that have been examined have proved to be legumes, mainly *Heliotropium*, *Indigofera* and *Trianthema*. These are all herbaceous plants that typically grow rapidly after rains, and their seeds provide a rich supply of high-protein foods for both breeding females and chicks (Kalchreuter 1980).

Social behaviour

Although it is known to be monogamous, no specific information is available on the social behaviour of this species.

Reproductive biology

The breeding season is fairly extended in northern Tanzania, lasting from May to November or December, with a probable peak between June and August. This span also generally corresponds to the breeding period of the chestnut-bellied sandgrouse in the same area of northern Tanzania, although the black-faced sandgrouse appeared to Kalchreuter (1980) to have a somewhat shorter and more precisely timed season. Neither species had a definite second breeding peak, although Kalchreuter believed that breeding begins seasonally with older birds in May and June, which produce the breeding peak. These are followed by renesting pairs, older juveniles, and possibly also second-brooders, although evidence for juvenile breeding and/or double-brooding was not directly obtained by Kalchreuter. Throughout the year, males of both the black-faced and the chestnut-bellied sandgrouse were found that were actively moulting their wing and tail feathers, but females evidently more or less confined their moulting to the time of minimum breeding activity between August and April.

The nest is placed on sandy or stony ground, and clutch sizes of two (in the Serengeti plain of Tanzania) or three (elsewhere) eggs have been reported. Nothing else of consequence has been reported on this species' nesting biology and behaviour (Urban et al. 1986).

Evolutionary relationships

Snow (1978) considered this to be a member of the *quadricinctus* species group (which also included *indicus*, *lichtensteinii* and *bicinctus*), although Urban et al. (1986) included *decoratus* only as a probable member of this same assemblage. Maclean (1984) included it in a separate group that otherwise comprised *coronatus* and *personatus*, and I agree that these species, especially *personatus*, are probably fairly closely related to *decoratus* (see Figure 3).

Status and conservation outlook

Although of limited distribution, this species is locally common through much of Kenya at elevations below 1800 m, and it also occurs rather commonly over much of southern Ethiopia, southern Somalia and northern and eastern Tanzania.

MADAGASCAN SANDGROUSE (Plate 44)

Pterocles personatus Gould 1843

Other vernacular names: Malagasi sandgrouse, masked sandgrouse; ganga masqué, ganga de Madagascar (French); Madagaskarflughuhn (German).

Distribution of species (Map 33)

Endemic to western and south-western Madagascar. No subspecies recognized.

Measurements (mm)

Wing, males 216–248, females 203–216; tail, males 94–127, females 86–108; tarsus, both sexes c. 25.4 (D.G. Elliot 1878, Ogilvie-Grant 1893). Egg, av. 44.8 × 32.9 (Appert and Etchécopar 1962).

Weights

No information. Egg, 18 g (estimated).

Description

Adult male. Similar to *coronatus*, but generally darker both above and below, the axillaries and upper surface of the shaft of the outermost long primary (tenth) dusky, never white. A wide black band surrounds the gape, forming a facial 'mask'; crown and upper back dark isabelline, shading to yellowish on the nape; lower back, rump and upper tail-coverts blackish-grey, thickly spotted with whitish-buff; primaries, primary coverts, and secondaries black; scapulars vinaceous brown, paler at the extremities; rest of the wing-coverts yellowish-buff, the secondary coverts each with a black shaft-stripe nearly reaching the tip, and some of the median wing-coverts with a brown terminal band; throat white, tinged with buff; chest vinaceous buff; sides of breast white, rest of breast and belly rufous buff, all closely barred with black; tarsi and under tail-coverts buff; tail of fourteen to sixteen rectrices, blackish-grey, irregularly barred and widely tipped with white. Axillaries black; underwing lining dark brown. Iris brown; bill and toes apparently bluish; colour of bare eye-ring not reported, but shown as red in Plate 44.

Adult female. Differs from the male in having the throat and sides of the face buffy white, the breast vinous, the top of the head spotted with blackish-brown; the nape, upper back, lesser and median wing-coverts regularly barred with the same colour; the scapulars blackish-brown irregularly barred with buff, and the greater secondary coverts buff irregularly barred with black; flanks and chest barred with black and white; the abdomen and vent with rufous and black; tail blackish-brown, barred and largely tipped with yellowish-white. Bill and toes bluish-black.

Immature. Similar to the female in having the breast vinous in the centre, pale buff on the sides, each feather with two narrow brown bars. Abdomen, flanks and vent all chestnut, barred sparingly with black. Wings very pale buff, barred with black (D. G. Elliot 1878). Presumably the outer primaries and most other juvenal feathers are tipped or edged with buff as in other sandgrouse species.

Identification

In the hand. Males of this large species have a black facial mask, a blackish shaft on the tenth primary, underparts that are rufous buff, barred with black, and tarsi that are uniformly buff. Females are similar but lack the black mask and are more heavily barred throughout, including the upper wing-coverts.

In the field. This is the only sandgrouse species occurring in Madagascar. Both sexes have rather strongly barred flanks, but lack pectoral banding. Its flight-call is a distinctive *catch-catcha-catcha* (Rand 1936).

General biology and ecology

Little has been written on this species, which has the most restricted range of any of the sandgrouse, but Rand (1936) found it to be common in southern Madagascar. There it occurred in open wooded country and the plains but avoided the more heavily wooded areas. Farther north (above Soalala) it was only fairly common on palm plains and savannahs. Generally Rand believed it to be a bird of the open brush, savannahs and plains of the greater portion of the western savannahs and subdesert habitats of Madagascar.

Rand found the species feeding in dry areas, but coming to rivers, ponds and lakes to drink and to rest on sandbars. It was usually seen in groups of from three or four to as many as twenty to thirty. It was often observed flying 30–100 m above ground, and usually came to water during the morning, sometimes remaining to rest on sandbars during the day. One large gathering was observed in late November, when 300–400 birds had arrived on a river's edge by 7 a.m., in flocks of up to thirty birds. By 8.30 a.m. all of them had left. Thus, like most sandgrouse, it is evidently an early-morning watering species.

Nothing specific has been written on this species' foods.

Social behaviour

Other than the comments on its relative gregariousness noted above, there is no specific information on the social behaviour of this species.

Reproductive biology

Rand (1936) found evidence of breeding from May 24, when a half-grown juvenile was collected, until late October, when a two-thirds-grown juvenile was obtained and a nest with three heavily incubated eggs was found. Birds in breeding condition were obtained in late June and early July. Rand thus believed that breeding probably occurs throughout the year. A nest with three eggs was also found in mid-June by Appert and Etchécopar (1962).

The nest found by Rand was located in an open, slight grassy area on a brushy, rocky hillside, and consisted of a slight scrape in gravelly soil, with a little grass at the bottom of the scrape. The nest found by Appert and Etchécopar was within about 10 m of a large specimen of *Tamarindus indica*. The location was in undulating steppe vegetation having a few trees and bushes, and the nest was mostly surrounded by the grass *Heteropogon contortus*. These authors judged the species to prefer a nesting habitat of flat terrain with little vegetation present, or only very short vegetation.

There is no information on the pattern of egg-laying, incubation behaviour or duration, or any other aspect of parental behaviour.

Evolutionary relationships

Maclean (1984) included this species in a group that also included *decoratus* and *coronatus*, which he characterized as all having black forehead colouration in males, and an absence of elongated central rectrices. The geographic distribution of *decoratus* in East Africa would make it a logical candidate for the nearest African relative of this species, and indeed the plumage patterns of the two are very similar. For these reasons I believe they are probably close relatives (see Figure 3).

Status and conservation outlook

There is no specific information on the status of this species, but Dee (1986) has stated that it is widespread in the western half of Madagascar, and is rather common. It is probably most common on the central grassland plateau area, but seemingly it can occupy a fairly broad range of open brush or sparsely wooded habitats, as might be found widely in western Madagascar.

LICHTENSTEIN'S SANDGROUSE (Plate 46)

Pterocles lichtensteinii Temminck 1825

Other vernacular names: Abyssinian sandgrouse, Arabian close-barred sandgrouse ('*arabicus*'), close-barred sandgrouse, Somaliland sandgrouse(*lichtensteinii*), Suk sandgrouse (*sukensis*); ganga de Lichtenstein (French); Wellenflughuhn (German).

Distribution of species (Map 36)

Africa, from Mauritania east through the Sahara to Ethiopia and Somalia, and south to northern Uganda and central Kenya; also the Arabian peninsula, Socotra Island, and from Israel, Jordan and Sinai east to Pakistan's North West Frontier Province and western Sind. Resident, but locally nomadic.

Distribution of subspecies

P. l. tarquis Geyr von Schweppenburg 1916: Sahara and the Sahel zone from Morocco and Mauritania east to Chad.

P. l. sukensis Neumann 1909: from southern Ethiopia and south-eastern Sudan south-west to north-eastern Uganda and northern Kenya.

P. l. lichtensteinii Temminck: south-eastern Egypt, Sudan (except south-east), northern Ethiopia, and northern Somalia, east through Israel, Jordan and Sinai, the Arabian peninsula, Socotra Island, and Pakistan east to North West Frontier Province and western Sind. Includes *arabicus* Neumann 1909, considered by Ali and Ripley (1983) to be part of *indicus*.

P. l. ingramsi Bates and Kinnear 1937: Hadhramaut (southern Yemen).

Measurements (races combined, mm)

Wing, males 187–195 (av. of 5, 191), females 176–187 (av. of 3, 180.7); tail, males 70–81 (av. of 7, 75.7), females 66–72 (av. of 3, 70); tarsus, males 26–28 (av. of 3, 27), two females 25; culmen, males 13.3–17.1 (av. of 7, 14.9), females 13.7–14.6 (av. of 3, 14). Egg, av. 42 × 26 (Cramp 1985).

Weights (g)

Males (*sukensis*), 175–250 (av. of 11, 217), females 190–230 (av. of 7, 207) (Urban *et al.* 1986). Egg, 16 (estimated).

Description

Adult male. Resembles *indicus*, including a black-and-white forehead pattern, but the ground colour of the upperparts pale whitish-buff and the throat and neck spotted with black; the rest of the upperparts and the chest closely barred with narrow black bands; upper breast yellowish-buff, divided in the middle by a narrow dark chestnut-to-black band, and separated posteriorly from the rest of the barred whitish and black underparts by a second black band (sometimes poorly developed), giving the lower breast a four-banded sequence of wide yellow and narrow blackish bands; outer web of the outer secondaries nearly white, tipped with yellowish-buff and barred with black; tarsi pale buff to white; rectrices (fourteen or sixteen) rufous buff and wedge-shaped, with black subterminal banding and yellow-ochre to deep rufous-buff tips; underwing lining light brown, axillaries and undersides of remiges more greyish, with brownish mottling at tips. Iris brown; eye-ring yellow; bill brownish-red; toes orange-yellow.

Adult female. Similar to the male, but lacking the black-and-white forehead markings, the breast-banding, and the heavy black barring on the sides and underparts. Closely resembling the female of

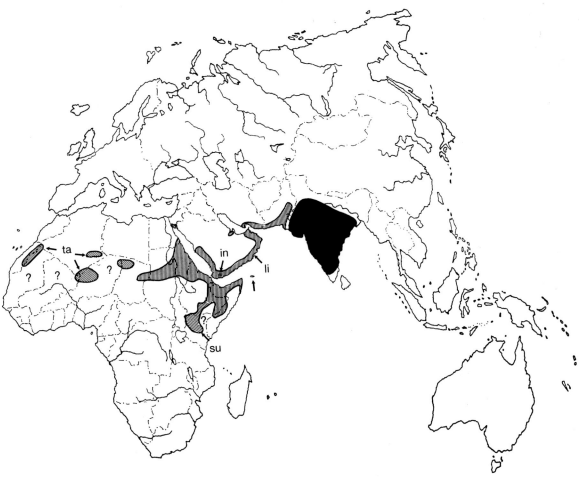

Map. 36. Distribution of the painted (solid) and Lichtenstein's (hatched) sandgrouse, including the locations of the latter's races *ingramsi* (in), *lichtensteinii* (li), *sukensis* (su) and *tarquis* (ta).

bicinctus, but generally less coarsely barred, especially above. Differs from the male in being generally browner throughout, having the throat, chest and lower breast uniformly and narrowly barred with black, the head lacking black and white patterning and instead being rufous buff with narrow black streaks, the barring on the upperparts much finer and closer, the abdomen narrowly barred with black like the upperparts, and the tips of the scapulars and larger wing-coverts whitish-buff.

Immature male. Differs from the adult male in having the spots on the throat and neck small and indistinct, the chest and upperpart barring pale and irregular, and only a few adult feathers on the wing-coverts and scapulars. The chestnut pectoral band and head patterning of the adult male is incomplete, and the terminal black band lacking; the juvenal primaries are tipped with buff and brownish-black vermiculations.

Immature female. Similar to adult female, but more closely barred, and the remiges margined with buff and mottled grey; older birds separable only as long as some juvenal outer primaries retained (Cramp 1985).

Identification

In the hand. Males have a uniformly white to buffy tarsus, a pectoral band of four bars (a repeated pattern of a broad yellow-buff band followed by a narrower black bar), wing-coverts white, narrowly tipped with black, and with buff tips, and a throat lightly spotted with black. Females have an unbarred buffy tarsus, lack pectoral bands, the throat is thickly spotted with blackish, and the wing-coverts have narrow and regular black barring.

In the field. Males of this extreme desert-inhabiting species are banded with black and white on the forehead and crown, as is the broadly sympatric four-banded sandgrouse, the latter also having a black and white band on the lower breast. The fine black barring on both back and underparts is distinctive for both sexes; in males this is set off on the lower breast by alternating bands of yellowish-buff and black. Females are more completely and uniformly barred throughout than other sandgrouse, and their whitish to buffy throats are spotted with blackish all the way to the chin. Their calls include a repeated liquid whistling *quitoo* flight-call, and they also utter musical chattering notes when at water.

General biology and ecology

This is a species highly adapted to extreme desert conditions. Thomas and Robin (1977) classified it as a true desert species; on the basis of the water-holding capacity of its feathers it ranked the highest of seven species tested, and its relative kidney weight was also ranked as perhaps the most desert-adapted of five sandgrouse species analysed. Snow (1978) also regarded it as the most desert-adapted of the four species he regarded as forming a single species-group. This species prefers rocky or scrubby habitats, especially such as bush-covered stony hillsides, bushy wadis, clearings in thornbush or scrub, and open patches of *Acacia*, or similar plants which may provide shade during the hottest parts of the day. It avoids open deserts wholly lacking in vegetation, and cultivated habitats, and is also not associated with uniformly sandy substrates. However, it does utilize partly sandy and partly stony sites, such as in ravines where there are a few thorny shrubs or tamarisk cover (Ali and Ripley 1983, Cramp 1985, Urban *et al.* 1986).

Unlike most other sandgrouse (all but those in its own species-group), it is nocturnal to crepuscular in its watering activities, flying considerable distances each dawn and dusk to water, although exact distances are unknown. As with its close relatives, these tend to be low-altitude flights rather than the high-altitude flights of the species that water during daylight hours. However, like all sandgrouse, the birds typically land some distance away from the water's edge and walk or run down to it after waiting a minute or two to make certain it is safe. After drinking quickly, the birds leave directly. Water of low salt content may be drunk, and similarly the major food plants are rather low in salts, an estimated 8–35 per cent of the major ions being derived directly from these foods (Thomas and Robin 1977).

Several studies have suggested that the seeds of *Acacia*, especially *A. sayal*, represent important food plants, those of asphodel *Asphodelus tenuifolius* and *Cassia* being used when acacia seeds become unavailable. Meinertzhagen (1954) found 3674 acacia seeds in the crops of ten birds. Two specimens obtained by Thomas and Robin (1977) contained only the seeds of acacia and asphodel, although others also included the flowers and leaves of *Mesembryanthemum* and the seeds of *Plantago* and *Reseda*. A few other food plants such as *Salsola* seeds and insects or their larvae have been mentioned as apparently minor foods (Cramp 1985).

Social behaviour

Almost nothing is known of the social behaviour of this species, which is usually not found in large groups except during watering, and is also a solitary nester. A few observations of display have been reported, including tail-raising and fanning of the rectrices, along with wing-drooping. Captive birds have been observed strutting and 'pouting', making turning and hopping movements, and raising the rear end and wings as a prelude to attack. Wing-clapping and calling were also associated with these activities (Heuglin 1873).

Reproductive biology

Few nests of this species have been found. In Sind (Pakistan) two nests were found on bare pebbly ground on which some thorn-bushes and tamarisk were growing. Both of these nests were found in May (Baker 1935). In north-eastern Africa, eggs have been found in June and July, in Somalia during February and July, and in East Africa during May, June and September. The few nests so far found have typically been among scattered trees or rocks, either in the open or under the shelter of a low shrub. Clutches of both two and three eggs have been found, but there is too little information available as to judge the typical clutch size, although Meinertzhagen (1954) judged that only rarely are three eggs laid. Similarly there is no information on nest-site selection, egg-laying, incubation behaviour or duration, or the care of the young.

As noted earlier, the breast feathers of this species are highly adapted to holding water (Cade and Maclean 1967), and have been found capable of holding more than did any others of the seven sandgrouse species tested (Thomas and Robin 1977). There is thus every reason to believe that the young are watered by the male parent in the usual sandgrouse manner, although direct observations of this are apparently still lacking.

Thomas and Robin (1983) observed a pair with a single chick in late May in Morocco, noting that for about an hour the female and chick sheltered in the shade of a rock, while the male stood nearby and performed gular fluttering. Later, as the birds moved off, the chick walked in the shade of the female, and hid immediately in the shade of a rock when followed. Like its near relatives *indicus*, *coronatus* and *senegallus*, the chick of this species is very pale-coloured and almost unpatterned, and apparently this reflects an adaptation for survival in extreme deserts.

Evolutionary relationships

This is clearly a very close relative of such forms as *bicinctus* and *indicus*; indeed, Meinertzhagen (1954) believed that all three should all be regarded as conspecific, and Vaurie (1961) regarded them as closely related replacement forms. Snow (1978) additionally included *quadricinctus* and *decoratus* in this phyletic assemblage, which he considered to represent a species-group. Snow, and later Urban *et al.* (1986), furthermore considered the two most closely related forms in this group to be *indicus* and *quadricinctus*; these same taxa were considered a distinct group by Maclean (1984) and were recognized as a full genus (*Nyctiperdix*) by Bowen (1927) as well as more recently by Wolters (1974). I have tentatively excluded *decoratus* from this group, but otherwise agree that this assemblage is a closely related array of largely allopatric forms (see Figure 3).

DOUBLE-BANDED SANDGROUSE (Plate 47)

Pterocles bicinctus Temminck 1815

Other vernacular names: ganga bibande (French); Doppelbandflughuhn, Nachtflughuhn (German).

Distribution of species (Map 37)

Resident from south-western Angola and Namibia through northern Cape Province, Transvaal, Botswana and southern Zambia to western Mozambique and extreme southern Malawi (Urban *et al.* 1986).

Distribution of subspecies

P. b. bicinctus Temminck: extreme southern Angola, Namibia, Botswana and north-western Cape Province. Includes *chobiensis* Roberts 1932.

P. b. ansorgei Benson 1947: south-western Angola.

P. b. multicolor Hartert 1908: Mozambique, Malawi and Zambia south to the Transvaal. Includes *usheri* Benson 1947.

Measurements (*bicinctus*, mm)

Wing, males 172–187 (av. of 10, 181), females 171–185 (av. of 10, 177); tail, males 73.5–86.5 (av. of

Map. 37. Distribution of the four-banded (solid) and double-banded (hatched) sandgrouse, including the locations of the latter's races *ansorgei* (an), *bicinctus* (bi) and *multicolor* (mu).

10, 80.6), females 76–82.5 (av. of 10, 79.1); tarsus, males 22–26.5 (av. of 10, 23.5), females 22–26 (av. of 10, 23.5). Egg, av. 37.3 × 26.7 (Urban *et al.* 1986).

Weights (g)

Males 215–250 (av. of 9, 234), females 210–280 (av. of 19, 239) (Urban *et al.* 1986). Egg, 14 (estimated).

Description

Adult male. Head, upper back and chest as in *indicus*; rest of back, rump, and upper tail-coverts blackish-brown, the feathers slightly tipped or spotted with white and irregularly barred and marked with rufous buff; scapulars the same, but the bars very irregular, and each feather has a rounded terminal white blotch, producing a distinctly spotted effect; wings as in *indicus*, but all the secondaries brownish-black on the outer web to the base, and the outer ones ornamented with one or two oblique white bands; the greater and some of the median wing-coverts blackish-brown, tipped with white; chest separated from the upper breast by a white and then black zone; rest of underparts as in *indicus*; rectrices fourteen (sixteen reported by Ogilvie-Grant), deep buff banded with brownish-black, and mostly tipped with buff; underwing lining greyish-brown; axillaries grey. Iris brown, eye-ring yellow, bill yellowish to reddish with darker tip; toes dull yellowish to brownish.

Adult female. Chin and fore-throat creamy buff, with the lower throat and upper breast becoming pinkish-buff, mottled and barred with sepia or blackish, and the sides becoming greyish-white, finely barred with black. There is no pectoral banding as in the male. The tibia feathers are buffy white, and the wing-coverts and scapulars are terminally spotted with white and transversely barred with dark brown. Generally very like the female of *indicus*, but the black barring on the upperparts and chest wider and less regularly barred, and the scapulars and smaller wing-coverts tipped with white; the terminal black bars on the upperparts and chest are more crescent-like or bridge-shaped, the latter producing a spotted appearance.

Immature (both sexes). Juveniles are similar to the adult female, but more pinkish-fawn and less barred with speckled and coarse vermiculations and sub-terminal bars of dark brown on breast, underparts finely barred with brown, primaries tipped with pinkish-buff, most feathers edged with buff. With the assumption of the first basic plumage both sexes are reportedly exactly like those of the respective adult plumages (Urban *et al.* 1986, Clancey 1967);

details of primary moult are still unstudied, but perhaps young birds might be recognized for a time by their worn and faded outer primaries.

Identification

In the hand. Males have a double pectoral band (white followed by a wider black bar), a white tarsus barred with blackish-brown, and a buffy throat. They are very similar to *indicus* but have white dorsal spotting that is lacking in that species. Females have no pectoral band, the throat is heavily spotted with black, the tarsus is similarly barred, and the underpart barring extends to the throat. They too are very similar to *indicus*, but the upperpart markings are less tranversely barred and more wavy or crescent-like, and the cheeks and throat are more distinctly spotted with blackish.

In the field. In its southern African range this species is the only sandgrouse that has a white forehead and is distinctly barred with black and white on the flanks (males), or has a short wedge-shaped tail and breast-barring extending up to the throat (females). The flight calls are a series of harsh *chuck-chuck* notes, and when on the ground the birds utter a complex phrase sounding like 'Oh NO, he's gone and done it AGAIN'.

General biology and ecology

This species mainly occupies acacia-veld and light woodland savannahs, thornveld or bushveld, where a supply of water is fairly nearby and the climate ranges from moderately moist to quite arid. Rocky rather than sandy substrates are preferred; these include barren and stony hill slopes, open woodlands where the soil is loose and stony, gravelly plains, and generally rocky ground. Desert edges are sometimes used, as are *Brachystegia* woodlands, but such habitats may be somewhat exceptional (Clancey 1967, Urban *et al.* 1986).

Watering is probably done on a daily basis, the birds leaving their foraging areas near sunset and flying in small groups and at low altitude to water. They often arrive well after sunset and may remain until after dark. Like other sandgrouse, they typically land near water, and then after a period of watchful waiting run to the water to drink. From there they fly back to their foraging areas, where nocturnal feeding may occur if moonlight allows. Otherwise they feed during early morning and late afternoon, spending the hottest part of the day in the shade of trees or rocks (Urban *et al.* 1986).

When foraging (at least in captivity), the birds were observed typically to search throughout for food in a limited area (compare the Namaqua sandgrouse), and generally spent the afternoon resting in the shade. At such times the birds often huddled in a group, either standing or sitting in the shade, fairly close together but not tightly packed. On the other hand, nocturnal huddling was done as a tightly packed group, often beside a bush or small stone, as many as seven birds being part of the group. Although in nocturnal huddling the Namaqua sandgrouse were often found to be organized into an arrangement of adjacent birds sitting in opposite directions, this was not found to be the case in the double-banded sandgrouse. Even at ambient daytime temperatures of up to 50°C, both of these species are evidently able to maintain rather constant internal body temperatures of 40–42°C, using such behavioural devices as feather erection, gular fluttering and wing-drooping (Thomas *et al.* 1981).

Foods include various seeds such as those of grasses, herbs, and weeds, including the introduced weed *Bidens bidentata*.

Social behaviour

Little is known of the social behaviour of this species, other than that it is monogamous and a solitary nester.

Reproductive biology

Breeding primarily occurs during the winter dry season over this species' range. In Namibia it mainly breeds between May and October, but it has also been reported nesting in November and December. The Botswana population ('*chobiensis*') also breeds during the cool and dry winter months from May to August, and the more easterly populations in Zimbabwe and Transvaal breed between April or May and September or October. There is also an October breeding record for Angola (Clancey 1967, Urban *et al.* 1986).

The nest is placed on the ground among thin grasses or under a shrub, or on rather bare ground underneath a tree. There are two or three eggs; the mean of seventeen clutches in Namibia was 2.6 eggs. Little is known of the incubation behaviour, but a captive pair incubated some infertile eggs for 33 days, the female incubating from 7 or 8 p.m. until about 1 p.m. the following day, and the male during the afternoon. A similar pattern of the female incubating until about 9.30 a.m. and the male the rest of the daylight hours has also been observed, which is nearly the opposite of the usual sandgrouse pattern (Urban *et al.* 1986).

Evolutionary relationships

This is certainly a member of the assemblage recognized by Snow (1978) as the *P. quadricinctus* species-group, although *decoratus* is only questionably a member of this group. Both sexes are remarkably similar in colouration to *indicus*, in spite of the fact that these two species represent the opposite geographic ends of this broadly distributed array of essentially geographic replacement forms. Like Maclean (1984), I have tentatively concluded that *bicinctus* might be regarded as a derivative of ancestral *lichtensteinii* stock, which also apparently produced both *quadricinctus* and *indicus* (see Figure 3).

Status and conservation outlook

This is apparently a fairly common species over much of its somewhat limited range. It is ecologically replaced to the north by *decoratus*, another African endemic sandgrouse with a comparable small range.

FOUR-BANDED SANDGROUSE (Plate 48)

Pterocles quadricinctus Temminck 1815

Other vernacular names: Lowe's four-banded sandgrouse; ganga quadribande (French); Buschflughuhn, Dreibandenflughuhn (German).

Distribution of species (Map 37)

Widespread in the sub-Saharan Sahel from Senegambia east to Ethiopia and north-western Kenya. Resident or locally migratory. No subspecies recognized by Urban *et al.* (1986).

Measurements (mm)

Wing, males 175–182, females 176–184; tail, males 75–80, females 72–78; tarsus, males 25–26, females 23–25; culmen, both sexes 13–15. Egg (East Africa, averaging slightly larger in west), av. 38.7 × 27.6 (Urban *et al.* 1986).

Weights (g)

No information. Egg, 16 (estimated).

Description

Adult male. Differs from *indicus* in the markings of the yellowish-buff greater secondary and inner median wing-coverts, which have a black bar, narrowly edged on either side with white, across the external third of each feather. Differs from *lichtensteinii* in having a white (not yellow) lower breast-band, which is bounded posteriorly by a narrower black band and anteriorly by a chestnut band (rather than having two yellow breastbands separated by black). Rectrices fourteen (sixteen reported by Ogilvie-Grant), banded with ochre-yellow and black; underwing lining barred greyish-brown and white to brownish-grey, axillaries grey. Iris dark brown, eye-ring yellow; bill dull yellow; toes yellow.

Adult female. Generally like the male, but lacking the black-and-white forehead markings and the breast-bands, these areas instead mostly rufous buff, with darker spotting and barring. Differs from *indicus* and *lichtensteinii* in having the front of the neck and the chest uniform buff without blackish bars; otherwise very similar to both except that the black bars on the scapulars are wider and less numerous. The greater secondary coverts and outer median wing-coverts also have a black bar across their terminal thirds, as in the male, but lack white margins. Bill yellowish, with a darker tip, or mostly black.

Immature male. After losing the female-like juvenal plumage, increasingly like the adult male, but the back and sides of the neck barred and the sides of the throat spotted like the female, whilst the pectoral band is incompletely formed and the primaries are widely tipped with buff and mottled or vermiculated with black.

Immature female. Like the adult female, but generally more rufous, the upperparts tinged with buff, and the primaries broadly tipped with rufous or whitish buff (Urban *et al.* 1986).

Identification

In the hand. Males have a pectoral band of three approximately equally wide bars (chestnut, buffy white and black), an unspotted buffy throat, an unpatterned whitish tarsus, and wing-coverts crossed by one or black bars and narrowly edged on either side with white. Females lack a pectoral band, and have an upper breast and throat of unspotted buff and a tarsus barred with black. Both sexes are most similar to *indicus*, but males differ in having black-barred wing-coverts, and females in their black of dark barring on the upper breast ('chest'). They also resemble *bicinctus*, but males of that species are much more strongly spotted and barred with white on the scapulars and back, and females are spotted with blackish from the lower breast right up to the throat or chin.

In the field. In its sub-Saharan range this wedge-tailed species is the only sandgrouse with a black-and-white forehead and usually tricoloured (chestnut, white and black) breast banding; the otherwise very similar and sympatric Lichtenstein's sandgrouse has bicoloured black and yellow breast-banding. Females are unique in having a plain buff upper breast ('chest') and throat contrasting with a distinctly barred lower breast and belly. Their calls include soft or shrill piping *wur-wulli* or *pirrou-ee* whistles in flight, and loud twittering noises at water.

General biology and ecology

This species occupies open or partly wooded savannahs, open and bushy grasslands, coastal dune scrub, and some cultivated areas, including dissected pastures. Stony, pebbly, or cotton (clay) soils are frequent substrates. It is more closely associated with somewhat wooded or cultivated vegetation than is the rock-associated Lichtenstein's sandgrouse, and is also apparently not associated with sandy substrates (Urban *et al.* 1986).

Like its close relatives, it is rather inactive during the day, but waters before dawn and after dusk, typically flying to water at around sunset and remaining there until dark. Although flocks going to water are typically small, fairly large numbers of birds may aggregate at favoured waterholes. Some drinking may also occur before sunrise (Lewis *et al.* 1984). Movements greater than daily flights to water may also occur seasonally; in particular there appears to be a northward movement to the sub-Saharan sahel zone during the rainy season (Urban *et al.* 1986).

Foods of this species are still unstudied, but probably consist of small, dry seeds like those of related species.

Social behaviour

There is no specific information on this, but the birds are known to be monogamous and solitary nesters.

Reproductive biology

Breeding records from the western edge of the species' range (Senegambia) are from November to June, but mainly during March, whilst in Cameroon there is a record of chicks in January. At the eastern edge in Sudan and Ethiopia there are breeding records for February and March, and in Ethiopia perhaps also for July, August and December. For the area of northern Uganda and western Kenya there is a January breeding record (Urban *et al.* 1986).

There are typically two or three eggs in the clutch

(two out of three clutches had three eggs), and the nest site is often placed among the dried leaves of *Bauhinia* trees, where the eggs are cryptically matched to the background. The nest may also be placed on bare stony soil, under a shrub, or among broken scrub and bushes. Almost nothing else is known of the species' breeding biology (Urban *et al.* 1986).

Evolutionary relationships

This species is certainly an extremely close relative of *indicus*, and as noted in earlier species accounts, it has been suggested by Snow (1978), Urban *et al.* (1986) and others that they might be easily regarded as a superspecies. Maclean (1984) has diagrammatically derived both of these forms, as well as *bicinctus*, from a *lichtensteinii*-like ancestral type, and I have made a very similar estimate of relationships in this assemblage (see Figure 3).

Status and conservation outlook

This is a fairly widespread and generally uncommon-to-common species throughout its range. Although little-studied, there is no reason to believe that it is particularly rare or threatened anywhere in its range.

PAINTED SANDGROUSE (Plate 49)

Pterocles indicus (Gmelin) 1786

Other vernacular names: Indian sandgrouse; ganga des Indes (French); Bindenflughuhn, Indisches Flughuhn (German).

Distribution of species (Map 36)

Resident in eastern Pakistan and peninsular India from Rajasthan and Gujarat east to Bihar and Orissa, and south to Mysore and Tamil Nadu (Ali and Ripley 1983).

Distribution of subspecies

None recognized by Peters (1937). Although Ali and Ripley (1983) considered *arabicus* Neumann 1909 (from western Pakistan to southern Arabia) to be a subspecies of *indicus*, Peters and other more recent authors consider *arabicus* to be part of *lichtensteinii*; indeed, Vaurie (1965) regards this form as inseparable from nominate *lichtensteinii*.

Measurements (mm)

Wing, both sexes 158–184; tail, both sexes 80–101; tarsus, both sexes c. 23–25; exposed culmen, both sexes c. 13–15 (Ali and Ripley 1983). Egg, av. 35.8 × 25.

Weights (g)

Males 166–208, females 166–176 (Ali and Ripley 1983). Egg, 12.4 (estimated).

Description

Adult male. Lores, forehead and superciliary stripes white, a wide black band crossing the forehead just in front of the eyes, and an oval black spot above each eye; feathers of crown buff, each with a wide black shaft-stripe; upper back, throat and chest yellowish-buff; rest of back, rump and upper and under tail-coverts black, barred with white or yellowish-white, narrowly tipped with buff; primaries, primary coverts and secondaries brownish-black; the primaries narrowly margined with white, and the secondaries with a basal patch of dirty white, this gradually increasing in extent outwardly to the outermost secondary, which is pale to the tip; other webs of the seventh and eighth secondaries crossed by one, and the ninth and tenth by two oblique black bands narrowly margined on both sides with white; a few of the outer median and lesser wing-coverts yellowish-buff; greater wing-coverts and rest of median coverts barred alternately with wide stripes of white and dark grey and widely tipped with yellowish-buff; scapulars and rectrices (sixteen) black, barred with rufous buff and widely tipped with yellowish-buff; chest bounded by a moderately wide chestnut band, followed by somewhat wider bands of pale yellowish-buff and black; the rest of the underparts with narrow bars of white and black; underwing lining and axillaries medium grey, the underside of the remiges a somewhat darker greyish-black. Iris brown; eye-ring yellow to yelowish-green; bill orange to orange-brown; toes dull yellow, orange-yellow, or greyish-green.

Adult female. Differs from the male in having no white and black markings on the head; sides of neck and throat spotted with black; the chest and upperparts rufous buff closely barred with black; the ends of the scapulars and most of the outer webs of the wing-coverts yellowish-buff. No pectoral band, and the outer web of the outermost primary bright buff. Generally more coarsely barred than the female of *lichtensteinii*, and the chest barred to the throat as in *bicinctus* but not *quadricinctus*, differing from the former mainly in having the barring of the wing-coverts more wavy and irregular.

Immature. Immature males differ from adult males in having the upperparts female-like, but the black bars are broken up into vermiculations on most feathers; the tips of the primaries are vermiculated with black and buff, and the chest has only traces of a chestnut band. Young females also have vermiculated primary tips, but otherwise resemble adult females. Probably the outer primaries are retained in the post-juvenal moult, as has been reported to be true for *lichtensteinii*, which would assist in identifying birds of the year once their adult body plumage has been attained.

Identification

In the hand. Males are identifiable by the presence of three differently coloured pectoral bars (chestnut, buffy white, and black), and each wing-covert having a broad bicoloured white and grey band towards the tip, or sometimes four alternating and narrower white and grey bars. Females have no pectoral bands, have only a few black spots at the base of the throat, the wing-coverts are narrowly barred with buff and black, and the tarsi are barred with blackish-brown. Both are very similar to *quadricinctus*, but males of *indicus* may be separated by their dark grey (not black) barred wing-covert pattern and females by their barred rather than uniformly buff upper breast and throat, although the lower breast is barred in both. Also very similar to *bicinctus*, but the males of that species have white-tipped scapulars and mantle feathers, and females are more distinctly spotted on the cheeks and throat. Females are more coarsely barred than those of *lichtensteinii*, and in that species the tarsi are pure white rather than barred with black.

In the field. In India this species is recognized by its wedge-shaped tail, darkly barred flanks and undersides, and dark underwing lining. In Pakistan and north-western India it may come into contact with *lichtensteinii*, males of which have a broad yellowish and closely barred breast with no lower chestnut band, and (at least in that region) the larger wing-coverts are not so broadly banded with black and white. Females of *lichtensteinii* likewise lack chestnut tones above, and the feathers have a generally more whitish ground colour, especially below. The flight call of *indicus* is a distinctive *chirik-chirik*.

General biology and ecology

This species is largely associated with bare, stony substrates rather than sandy ones, and also apparently avoids coastal areas. It is largely found in dry, broken country, including stony hillsides and rocky ravines more or less covered by scrub-jungle. It also occurs on rocky ground in thinly scrubbed woodlands, and is especially associated with open stony burnt-grass jungle (Baker 1935, Ali and Ripley 1983).

The species is distinctly crepuscular rather than diurnal in its drinking schedule, the birds often flying considerable distances before dawn to reach water, and making a second flight in the evening hours when it is almost too dark to see. Large numbers of birds typically arrive at jungle pools, the flocks usually keeping well below the skyline and thus being particularly hard to observe, although constantly calling as they approach and depart from water. Like its foraging behaviour, its foods are likewise still not known in any detail, but they reportedly include the hard seeds of various weeds and grasses, some berries, and apparently also small insects such as termites (Ali and Ripley 1983).

Social behaviour

Little has been reported on this species' sociality, other than that the birds often are found in groups of two or three, or in small packs of about eight to ten (Ali and Ripley 1983). The birds are believed to be generally resident, and there is no reason to doubt that they form strong, perhaps long-term pair-bonds.

Reproductive biology

The breeding season of this species is not well defined, and although it may cover practically the entire year, it may be concentrated between March and June (Ali and Ripley 1983) or between April and June, with scattered records for all other months except July and September (Baker 1935).

The nest is normally only a simple scrape among stones and earth with little or no vegetation, although at least one rather 'abnormal' nest was found on sandy soil, under the shadow of a small tree, and was described as being a well-assembled saucer-like structure of dried grass and other dry vegetation. However, it is common for the eggs to be placed under the shade of a bush, tree, or some other vegetation, or where a stone or boulder provides some shade during the hottest part of the day. There are typically three eggs, but sometimes only two (Baker 1935).

Both sexes incubate, and Baker (1935) estimated the period to be 21–23 days, a duration apparently based on data from other species. He also believed that the eggs are kept damp by the bringing of water to the nest in the adults' feathers, although this too was evidently not based on direct evidence. He did note that incubating birds have been observed panting and spreading their wings out while sitting over

C. Sandgrouse (Family Pteroclidae)   255

the eggs dring extremely hot periods. Ali and Ripley (1983) stated that the incubation period is still undetermined, and made no comments as to possible parental watering of the eggs or young.

Evolutionary relationships

As Snow (1978) has observed, *quadricinctus* is so strikingly similar to *indicus*, and their ranges so clearly represent geographic replacement types, that it is evident that *indicus* must be the result of a fairly recent eastward spread and colonization of the Indian region by an ancestral population of *quadricinctus*. As such, the two forms represent a superspecies, and both are part of a larger species group that also includes the African forms *bicinctus* and *lichtensteinii*. Maclean recognized these same taxa as constituting a single group, and indeed Bowen (1927) proposed that they all be included within the genus *Nyctiperdix*. Wolters (1974, 1975) also recognized these forms as belonging to the subgenus *Nyctiperdix*, but included several additional species within a larger genus of the same name.

Status and conservation outlook

This is evidently a moderately common species over some parts of India, although it is locally nomadic as well as resident, and tends to concentrate in some areas near the end of the rains. In the absence of more information, it seems unlikely that this species requires any special conservation attention.

YELLOW-THROATED SANDGROUSE (Plate 50)

Pterocles gutturalis Smith 1836

Other vernacular names: Kenya yellow-throated sandgrouse (*saturatior*), Smith's chestnut-vented sandgrouse; ganga à gorge jaune (French); Gelbkehlflughuhn (German).

Distribution of species (Map 38)

Disjunctive; one more-northerly subspecies resident from northern Ethiopia south through Kenya and Tanzania to north-eastern Zambia, the other somewhat migratory, breeding in southern Zambia and northern Botswana, and wintering to western Zimbabwe, south-eastern Botswana, Transvaal and northern Cape Province.

Distribution of subspecies

P. g. saturatior Hartert 1900: Ethiopia south to northern Zambia.

Map. 38. Distribution of the yellow-throated sandgrouse, including the locations of its races *gutturalis* (gu) and *saturatior* (sa).

P. g. gutturalis Smith 1836: southern Zambia and Botswana south to Transvaal and Cape Province.

Measurements (*gutturalis*, mm)

Wing, males 208–228 (av. of 10, 217), females 205–220 (av. of 10, 213); tail, males 83.5–94 (av. of 10, 91.1), females 75–88 (av. of 10, 81.1); tarsus, males 28–32 (av. of 10, 29.9), females 28–31 (av. of 10, 29.2); culmen, males 13.5–15 (av. of 10, 14.6), females 13.5–16 (av. of 10, 14.9), (Urban *et al.* 1986). Egg, av. (of 9) 45.6 × 33.9 (W. Tarboton, personal communication).

Weights (g)

Males 340–345 (av. of 3, 342), females 285–400 (av. of 6, 336) (Urban *et al.* 1986). Egg, 14.7 (estimated).

Description

Adult male. A black band from the nostril to the eye; a narrow buff stripe from above the nostril to the back of the eye; crown dull olive-buff, shading to dull yellowish-buff on the neck; chin, cheeks and throat yellowish-buff, separated from the neck by a black band; back, rump, and upper tail-coverts olive-grey, scapulars blackish, with greyish-buff tips; primaries, primary coverts and secondaries black, slightly margined with white; rest of the wing-coverts grey, with bright rusty tips; chest grey, tinged with vinaceous and becoming dark chestnut on the breast, belly and under tail-coverts; tarsi rufous buff; rectrices mostly black, barred and tipped with chestnut, the central pair like the upper tail-coverts; underwing lining and undersurface of remiges brownish-black, axillaries black. Iris dark brown; eye-ring grey; bill pale bluish-grey; toes pinkish-brown.

Adult female. A brownish-black band from the nostril to the eye, and a narrow black stripe from the lores to the back of the eye, and the chin and throat yellowish-buff; feathers of the crown, back, rump, upper tail-coverts, and chest yellowish-buff, each with a wide black shaft-stripe widening terminally; the lower tail-coverts also barred with black; primaries, secondaries and primary coverts as in the male, but the outer secondaries barred outwardly with buff; scapulars black, tipped and toothed on both webs with bright yellowish-buff, and with a wide W-shaped subterminal band of the same colour; rest of wing-coverts the same, but with the dark parts much reduced; breast, belly, and flanks dull rufous buff closely barred with black, under tail-coverts dark chestnut; central pair of rectrices yellowish-buff, barred with black like the longer upper tail-coverts; rest like those of the male.

Immature (both sexes). Similar to the adult female, but the upperparts with smaller spots and narrower olive-buff bars; the primaries broadly tipped and edged with olive-buff.

Identification

In the hand. Males have a uniformly rufous-brown tarsus, a yellowish face terminated above and below by black bands, and deep uniform chestnut underparts. Females have the yellow throat not terminated by black banding (but a brown loral stripe) and the chestnut underparts barred with black.

In the field. The presence of a wedge-shaped tail, a yellowish face and throat, and dark chestnut underparts identifies this species of sandgrouse, which also has dark underwing linings. Its primary call is a distinctive, far-carrying and four-syllable *ipi, aw-aw*. These last two notes are easily mimicked, the first being slightly higher pitched than the second. At drinking locations on the ground, the last two notes only are uttered. These notes have sometimes been described as *glock, glock* calls. The species is not a 'twittering' form (W. Tarboton, personal communication). When irritated, the birds make a growling sound.

General biology and ecology

The habitat of this species is perhaps the most unusual of all the African sandgrouse, consisting of shortgrass plains and clay-like, often relatively moist, soils, usually near rivers or swamps or on floodplains. It also occupies recently burnt ground, as well as ploughed or fallow fields in cultivated areas. Also unlike the other sandgrouse, its range extends from the hot and semi-arid parts of southern Africa northward through equatorial areas and beyond to the Red Sea. Like some other sandgrouse it is distinctly nomadic to migratory in some regions, tending to move with the wet and dry seasons. It is the largest of the African sandgrouse, and normally is less common than the smaller species with which it sometimes associates (Clancey 1967, Urban *et al.* 1986).

In contrast to the double-banded, four-banded and Lichtenstein's sandgrouse, it is a daytime waterer; groups of from about ten to fifty or more birds often gather at drinking areas in post-sunrise hours from about 7 to 10 a.m., or even later. According to W. Tarboton (personal communication) it has two peak drinking times, during early to mid-morning and again from mid- to late afternoon. At such times it may associate with chestnut-bellied and Burchell's sandgrouse, although the species tend to displace one another at drinking sites (Mungure 1974). Among grouped birds at drinking sites, wing-raising is a commonly observed behaviour that is of unknown significance but perhaps functions as an intention movement (W. Tarboton, personal communication).

At least in captivity, this species also has a strong tendency to wallow in mud, and apparently also needs green food, both of these being goose-like tendencies that perhaps make the yellow-throated sandgrouse unique among the sandgrouse assemblage. However, the presence of dry locations and access to sunshine is apparently important for keeping the birds successfully. Under such conditions both adults and chicks were found to consume not only grass, beets, and chickweeds but also such seeds as those of milo, flax, hemp, poppy, safflower, Mung beans, peanut hearts, and a very small amount of rape (Grueber 1987). Wild birds have been found to consume the seeds of the weeds *Achyramthes* and *Bidens*, the legumes *Cassia*, *Crotolaria* and *Sesbania*, and the grasses *Leersia* and *Rottboelia* (Clancey 1967).

Social behaviour

This is a monogamous species, with paired birds occurring together throughout the entire year, although little has been published on its behaviour. Presumptive courtship in the form of bobbing movements by a pair while facing one another has been mentioned (Urban *et al.* 1986), but head-bobbing in at least some other sandgrouse is used in a conflict depart–approach situation. According to Grueber (1987), who has not observed head-bobbing, the male's courtship display consists in erecting the feathers of the black neck-band and jumping about 15 cm vertically while making a quarter-turn with the tail erected and fanned. He believes that this display may serve to identify unmated males, as he has seen it most frequently performed by birds in groups. In one apparent early pair-bonding display, the male bowed toward an approaching female, raised his tail slightly, and uttered a low call. During later pair-bonding behaviour the birds slowly walk side by side, making somewhat rhythmical pecking movements. This behaviour continues until the birds have moved about 2–3 m together.

The female's invitation to copulation (Figure 53b) consists in a submissive sitting or crouching posture, with her head held low and stretched forward. No specific precopulatory display of the male was evident. The post-copulatory display of the male consists in holding the head arched downward, with the wings held away and downwards from the body, a posture that may be held for about 10–15 s. This posture is somewhat similar to that assumed during the pair-bonding display, but is performed only by the male (L. Grueber, personal communication).

Grueber has observed copulation in his captive birds (in southern California) during nearly all months, but with a peak (nine out of thirty-six observations) during May. He believes that vocalizations may play a very important role in pair-bonding, although they are so weak as to be almost inaudible to humans when more than a metre or two away, and only by watching their throat movements is it evident that they are actually vocalizing.

Reproductive biology

Because of its broad north–south distribution, the breeding season might well be highly diverse in various regions, but it mainly occurs during the dry season between April and September throughout the species' range. Thus in the Transvaal, breeding occurs from April to August, with most of the records for April and May (W. Tarboton, personal communication). In Zambia it breeds between May and September, with a peak in July, in the Serengeti of Tanzania from June to August, and elsewhere in Tanzania generally from April to September, with a peak in July. To the north in interior Kenya and north-eastern Tanzania there is a somewhat later peak, in August. Finally, in Ethiopia the breeding season is similarly from July to September (Urban *et al.* 1986).

The nest scrape, like that of other sandgrouse, is often placed in a natural depression of the hoof-print of an ungulate, or is partly hidden in a grass tuft or stubble. There are usually three eggs, but sometimes only two. Loren Grueber has consistently (four clutches) found clutches of three eggs in captive birds, the eggs being laid at two-day intervals. These clutches have been laid during four different months, with three 1985 clutches laid in January, May and December. In one instance incubation began the day after the laying of the last egg in the clutch, and the male typically incubated during the night and the female during the day. In one instance the female took over all the incubation after 5 days, and three chicks were hatched within a two-hour period after some 25 days of incubation (Grueber 1987).

Like other sandgrouse, the male alone apparently normally brings water to the chicks in his breast feathers. In one instance, after the eggs hatched, the female persistently chased the male several times with a hatched eggshell in her beak, presumably thus informing him that the chicks had hatched and that he should begin the watering process. However, the male did not begin this activity, and the young had to be hand-watered with a soaked cotton swab for 6 days, after which they began to drink on their own. Three days after hatching, the female also led the young on a very detailed and orderly tour of the aviary, apparently so that the young could familiarize themselves with it. This female began a new clutch less than 7 months later (L. Grueber, personal communication).

Evolutionary relationships

According to Snow (1978) and Urban *et al.* (1986), this is a rather isolated form, with no close relatives. Maclean (1984) also considered it to represent a separate monospecific phyletic lineage, but one that he thought might be more closely related to *orientalis* than is suggested by the plumage. I have likewise tentatively (see Figure 3) placed it in a rather isolated position.

Status and conservation outlook

This is a generally common and sometimes abundant species, especially on the grasslands of East Africa, and evidently is not presently a species of concern for conservationists of most regions, although in South Africa it has been listed as a threatened species (Brooke 1984).

BURCHELL'S SANDGROUSE (Plate 51)

Pterocles burchelli Sclater 1922

Other vernacular names: spotted sandgrouse, variegated sandgrouse; ganga de Burchell (French); Fleckenflughuhn, Tupfelflughuhn (German).

Distribution of species (Map 39)

Resident from extreme south-eastern Angola south to northern and eastern Namibia and northern Cape Province, and east to Botswana, western Transvaal and western Orange Free State.

Fig. 53. Social behaviour of yellow-throated sandgrouse: (a) threat posture; (b, c) copulation sequence, with an additional bird nearby. After photos by L. Grueber.

Distribution of subspecies
None recognized by Urban *et al.* (1986); Clancey (1967) recognized *makarikari* Roberts 1932.

Measurements (mm)

Wing, males 163–175 (av. of 10, 169), females 163–173 (av. of 10, 167); tail, males 66–75 (av. of 10, 70.8), females 63–72 (av. of 10, 66.7); tarsus, males 23–29 (av. of 10, 27.3), females 25–29 (av. of 10, 26.7); culmen, males 11–12.5 (av. of 10, 11.8), females 9.5–12 (av. of 10, 10.9). Egg, av. 36.7 × 25.7 (Urban *et al.* 1986).

Weights (g)

Males 180–200 (av. of 3, 192), females 160–185 (av. of 4, 171) (Urban *et al.* 1986). Egg, 13.4 (estimated).

Description

Adult male. Throat, lores and superciliary stripes pale grey; ear-coverts buff; crown feathers dark brown, with the outer parts of each web buff; back, rump and upper tail-coverts blackish, tipped with dull olive; scapulars, lesser and median wing-coverts somewhat brighter, tipped with yellowish-olive, and each feather with one or two round white spots on lateral margins; greater secondary coverts grey, with a rufous stripe down the shaft, a white spot on each web, and a wide white margin; primaries, primary coverts and secondaries black, the shafts of the outer primaries and the margins of the inner primaries and outer secondaries white; chest, breast and belly dull rufous, becoming buff on the tarsi and under tail-coverts; the chest and breast feathers also have round white spots near the tips of both webs, and the feathers of the belly are white basally; rectrices (fourteen) black, barred and tipped with buffy white. Axillaries and underwing lining dull rufous; under-surfaces of the remiges brownish-black. Iris brown, eye-ring yellow, bill blackish, toes yellowish-pink.

Adult female. Differs from the male in having the chin, throat and superciliary stripe pale buff; the belly and the thighs are pale rufous buff, indistinctly barred with white, and the axillaries and under wing-coverts are grey, tinged with rufous-buff.

Immature. Juvenile males have the throat and eye-stripe greyish-buff, and are generally similar to adult

Map. 39. Distribution of Burchell's sandgrouse.

female; the feathers with pale tips and marked with chevrons of light brown; the white wing spotting duller and more diffuse; the tail as in the adult female. Most juvenal feathers of both sexes are barred with buff, some innermost secondaries are barred on the outer vane with buff, and the tips of the juvenal primaries are white (Clancey 1967). Presumably some of the outermost juvenal primaries persist after most other adult feathers have grown in, as occurs in at least most other sandgrouse.

Identification

In the hand. Males have uniformly buff tarsi, the upper and lower body extensively spotted with white, and a greyish throat. Females are very similar but have a buff rather than greyish throat and some white barring on the belly.

In the field. This southern African species is the only sandgrouse of that region with extensive white spotting on the upperparts combined with unbarred flanks. Although the sympatric *bicinctus* has similar but somewhat larger dorsal spotting in males, its flanks are strongly barred, and it lacks the greyish throat of this species. Females have spotted (not barred) breasts, and both sexes exhibit deep buff to rufous underwing linings in flight. Their calls include mellow double-note *kowk-wok* or *chok-lit* flight calls, the second syllable higher and more accented, and they sometimes also utter cheeping notes when on the ground and sharp, repeated *quip* notes on take-off. However, the species calls only rather infrequently in flight compared with other sandgrouse, and sometimes a flock will arrive, drink, and depart from a waterhole in total silence (Maclean 1968).

General biology and ecology

This species is associated with grassy cover (such as *Aristida, Stipagrostis* and *Asthenatherum* about 30–50 cm high) in sandy areas, and rather dry and lightly wooded savannah habitats in semidesert climates, rather than true deserts. It is especially characteristic of the red sands of the Kalahari, a colour that well matches its bright rufous underparts. It is most abundant on open, rolling and grassy plains, but also extends into areas of open acaciaveld, as well as scrub and shrubby areas. It is shy, hard to locate and highly secretive, and when frightened the birds tend to crouch or sneak away in the grassy cover. Compared with some other sandgrouse they are fairly long-legged, and run quite easily (Clancey 1967, Urban *et al.* 1986).

Sometimes found breeding far from the nearest surface water, this species drinks during morning hours. Drinking usually occurs from 2 to 3 or 4 hours after sunrise, the birds apparently flying daily to available waterholes from as far as 70–80 km away. Flocks going to water usually fly at a considerable altitude, up to about 300 m, and probably cruise at speeds of about 80 km h^{-1}, so that perhaps as much as an hour is required for flying a maximum distance. Upon arrival, the birds sometimes land directly at the water's edge or even in it. Drinking is done very rapidly, after which the birds promptly return to feeding areas; indeed some birds are in such a haste to depart that they might not drink at all that day. After returning from their drink they probably forage for a time, but they spend most of the hot midday hours inactive and roosting. Drinking during the winter period is of shorter duration than in the summer, and on cloudy days the drinking period is also shortened. Afternoon or evening drinking by this species is apparently rare or absent, since it was never observed by Maclean, and brackish waters are avoided by the birds (Maclean 1968).

Foods consist of the usual array of seeds of grasses, forbs, and perhaps other plants. In the Kalahari desert the most important food plant is probably the annual chenopod forb *Lophiocarpus burchelli*, and one chick less than a week old was found to have 1400 seeds in its crop (Maclean 1968).

Social behaviour

Maclean (1968) was unable to observe any definite courtship in this species during his 19 months of study, but attributed this to the species' unusual shyness and the difficulty of seeing it in its dune habitat. However, he observed that head-bobbing was the typical approach–retreat agonistic display performed by birds in conflict at a water-hole. When in a defensive threat situation, the bird faces away from the opponent, with the breast lowered, the head raised, the wings flicked open, and the tail raised and spread. An upright-alert posture with accompanying neck-stretching was the initial response to possible danger, followed by crouching. Although no courtship behaviour was observed, Maclean noted that water-soaking behaviour by males increased in the winter, as was also the case with the Namaqua sandgrouse, suggesting to him that winter breeding was simultaneously occurring in both species.

Reproductive biology

Maclean (1968) found only a single nest (in April) of this species, compared with thirty-six nests of the

Namaqua sandgrouse in the same area, during his studies in the Kalahari desert. Nonetheless he believed that the birds probably breed in loose assemblages, since pairs semed to be scattered over areas of about 8 ha each, and between such areas there were no birds present at all.

Data from throughout the species' range indicate that it breeds from April to October, after the rains and during the subsequent dry winter season, by which time rapidly growing annual plants have matured and set their seeds. The nest is often placed in short grasses or among stunted shrubs, with little surrounding cover. There are typically three eggs present, but rarely only two. The incubation period is unknown, but certainly considerably longer than the 16-day estimate of Clancey (1967). The incubating birds are very shy when on the nest, and often walk away from it before taking off when approached by a possible predator or even a vehicle. It is known that both sexes share incubation, and that the male gathers water in his breast feathers to pass on to the chicks. Both parents tend the young, which become capable of fight when they are about half the size of the adults. Moulting probably occurs after the breeding period, or from August to early December (Clancey 1967).

Evolutionary relationships

This is apparently a rather isolated species of *Pterocles* in the opinion of Snow (1978), Maclean (1984), Urban *et al.* (1986), and Wolters (1974, 1975), the last-named author even placing it in a monotypic genus (*Calopterocles*). Maclean judged it to be the result of a relatively old invasion of southern Africa by ancestral sandgrouse stock, and both its plumage and behaviour, as well as some aspects of its anatomy (such as its relatively long legs), appear to have been influenced by its adaptations to a sandy, grassy environment.

Status and conservation outlook

The abundance of this species is probably associated more with the distribution of its available habitat than with any other single factor. In contrast to the more water-dependent Namaqua sandgrouse, it has not with any certainty spread its breeding range in the Kalahari as farming has developed there since the 1930s, although that remains a distinct possibility (Maclean 1968). In any case, it is apparently a relatively common species over much of its limited geographic range, and is likely to remain so, given its ability to survive in such environments as it does.

References

Unfortunately, the literature on bustards, hemipodes, and sandgrouse of the world is relatively scanty and greatly scattered through the ornithological literature. Bibliographies for the hemipodes and sandgrouse are still wholly lacking, but Downes (1975) has compiled a partial and now rather outdated bibliography for the bustards. For the reader's further information, *Bustard Studies* is a journal published irregularly by the Bustard Group of the International Council for Bird Protection (ICBP), Cambridge. This group also provides a bustard literature reference collection and is a repository for various translations on bustard biology, including several cited below.

Ahmed, S.N. (1985). Protection and conservation of houbara bustards in the Punjab. *Bustard Studies* 3, 39–42.

Aldrich, H.C. (1943). Some notes on the common sandgrouse (*Pterocles exustus* Temminck) in the Kaira district. *Journal of the Bombay Natural History Society* 44, 123–5.

Alekseev, A.F. (1985). The houbara bustard in the north west Kyzylkum. *Bustard Studies* 3, 87–92.

Ali, S. and Rahmani, A.R. (ed.) (1982). *Study of ecology of certain endangered species of wildlife and their habitats. The great Indian bustard*, Bombay Natural History Society Annual Report I, 1981–82; 154 pp.

—— and —— (1984). *Study of ecology of certain endangered species of wildlife and their habitats. The great Indian bustard*, Bombay Natural History Society Annual Report II, 1982–84; 100 pp. (See also Manakadan and Rahmani 1986.)

—— and Ripley, S.D. (1983). *Handbook of the birds of India and Pakistan*, Compact Edition. Oxford University Press.

—— Daniel, J.C. and Rahmani, A.R. (ed.) (1985). *Study of ecology of certain endangered species of wildlife and their habitats. The floricans*, Bombay Natural History Society Annual Report I, 1984–85; 99 pp. (See also Rahmani *et al.* 1988b.)

Allan, D. (1988). Bustard alert. *Palea* 7, 86–9.

Allen, P.M. and Clifton, M.P. (1972). Aggressive behaviour of kori bustard *Otis kori*. *Bulletin of the East Africa Natural History Society (EANHS Bulletin)* (Nov.), 188–9.

Amadon, D. and du Pont, J.E. (1970). Notes on Philippine birds. *Nemouria* 1, 1–14.

André, R. (1985). Some aspects of the biological reproduction of the little bustard and a contribution to an estimate of its population in France 1978/1979. *Bustard Studies* 2, 153–60.

Appayya, M.K. (1982). Breeding of bustards—an observation in Australia. *Journal of the Bombay Natural History Society* 79, 195–7.

Appert, O. and Etchécopar, R.D. (1962). Note sur la nidification de *Pterocles personatus*. *Oiseau* 32, 179–80.

Archer, C.F. and Godman, E.M. (1937). *The birds of British Somaliland and the Gulf of Aden*, Vol. 2. Gurney and Jackson, London.

Ash, J.S. and Miskell, J.E. (1983). Birds of Somalia, their habitat, status and distribution. *Scopus* (Spec. Suppl. 1), 1–97.

Astley-Maberly, C.T. (1967). An unusual display of the red-crested korhaan. *Bokmakierie* 19, 41.

Baker, E.C.S. (1921). *Game-birds of India, Burma and Ceylon*, Vol. II. Bombay Natural History Society, Bombay.

—— (1928). *The fauna of British India including Ceylon and Burma*, Vol. 5. Taylor and Francis, London.

—— (1930). *Game-birds of India, Burma and Ceylon*, Vol. 3. John Bale and Son, London.

—— (1935). *The nidification of birds of the Indian Empire*, Vol. 4. Taylor and Francis, London.

Bannerman, D. (1931). *Birds of tropical West Africa*, Vol. 2. Oliver and Boyd, Edinburgh.

—— (1959). *Birds of the British Isles*, Vol. 8. Oliver and Boyd, Edinburgh.

—— (1963). *Birds of the Atlantic islands: 1, History of the birds of the Canary Islands and Salvages*. Oliver and Boyd, Edinburgh.

Beddard, F.E. (1898). *The structure and classification of birds*. Longmans, Green, London.

Bell, J. (1970). The white-quilled black bustard. *Animal Kingdom* 73 (6), 25–8.

—— and Bruning, D. (1974). Hand-rearing hemipodes at the New York Zoological Park. *International Zoo Yearbook* 14, 196–8.

Bennett, S. (1985). The distribution and status of the black-breasted buttonquail *Turnix melanogaster* (Gould, 1837). *Emu* 85, 57–62.

Benson, C.V. and Irwin, M.P.S. (1972). Variation in tarsal and other measurements in *Otis denhami*, with some distributional notes. *Bulletin of the British Ornithologists' Club* 92, 70–7.

——, Brooke, R.K., Dowsett, R.J. and Irwin, M.P.S. (1972). *The birds of Zambia*. Collins, London.

——, Beamish, H.H., Jourdain, C., Salvin, C. and Watson, G.E. (1975). The birds of the Isles Glorieuses. *Atoll Research Bulletin* 176, 1–30.

——, Colebrook-Robjent, J.F.R. and Williams, A. (1976). Contribution à l'ornithologie de Madagascar. *Oiseau* 46, 103–34, 209–42, 367–86.

Blakers, M., Davies, S.J.J.F. and Reilly, P.N. (1984). *The atlas of Australian birds*. RAOU and Melbourne University Press, Victoria.

Bock, W.J. and McEvey, A. (1969). Osteology of *Pedionomus torquatus* (Aves, Pedionomidae) and its allies. *Proceedings Royal Society of Victoria* **82**, 187–232.

Böhm, D. (1985). Zur Brutbiologie der Flughühner (Pteroclididae). Beobachtungen aus Anlass einer gelungen Zucht des Braunbauchflughühner (*Pterocles exustus exustus* Temminck). *Gefiederte Welt* **109**, 257–60.

Boobyer, M.G. (1989). The eco-ethology of the Karoo korhaan (*Eupodotis vigorsii*). M.S. Thesis, University of Cape Town.

Borisenko, V.A. (1977). On the numbers of *Otis tetrix* and *Otis tarda* in some regions of Kazakhstan. In *Rare and vanishing animals and birds of Kazakhstan* (ed. A.A. Sludskii, Alma Ata). (In Russian.)

Bowen, W.W. (1927). Remarks on the classification of the Pteroclididae. *American Museum Novitates* **272**, 1–12.

Bradbury, J.W. and Gibson, R.M. (1983). Lek and mate choice. In *Mate choice* (ed. P. Bateson), pp. 109–38. Cambridge University Press, Cambridge.

Brooke, R.K. (1984). *South African red data book: birds.* CSIR, Pretoria.

Brosset, A. (1961). Ecologie des oiseaux du Maroc oriental. *Travaux de l'Institut Scientifique Chérifien (Zoologie)* **22**, 1–155.

Bruning, D. (1971). Hemipodes, *Turnix sylvatica. Game Bird Breeders' Gazette* **22** (5), 7–10.

Butler, A.G. (1905a). On breeding *Turnix nigricollis* in German birdrooms. *Avicultural Magazine* **3**, 195–203.

— (1905b). On breeding *Turnix lepurana* in German birdrooms. *Avicultural Magazine* **3**, 217–22.

Cade, T.J. (1965). Relations between raptors and columbiform birds at a desert water hole. *Wilson Bulletin* **77**, 340–5.

— and Maclean, G.L. (1967). Transport of water by adult sandgrouse to their young. *Condor* **69**, 323–43.

—, Willoughby, E.J. and Maclean, G.L. (1966). Drinking behavior of sandgrouse in the Namib and Kalahari deserts, Africa. *Auk* **83**, 124–6.

Cardosa, J.J.V. (1985). A project to halt the decline of the great bustard on the Extremadura plains. *Bustard Studies* **2**, 73–4.

Casado, M.A., Levassor, C. and Parra, F. (1983). Régime alimentaire estival du ganga cata *Pterocles alchata* (L.) dans le centre de l'Espagne. *Alauda* **51**, 203–9.

Cassels, K.A.H. and Elliot, H.K.I. (1975). An undescribed display of the red-crested korhaan *Lophotis ruficrista*. *Bulletin of the British Ornithologists' Club* **95**, 116–7.

Chandler, A.C. (1916). A study of the structure of feathers, with reference to their taxonomic significance. *University of California Publications in Zoology* **13**, 243–446.

Chapin, J. (1939). Birds of the Belgian Congo. *Bulletin of the American Museum of Natural History* **75**, 1–632.

Chappuis, G., Erard, C. and Morel, G.J. (1979). Données comparatives sur la morphologie et les vocalisations des diverses formes d'*Eupodotis ruficrista* (Smith). *Malimbus* **1**, 74–89.

Cheng Tso-hsin (ed.) (1963). *(China's economic fauna: birds).* Beijing. (In Chinese.)

Cheylan, G. (1975). Esquisse écologique d'une zone semi-aride: La Crau (Bouches-du-Rhone). *Alauda* **43**, 23–54.

Christensen, G.C. (1963). The imperial sandgrouse in the Thar Desert of India. *Occasional Papers of the Biological Society of Nevada*, No.2; 7 pp.

—, Bohl, W.H. and Bump, G. (1964). A study and review of the common Indian sandgrouse and the imperial sandgrouse. U.S. Fish and Wildlife Service, *Special Scientific Report (Wildlife)* **84**, 1–71.

Clancey, P.A. (1967). *Gamebirds of southern Africa.* Purnell, Cape Town.

— (1972–3). The magnificent bustards. *Bokmakierie* **24**, 74–9; **25**, 10–4.

— (1986). Endemicity in the southern African avifauna. *Durban Museum Novitates* **13**, 246–84.

— (1989). Four additional species of southern African endemic birds. *Durban Museum Novitates* **14**, 140–52.

Clements, J. *Birds of the world: A checklist.* Facts on File, New York.

Collar, N.J. (1979). Bustard Group general report. *ICBP Bulletin* **XIII**, 129–34.

— (1980). The world status of the houbara: a preliminary review. Symposium papers on the great bustard and houbara bustard, Athens, Greece, May 24, 1979; 12 pp. Fondation Internationale pour le Sauvegarde du Gibier; Conseil Internationale de la Chasse/Game Conservancy.

— (1985). The world status of the great bustard. *Bustard Studies* **2**, 1–20.

— and Andrew, P. (1988). *Birds to watch. The ICBP world checklist of threatened birds*, ICBP Technical Publication No.8. Smithsonian Institution Press, Washington, D.C.

— and Stuart, S.N. (1985). *Threatened birds of Africa and related islands*, ICBP/IUCN Red Data Book, 3rd ed. (Pt.1). ICBP, Cambridge.

— et al. (1983). Report of the ICBP Fuerteventura houbara expedition, 1979. *Bustard Studies* **1**, 1–93.

Collins, D.R. (1984). A study of the Canarian houbara (*Chlamydotis undulata fuertaventurae*), with special reference to its behaviour and ecology. M.Phil. thesis, University of London.

Cornwallis, L. (1983). A review of the bustard situation in Iran. In Goriup and Vardhan (1983), pp. 81–8.

Coverdale, M.A.C. (1987). Display of black-bellied bustard *Eupodotis melanogaster*. *Scopus* **11**, 52.

Cracraft, J. (1981). Toward a phylogenetic classification of the Recent birds of the world (class Aves). *Auk* **98**, 681–714.

Cramp, S. (ed.) (1985). *Birds of the western Palearctic*, Vol.4. Oxford University Press, Oxford.

— and Simmons, K.E.L. (eds.) (1980). *Birds of the western Palearctic*, Vol.2. Oxford University Press, Oxford.

Dee, T.J. (1986). *The endemic birds of Madagascar.* ICBP, Cambridge.

Deignan, H.G. (1945). The birds of northern Thailand. *Bulletin of the US National Museum* **186**, 1–615.

Delacour, J. (1929). On the birds collected during the fourth expedition to French Indo-China. *Ibis*, 12th ser., **5**, 193–220.

Dementiev, G.P. and Gladkov, N.A. (1951). *Birds of the Soviet Union*, Vol.2 (translated 1968 from Russian by the Israel Program for Scientific Translations, Jerusalem). U.S. Dept. of the Interior and National Science Foundation, Washington, D.C.

Dharmakumarsinhji, K.S. (1945). The bustard-quail at home. *Avicultural Magazine* **10**, 58–60.

—— (1950). The lesser florican [*Sypheotides indica* (Miller)]: its courtship display, behaviour and habits. *Journal of the Bombay National History Society* **49**, 201–16.

—— (1957). Ecological study of the great Indian bustard *Ardeotis nigriceps* (Vigors) in Kathiawar Peninsula, western India. *Journal of the Zoological Society of India* **9**, 140–55.

—— (1966). Display, posturing and behaviour of the great Indian bustard *Choriotis nigriceps* Vigors, Aves, Otididae. Proceedings of the All-India Congress of Zoology, 1962, pp. 277–83.

Dixon, J.E.W. (1978). Animal remains recovered from sandgrouse (Aves, Pteroclididae) crops in the Etosha National Park. *Madoqua* **11**, 75–6.

—— and Louw, G. (1978). Seasonal effects on nutrition, reproduction and aspects of thermoregulation in the Namaqua sandgrouse (*Pterocles namaqua*). *Madoqua* **11**, 19–29.

Dobai, C. (1983). Breeding, protection and management of the great bustard in Czechoslovakia and Europe. In Goriup and Vardhan (1983), pp. 104–13.

Dornbusch, M. (1983). Status, ecology, and conservation of the great bustard in GDR. In Goriup and Vardhan (1983), pp. 89–90.

Downes, M. (1975). *A bibliography of the bustards: first working draft*. Wildlife Branch, Dept. of Agriculture, Stock and Fisheries, Papua New Guinea; 209 pp.

—— (1982). Re-establishment of the bustard in Victoria. In *Wildlife management in the '80s* Proceedings of a Conference, Field and Game Federation of Australia and Graduate School of Environmental Sciences, Monash University, November 27–29, 1981 (ed. T. Riney), pp. 227–39.

Du Pont, J.E. (1971). *Philippine birds*, Delaware Museum of Natural History, Monograph Series No.2.

—— (1976). Notes on Philippine birds (No.4). *Nemouria* **17**, 1–13.

Earlé, R.A., Louw, S. and Herholdt, J.J. (1989). Notes on the measurements and diet of Ludwig's bustard. *Ostrich* **59**, 178–9.

Elgood, J.H., Fry, C.H. and Dowsett, R.J. (1973). African migrants in Nigeria. *Ibis* **115**, 1–45, 374–411.

Elliot, D.G. (1878). A study of the Pteroclidae or family of the sand-grouse. *Proceedings of the Zoological Society of London*, 223–64.

—— (1885). Opisthocomi, Gallinae, Pterocles, Columbae. In *The standard natural history*, Vol.4: *Birds* (ed. J.S. Kinsley), pp. 196–259. S.E. Cassino, Boston.

Elliot, W. (1880). Notes on the great Indian bustard with special reference to its gular pouch. *Proceedings of the Zoological Society of London*, 486–9.

Emlen, S.T. and Oring, L.W. (1977). Ecology, sexual selection and the evolution of mating systems. *Science (Washington, D.C.)* **197**, 215–23.

Ena, V., Lucio, A. and Purroy, F.J. (1985). The great bustard in Leon, Spain. *Bustard Studies* **2**, 35–52.

——, Martinez, A. and Thomas, D.H. (1987). Breeding success of the great bustard *Otis tarda*, in Zamora Province, Spain, in 1984. *Ibis* **129**, 364–70.

Etchécopar, R.D. and Hüe, F. (1967). *The birds of North Africa from the Canary Islands to the Red Sea*. Oliver and Boyd, Edinburgh.

—— and —— (1978). *Les oiseaux de Chine, non-passereaux*. N. Boubée, Paris.

Faruqi, S.A., Bump, G., Nanda, P.C. and Christensen, G.C. (1960). A study of the seasonal foods of the black francolin (*Francolinus francolinus* Linnaeus), the grey francolin [*F. pondicerianus* (Gmelin)], and the common sandgrouse (*Pterocles extusus* Temminck) in India and Pakistan. *Journal of the Bombay Natural History Society* **57**, 354–61.

Ferguson-Lees, I.J. (1967). Studies of less-familiar birds. Little bustard. *British Birds* **60**, 80–4.

—— (1969). Studies of less-familiar birds. Pin-tailed sandgrouse. *British Birds* **62**, 533–41.

Finn, F. (1915). *Indian sporting birds*. F. Edwards, London.

Fitzherbert, K. (1978). Observations on breeding and display in a colony of captive Australian bustards (*Ardeotis australis*). B.Sc. thesis, Monash University, Victoria.

—— (1983). Seasonal weight changes and display in captivity of the Australian bustard. In Goriup and Vardhan (1983), pp. 210–26.

—— and Baker-Gabb, D.J. (1988). Australasian grasslands and their threatened avifauna. In Goriup (1988a), pp. 227–50.

Fjeldså, J. (1976). The systematic affinities of sandgrouse, Pteroclididae. *Vidensk, Meddelelser dansk naturh. Foren.* **139**, 179–243.

—— (1977). *Guide to the young of European precocial birds*. Scarv Nature Publications, Strandgarden.

Flieg, G.M. (1973). Breeding biology and behaviour of the South African hemipode in captivity. *Avicultural Magazine* **79**, 55–9.

Flower, W.H. (1865). On the gular pouch of the great bustard (*Otis tarda*). *Proceedings of the Zoological Society of London*, 747–8.

Fodor, T. (1966). (Studies of bustards under artificial conditions.) In *Különlenyomat az Allattani Közlemények* **53**, Nos.1–4, pp. 59–62. Akademiai Nyomda, Budapest. (In Hungarian, translated for ICBP Bustard Group, Cambridge.)

Friedmann, H. (1930). Birds collected on the Childs Frick expedition to Ethiopia and Kenya Colony, Pt.I. *Bulletin of the U.S. National Museum* **153**, 1–516

Frisch, O. von. (1969a). Zur Jugendentwicklung und Ethologie des Spiessflughuhns (*Pterocles alchata*). *Bonner Zoologische Beiträge* **20**, 130–44.

—— (1969b). Aufzucht von Zwergtrappen (*Tetrax tetrax*). *Gefiederte Welt* **93**, 204–6.

—— (1970). Zur Brutbiologie und Zucht des Spiessflughuhns (*Pterocles alchata*) in Gefangenschaft. *Journal für Ornithologie* **111**, 189–95.

—— (1976). Zur Biologie der Zwergtrappe (*Tetrax tetrax*). *Bonner Zoologische Beiträge* **27**, 21–38.

Frith, H.J. (ed.) (1969). *Birds of the Australian high country*. A.H. and A.W. Reed, Sydney.

—— (1976). *Reader's Digest complete book of Australian birds.* Reader's Digest Service Pty Ltd, Sydney.

Fulgenhauer, J. (1980). Behaviour of button quail. *Sunbird* **11**, 25–39.

Gadow, H. (1882). On some points in the anatomy of *Pterocles*, with remarks on its systematic position. *Proceedings of the Zoological Society of London*, 312–32.

—— (1892). On the classification of birds. *Proceedings of the Zoological Society of London*, 229–56.

Garrod, A.H. (1874a). On certain muscles in birds and their value in classification. *Proceedings of the Zoological Society of London*, 111–23.

—— (1874b). On some points in the anatomy of the Columbae. *Proceedings of the Zoological Society of London*, 249–59.

—— (1874c). On the 'showing-off' of the Australian bustard (*Eupodotis australis*). *Proceedings of the Zoological Society of London*, 471–3.

Gasperetti, J. and Gasperetti, P. (1981). Birds of Saudi Arabia. A note on Arabian ornithology—two endangered species. *Fauna of Saudi Arabia* **3**, 435–40.

Gavrin, V.F., Dolguschin, I.A., Korelov, M.N. and Kuzhmina, M.A. (1962). (Birds of Kazakhstan), Vol.2. Alma Ata. (English translation for ICBP Bustard Group by D.F. Vincent; not seen.)

George, U. (1969). Über das Tränken der Jungen andere Lebensäusserungen des Senegal-Flughuhns, *Pterocles senegallus*, in Marokko. *Journal für Ornithologie* **110**, 181–91.

—— (1970). Beobachtungen an *Pterocles senegallus* und *Pterocles coronatus* in der Nordwest-Sahara. *Journal für Ornithologie* **111**, 175–88.

—— (1977). *In the deserts of the earth.* Harcourt Brace Jovanovich, New York.

Gérodet, P. (1974). Notes marocaines sur la parade nuptiale de l'outarde houbara *Chlamydotis undulata. Oiseau* **44**, 149–52.

Gewalt, W. (1959). *Die Grosstrappe*, Neue Brehm Bücherei. Wittenburg-Lutherstadt, Leipzig.

—— (1964). The first success in zoo-breeding great bustards (*Otis tarda*). *Avicultural Magazine* **70**, 218–9.

—— (1965). Formveränderande Strukturen am Halse der männlichen Grosstrappe (*Otis tarda* L.). *Bonner Zoologische Beiträge* **161**, 288–300.

—— and Gewalt, I. (1966). Über Haltung und Zucht der Grosstrappe *Otis tarda* L. *Zoologische Garten*, new series, **32**, 265–322.

Glutz von Blotzheim, U.N. (ed.) (1973). *Handbuch der Vögel Mitteleuropas*, Vol.5; *Galliformes und Gruiformes.* Akademische Verlag, Frankfurt.

Goodman, S.M. and Meininger, P.L. (eds.) (1989). *The birds of Egypt.* Oxford University Press, Oxford.

Goodwin, D. (1965). Remarks on the drinking methods of some birds. *Avicultural Magazine* **71**, 76–80.

—— (1974). Birds of the Harold Hall Australian expeditions 1962–70. Gruiformes—bustard quails, cranes, rails, crakes, bustards. British Museum (Natural History) Publication No.745, 62–6.

Gore, R. (1976). The desert: an age-old challenge grows. *National Geographic* **156**, 586–640.

Goriup, P.A. (1983a). Houbara bustard: research and conservation in Pakistan. In Goriup and Vardhan (1983), pp. 267–72.

—— (ed.) (1983b). *The houbara bustard in Morocco. Report of a preliminary survey by the ICBP;* 23 pp. ICBP, Cambridge.

—— (1983c). Decline of the great Indian bustard. A literary review. In Goriup and Vardhan (1983), pp. 20–38.

—— (1985a). The 1980 breeding season at the Great Bustard Trust (U.K.). *Bustard Studies* **2**, 103–18.

—— (1985b). A note on the minimum required captive stocks of great bustards for augmenting wild populations. *Bustard Studies* **2**, 119–21.

—— (1987). Bustards in Meru National Park, Kenya. *British Ecological Society Bulletin* **18**, 189–91.

—— (ed.) (1988a). *Ecology and conservation of grassland birds*, Technical Publication No.7; 250 pp. ICBP, Cambridge.

—— (1988b). The avifauna and conservation of steppic habitats in western Europe, North Africa, and the Middle East. In Goriup (1988a), pp. 145–58.

—— and Karpowicz, Z.J. (1985). A review of the past and recent status of the lesser florican. *Bustard Studies* **3**, 163–82.

—— and Parr, D.F. (1985). Results of the ICBP bustard survey of Turkey, 1981. *Bustard Series* **2**, 77–98.

—— and Vardhan, H. (eds.) (1983). *Bustards in decline*, Proceedings of the International Symposium on Bustards, Jaipur, India, 1980. Natraj Publishers, Dehra Dun.

——, Osborne, P.E. and Everett, S.J. (1989). Bustards in Meru National Park, Kenya: a preliminary survey. *Bustard Studies* **4**, 52–67.

Gray, G. (1844–9). *The genera of birds: comprising their generic characters, a notice of the habits of the genus, and an extensive list of species referred to their several genera.* Longman, Brown, Green and Longmans, London.

Gregson, J. (1986). Breeding the little black bustard *Eupodotis afra* at the Paignton Zoological and Botanical Gardens. *Avicultural Magazine* **92**, 61–3.

Grice, D., Gaughley, G. and Short, J. (1986). Density and distribution of the Australian bustard *Ardeotis australis. Biological Conservation* **35**, 259–67.

Grote, H. (1936). Beiträge zur Biologie südostrussicher Steppenvögel, *Otis tetrax orientalis. Beiträge zur Fortpflanzungsbiologie der Vögel mit Berücksichtigung der Oölogie* **12**, 195–8.

Grueber, L.W. (1987). Raising sandgrouse. *Game Bird Breeders' Gazette* **36** (3), 16–9.

Grummt, W. (1985). Beobachtungen zur Haltung und zur Fortpflanzungsbiologie des Steppenhühner *Syrrhaptes paradoxus. Milu* **6**, 44–52.

Gruson, E.S. (1976). *Checklist of the world's birds.* New York Times Book Co., New York.

Grzimek, B. (1972). *Grzimek's animal life encyclopedia*, Vol.8: *Birds*, II. Van Nostrand Reinhold, New York.

Guichard, G. (1961). Note sur la biologie du ganga cata (*Pterocles a. alchata* L.). *Oiseau* **31**, 1–9.

Gupta, P.D. (1975). Stomach contents of the great Indian bustard, *Choriotis nigriceps* (Vigors). *Journal of the Bombay Natural History Society* **71**, 303–4.

Hachisuka, M. (1932). *The birds of the Philippine Islands.* H.F. and G. Witherby, London.
Haddane, B. (1985). The houbara bustard in Morocco: a brief review. *Bustard Studies* **3**, 109–12.
Hall, B.P. (1974). *Birds of the Harold Hall Australian expeditions, 1962–1970.* British Museum (Natural History), London.
Hanby, J. (1982). *Lion's share.* Houghton Mifflin, Boston.
Haribal, M., Lachungpa, U.G. and Rahmani, A.R. (1985). The lesser florican survey in Madhya Pradesh and Rajasthan. In Ali *et al.* (1985), pp. 42–50.
Hartert, E. (1898). On the birds of Lomblen, Pantar and Alor. *Novitates Zoologicae* **5**, 455–76.
Hassan, S.M. (1983). Status of the great Indian bustard (*Choriotis nigriceps*) in Madhya Pradesh. In Goriup and Vardhan (1983), pp. 44–50.
Heim de Balsac, H. and Mayaud, N. (1962). *Les oiseaux du nord-ouest de l'Afrique.* Editions Paul Lechavalier, Paris.
Heinroth, O. and Heinroth, M. (1927–8). *Die Vögel Mitteleuropas*, Vol.3. Lichterfelda, Berlin.
Hellmich, J. (1988). (On the mating behaviour of the kori bustard.) *Zoologische Garten* **58**, 345–52. (In German.)
Hendrickson, H.T. (1969). A comparative study of the egg white proteins of some species of the avian order Gruiformes. *Ibis* **111**, 80–91.
Herholdt, J.J. (1987). Some notes on the behaviour of the Ludwig's bustard *Neotis ludwigi* in the Orange Free State and northern Cape Province. *Mirafra* **4**, 34–5.
— (1988). The distribution of Stanley's and Ludwig's bustards in southern Africa: a review. *Ostrich* **59**, 8–13.
Heuglin, M.T. von (1873). *Ornithologie Nordost-Afrika's*, Vol.2. Kassel.
Hills, E.S. (1966). *Arid lands: a geographic appraisal.* Methuen, London.
Hoesch, W. (1959). Zur Biologie des südafrikanischen Laufhühnchens *Turnix sylvatica lepurana. Journal für Ornithologie* **100**, 341–9.
— (1960). Zum Brutverhalten des Laufhühnchens *Turnix sylvatica lepurana. Journal für Ornithologie* **101**, 265–75.
— and Niethammer, G. (1940). Die Vogelwelt Deutsch-Südwestafrikas. *Journal für Ornithologie* **88**, (Sonderheft), 1–404.
Höglund, J. (1989). Size and plumage dimorphism in lek-breeding birds: a comparative analysis. *American Naturalist* **89**, 72–87.
Hoogerwerf, A. (1962). Some ornithological notes on the smaller islands around Java. *Ardea* **50**, 199.
— (1964). On birds new for New Guinea or with a larger range than previously known. *Bulletin of the British Ornithologists' Club* **84**, 70–7, 94–6, 118–24, 142–8, 153–61.
Hopkins, G.H.E. (1942). The Mallophaga as an aid to the classification of birds. *Ibis*, 14th series, **6**, 94–106.
Howells, N.W. and Fynn, K.J. (1979). The occurrence of Denham's bustard at Wankie National Park and in north west Rhodesia with notes on movement and behaviour. *Honeyguide* **97**, 4–12.
Hsu, W. (1988). Steppe birds and conservation in China. In Goriup (1988a), pp. 221–26.

Hudson, R. (ed.) (1975). *Threatened birds of Europe.* Macmillan, London.
Hüe, F. and Etchécopar, R.D. (1957). Les ptéroclididés. *Oiseau* **27**, 35–58.
Hummel, D. (1985). A note on the invasions of western Europe by the great bustard in the winter seasons 1969/70 and 1978/79. *Bustard Studies* **2**, 75–6.
Huxley, T.H. (1867). On the classification of birds; and on the taxonomic value of the modifications of certain of the cranial bones observable in that class. *Proceedings of the Zoological Society of London*, 415–72.
— (1868). On the classification and distribution of the Alecteromorphae and Heteromorphae. *Proceedings of the Zoological Society of London*, 294–319.
Ilicek, V. and Flint, V.E. (1989). *Handbuch der Vögel der Sowjetunion*, Band 4: *Hühner und Kranichvögel.* A. Ziemsen Verlag, Wittenberg.
Inskipp, C. and Collar, N.J. (1984). The Bengal florican: its conservation in Nepal. *Oryx* **18**, 30–5.
— and Inskipp, T.P. (1983). *Report on a survey of Bengal floricans* Houbaropsis bengalensis *in Nepal and India, 1982*, Study Report No.2. ICBP, Cambridge.
— and — (1985a). A survey of Bengal floricans in Nepal and India, 1982. *Bustard Studies* **3**, 141–60.
— and — (1985b). *A guide to the birds of Nepal.* Tanager Books, Dover.
Isakov, Y.A. (1974). Present distribution and population status of the great bustard *Otis tarda* Linnaeus. *Journal of the Bombay Natural History Society* **71**, 433–44.
— (1982). Status of great bustard and little bustard populations in the USSR and perspectives on their conservation. In Abstracts of symposia, XVIII International Ornithological Congress, Moscow.
Istvan, S. (1983). Present status of great bustard (*Otis tarda*) in Hungary. In Goriup and Vardhan (1983), pp. 114–7.
Jackson, F.J. (1926). *Game-birds of Kenya and Uganda.* Williams and Norgate, London.
Jehl, Jr. J.R., and Murray, Jr. B.G. (1986). The evolution of normal and reverse sexual dimorphism in shorebirds and other birds. *Current Ornithology* **3**, 1–86.
Johansen, H. (1959). Die Vogelfauna Westsibiriens, III (Non-Passeres). *Journal für Ornithologie* **100**, 417–32.
Johnsgard, P.A. (1973). *Grouse and quails of North America.* University of Nebraska Press, Lincoln.
— (1981). *Plovers, sandpipers and snipes of the world.* University of Nebraska Press, Lincoln.
— (1983a). *Grouse of the world.* University of Nebraska Press, Lincoln.
— (1983b). *Cranes of the world.* University of Indiana Press, Bloomington.
— (1988). *Quails, partridges and francolins of the world.* Oxford University Press, Oxford.
Johst, E. (1972). Die Haltung und künstliche Aufzucht der Senegal-Trappe (*Eupodotis senegalensis* Vieillot). *Gefiederte Welt* **96**, 61–4.
Juana, E. de, Santos, T., Suarez, F. and Telleria, J.L. (1988). Status and conservation of steppe birds and their habitats in Spain. In Goriup (1988a), pp. 113–24.
Kalchreuter, H. (1979). Zur Mauser der äquatorialen

Flughühner *Pterocles exustus* und *P. decoratus*. *Bonner Zoologische Beiträge* **30**, 102–16.

— (1980). The breeding season of the chestnut-bellied sandgrouse *Pterocles exustus* and the black-faced sandgrouse *P. decoratus* in northern Tanzania and its relation to rainfall. *Proceedings, IV Pan African Ornithological Congress*, pp. 277–82.

Karim, F. (1985). A note on the disappearance of the Bengal florican from Bangladesh. *Bustard Studies* **3**, 161–2.

Karpowicz, Z.J. and Goriup, P.D. (1985). Occurrence of the great Indian bustard in the Okha Rann (Gujarat). *Bustard Studies* **3**, 133–8.

Kasparek, M. (1989). Status and distribution of the great bustard and little bustard in Turkey. *Bustard Studies* **4**, 80–113.

Kemp, A. and Tarboton, W. (1976). Small South African bustards. *Bokmakierie* **28**, 40–3.

King, W.B. (ed.) (1981). *Endangered birds of the world: the ICBP red data book*. Smithsonian Institution Press and ICBP, Washington, D.C.

Klös, H.G. (1977). News from the Berlin zoo. *Avicultural Magazine* **83**, 232–3.

Koenig A. (1896). Beiträge zur Ornis Algeriens. *Journal für Ornithologie* **43**, 113–238.

Kostin, Y.V. (1978). (Is the little bustard doomed?) *Bulletin of the Moscow Naturalists' Society, Biological Section* **83** (3), 67–71. (In Russian; translated for ICBP Bustard Group by M.G. Wilson.)

Kuleshova, L.V., Potorcha, V.I., Shilaev, Y.V. and Yakhontov, V.D. (1968). (Invasion of the Pallas' sandgrouse to the Far East.) *Ornithologiya* **9**, 354–5. (In Russian.)

Kumar, P. (1983). Great Indian bustard (*Choriotis nigriceps*) in Andhra Pradesh. In Goriup and Vardhan (1983), pp. 164–6.

Labitte, A. (1955). La reproduction d'*Otis tetrax* (L.) dans la partie nord du Département d'Eure-et-Loir. *Oiseau* **25**, 144–7.

Lachungpa, U.G. and Lachungpa, G. (1985). Lesser florican survey in Andhra Pradesh and Karnataka. In Ali *et al.* (1985), pp. 61–78.

— and Rahmani, A.R. (1985). Distribution of the lesser florican in India: a literature review. In Ali *et al.* (1985), pp. 5–19.

—, Rahmani, A.R., Lachungpa, G. and Sankaran, R. (1985). A preliminary survey of the Bengal florican *Eupodotis (Houbaropsis) bengalensis bengalensis* (Gmelin 1789) in Uttar Pradesh, West Bengal and Assam. In Ali *et al.* (1985), pp. 76–85.

Lack, P.C. (1975). Range expansion of the quail plover. *Bulletin of the East Africa Natural History Society (EANHS Bulletin)*, 110.

— (1983). The Canarian houbara: survey results, 1979. *Bustard Studies* **1**, 45–50.

Lamarche, B. (1980). Liste commentée des oiseaux du Mali: non-passereaux. *Malimbus* **2**, 121–58.

Launay, F. and Paillat, P. (1990). A behavioural repertoire of the adult houbara bustard (*Chlamydotis undulata maqueenii*). *Revue d'Ecologie (Terre Vie)* **45**, 65–87.

Lavee, D. (1985). The influence of grazing and intensive agriculture on the population size of the houbara bustard in the northern Negev, Israel. *Bustard Studies* **3**, 103–8.

— (1988). Why is the houbara *Chlamydotis undulata macqueenii* still an endangered species in Israel? *Biological Conservation* **45**, 47–54.

Lehmann, H. (1971). Vögel (Non-Passeriformes) eines bisher unbekannten Seegebietes in Zentral-Anatolien. *Vogelwelt* **92**, 161–81.

Lendon, A. (1938). The breeding in captivity of the little bustard-quail (*Turnix velox*). *Avicultural Magazine* 3 (5), 78–9.

Lewis, A.D., Loefler, I.J.P. and Pearson, D.J. (1984). Four-banded sandgrouse *Pterocles quadricinctus* in northwest Kenya. *Scopus* **8**, 46–8.

Little, J. de V. (1964). Observations on the call of the black-bellied korhaan. *Bokmakierie* **16**, 15.

Long, J.L. (1981). *Introduced birds of the world*. A.H. and A.W. Reed, Sydney and Wellington.

Longhurst, A. and Silvert, W. (1985). A management model for the great bustard in Iberia. *Bustard Studies* **2**, 57–72.

Lowe, P.R. (1923). Notes on the systematic position of *Ortyxelos*, together with remarks on the position of the Turnicomorphs and the position of the seed-snipe (Thinocoridae) and sand-grouse. *Ibis*, 11th series, **5**, 276–99.

— (1931). On the relations of the Gruimorphae to the Charadriimorphae and Rallimorphae, with special reference to the taxonomic position of Rostratulidae, Jacanidae, and Burhinidae (Oedicnemidae *olim*); with a suggested new order (Telmatormorphae). *Ibis* **73**, 491–534.

Lynes, H. (1925). On the birds of north and central Darfur, with notes on the west-central Kordofan and north Nuba provinces of British Sudan, *Ibis*, 12th series, **1**, 541–90.

Lynn-Allen, B.G. (1951). *Shot-gun and sunlight. The game birds of East Africa*. Batchworth Press, London.

Macdonald, J.D. (1957). *Contribution to the ornithology of western South Africa*. Trustees of the British Museum, London.

— (1971). Validity of the buff-breasted quail. *Sunbird* **2**, 1–5.

Maclean, G.L. (1967). Die systematische Stellung der Flughühner (Pteroclididae). *Journal für Ornithologie* **108**, 203–17.

— (1968). Field studies on the sandgrouse of the Kalahari Desert. *Living Bird* **7**, 209–35.

— (1969). The sandgrouse—doves or plovers. *Journal für Ornithologie* **110**, 104–7.

— (1976). Adaptations of sandgrouse for life in arid lands. *Proceedings, XVI Ornithological Congress*, 1974, pp. 502–16.

— (1983). Water transport in sandgrouse. *BioScience* **33**, 365–9.

— (1984). Evolutionary trends in the sandgrouse (Pteroclidae). *Malimbus* **6**, 75–8.

— (1985). Sandgrouse: models of adaptive compromise. *South African Journal of Wildlife Research* **15**, 1–6.

—, Maclean, C.M., Geldenhuys, J.N. and Allan, D.G.

(1983). Group size in the blue korhaan. *Ostrich* **54**, 244–5.
Mackworth-Praed, C.W. and Grant, C.H.B. (1952). *Birds of the southern third of Africa* (2 vols.). Longmans, London.
— and — (1962). *Birds of eastern and north eastern Africa* (2 vols.). Longmans, London.
Magrath, R.D., Ridley, M.W. and Woinarski, J.Z. (1985). Status and habitat requirements of lesser floricans in Kathiawar, western India. *Bustard Studies* **3**, 185–93.
Makatsch, W. (1976). *Die Eier der Vögel Europas*, Vol.2. Radebeul, Leipzig.
Malik, M.M. (1985). The distribution and conservation of houbara bustards in North West Frontier Province. *Bustard Studies* **3**, 81–8.
Malzy, P. (1962). La faune avienne du Mali (Basin du Niger). *Oiseau* **2** (suppl.), 1–81.
Manakadan, R. and Rahmani, A.R. (ed.) (1986). *Study of ecology of certain endangered species of wildlife and their habitats. Great Indian bustard*, Annual Report No.3, 1985–86. Bombay Natural History Society, Bombay. 50 pp.
Mansoori, J. (1985). The status of the houbara bustard in Iran. *Bustard Studies* **3**, 97–100.
Marchant, S. (1961). Observations on the breeding of the sandgrouse *Pterocles alchata* and *senegallus*. *Bulletin of the British Ornithologists' Club* **81**, 134–41.
— (1962). Watering of young in *Pterocles alchata*. *Bulletin of the British Ornithologists' Club* **82**, 123–24.
— (1963). The breeding of some Iraqi birds. *Ibis* **105**, 516–17.
Masterson, A.N.B. (1972). Notes on the Hottentot buttonquail. *Honeyguide* **74**, 12–16.
Mattingley, A.H.E. (1929). The love-display of the Australian bustard. *Emu* **28**, 198.
Mayaud, M. (1982). Les oiseaux du nord-ouest de l'Afrique. Notes comple'mentaires. *Alauda* **50**, 45–67, 116–45, 286–309.
McLachlan, G.R. (1985). The breeding season of the Namaqua sandgrouse. *Ostrich* **56**, 210–12.
McGinnes, W.G., Goldman, B.J. and Paylore, P. (ed.) (1968). *Deserts of the world*. University of Arizona Press, Tucson.
McGregor, R.C. (1909). *A manual of Philippine birds*, Part 1. Bureau of Printing, Manilla.
Meade-Waldo, E.G.B. (1896). The sandgrouse breeding in captivity. *Zoologist*, 298–9.
— (1897). Sandgrouse. *Avicultural Magazine* **3**, 177–80.
— (1905). A trip to the forest of Marmora, Morocco. *Ibis* (6th series), **1**, 161–4.
— (1906). Sandgrouse. *Avicultural Magazine* (new series), **4**, 219–22.
— (1922). Sandgrouse. *Bulletin of the British Ornithologists' Club* **42**, 69–70.
Mees, G.F. (1986). A list of the birds recorded from Bangka Island, Indonesia. *Zool. Verh.* (Leiden) **232**, 1–176.
Meinertzhagen, R. (1927). Systematic results of birds collected at high altitudes in Ladak and Sikkim. (Part II). *Ibis* (12th series), **3**, 573–633.
— (1954). *The birds of Arabia*. Oliver and Boyd, Edinburgh.

Mendelssohn, H. (1983). Observations on the houbara (*Chlamydotis undulata*) in Israel. In Goriup and Vardhan (1983), pp. 91–6.
—, Marder, U. and Stavy, M. (1979). Captive breeding of the houbara (*Chlamydotis undulata macqueenii*) and a description of its display. *ICBP Bulletin* **13**, 134–9.
—, — and — (1983). Captive breeding of the houbara (*Chlamydotis undulata macqueenii*) and the development of the young bird. In Goriup and Vardhan (1983). pp. 288–92.
Meyer de Schauensee, R. (1984). *The birds of China*. Smithsonian Institution Press, Washington, D.C.
Mian, A. (1986a). Ecological impact of Arab falconry on houbara bustard in Baluchistan. *Environmental Conservation* **13**, 41–6.
— (1986b). A contribution to the biology of the houbara: some evidence of breeding in Baluchistan. *Proceedings of the Pakistan Congress of Zoology* **5**, 261–9.
— (1988). A contribution to the biology of the houbara (*Chlamydotis undulata macqueenii*): some observations on 1983–84 wintering population in Baluchistan. *Journal of the Bombay Natural History Society* **81**, 9–25.
— and Dasti, A.A. (1985). The houbara bustard in Baluchistan, 1982–83: a preliminary review. *Bustard Studies* **3**, 45–50.
— and Surahio, M.I. (1983). Biology of houbara bustard (*Chlamydotis undulata macqueenii*) with reference to western Baluchistan. *Journal of the Bombay Natural History Society* **80**, 111–8.
Minchin, R.R. (1931). The display of the Australian bustard. *Avicultural Magazine* **4** (9), 147–8.
Mirza, Z.B. (1971). Preliminary report on the study of the population, mortality, sex ratio, age structure and food habits of the houbara bustard in Cholistan desert. *Outdoorsman (Karachi)* **1** (8), 40–2.
— (1985). A note on houbara bustards in Cholisan, Punjab. *Bustard Studies* **3**, 43–4.
Morris, A.K. (1971). The red-chested quail in New South Wales. *Emu* **71**, 178–80.
— and Kurz, N. (1977). Red-chested and little button-quail in the Mudgee district of New South Wales. *Corella* **1**, 77–9.
Mörs, F.E.O. (1915). On the breeding of the Kurrichaine button-quail (*Turnix lepurana*) in captivity. *Journal of the South African Ornithological Union* **10**, 19–21.
Mungure, S.A. (1974). A brief interesting observation on sandgrouse at Seronera River pool. *Bulletin of the East Africa Natural History Society (EANHS Bulletin)*, 52–3.
Murie, J. (1868). Observations concerning the presence and function of the gular pouch in *Otis kori* and *Otis australis*. *Proceedings of the Zoological Society of London*, 471–7.
Murray, B.G., Jr (1984). A demographic theory on the evolution of mating systems as exemplified by birds. In *Evolutionary Biology*, vol. 18 (ed. M. Hecht *et al.*), pp. 71–140. Plenum Press, New York.
Mwangli, E.M. (1988). The ecology of bustards in Nairobi National Park and the Kitelgela Conservation Area, Kenya. M.Sc. thesis, University of Nairobi.

— and Karanja, W.K. (1989). Home range, group size and sex composition of white-bellied and kori bustards. *Bustard Studies* **4**, 114–22.

Nader, I. (1983). Bustards in Saudi Arabia. In Goriup and Vardhan (1983), pp. 336–7.

Narayan, G. and Rosalind, L. (1988). Ecology of Bengal florican in Manos Wildlife Sanctuary. In Rahmani *et al.* (1988b), pp. 9–42.

Narayan, G., Sankaran, R., Rosalind, L. and Rahmani, A.R. (1989). Study of the ecology of some endangered birds and their habitats. Annual Report No. 4, 1988–1989. The floricans: *Houbaropsis bengalensis* and *Sypheotides indica*; 39 pp. Bombay Natural History Society, Bombay.

Neginhal, S.G. (1983). Status and distribution of the great Indian bustard in Karnataka. In Goriup and Vardhan (1983), pp. 76–80.

Neumann, O. (1934). Zwei neue geographische Rassen aus dem Süden des paläarktischen Gebiets. *Verhandlungen Ornithol. Gesellschaft Bayern* **20**, 470–2.

Newman, K. (1983). *Newman's birds of southern Africa*. Macmillan South Africa, Johannesburg.

Niethammer, G. (1934). Morphologische und histologische Untersuchungen am Kropf von *Pterocles orientalis* (L.) im Hinblick auf die systematische Stellung der Pterocliden. *Zoologischer Anzeiger* **107**, 199–202.

— (1937). Ueber den Kropf der männlichen Grosstrappe. *Ornithologische Monatsberichte* (supplement to *Journal für Ornithologie*), **45**, 189–92.

— (1940). Beobachtungen über die Balz und Untersuchungen über den Oesophagus südafrikanischer Trappen. *Ornithologische Monatsberichte* (supplement to *Journal für Ornithologie*), **48**, 29–33.

— (1961). Sonderbildungen an Ösophagus und Trachea beim Weibchen von *Turnix sylvatica lepurana*. *Journal for Ornithologie* **102**, 75–9.

Nishida, C., Sasaki, M. and Hori, H. (1981). Banding patterns and nucleolus organizing regions in somatic chromosomes of the Siberian great bustard, *Otis tarda*, with a note on the karyotypic similarities to the crane. *Chromosome Information Service* (31), 28–30.

Nitzsch, C.L. (1840). *System der Pterylographie*. Eduard Anton, Halle. (English translation 1867, by P. Sclater, Ray Society, London).

Ogilvie-Grant, W.R.O. (1889). On the genus *Turnix*. *Ibis* (6th series), **1**, 446–75.

— (1893). Catalogue of the game birds (Pterocletes, Gallinae, Opisthocomi, Hemipodii) in the collection of the British Museum. *Catalogue of the birds in the British Museum*, Vol.22, 1–585. British Museum (Natural History), London.

— (1896–7). *A hand-book to the game birds*, 2 vols. (Lloyd's Natural History). E. Lloyd Ltd, London.

Olson, S. (1970). [Review of Maclean's study of seedsnipe.] *Bird-banding* **41**, 258–59.

— and Steadman, D.W. (1981). The relationships of the Pedionomidae (Aves, Charadriiformes). *Smithsonian Contributions to Zoology* **337**, 1–25.

Oring, L. (1986). Avian polyandry. *Current Ornithology* **3**, 309–51.

Osborne, L. (1985). Progress toward the captive rearing of great bustards. *Bustard Studies* **2**, 123–30.

Osborne, P. (1986). Survey of the birds of Fuerteventura, Canary Islands, with special reference to the status of the Canarian houbara bustard *Chlamydotis undulata*. *ICBP Study Report* **10**, 1–76.

— (1989). The bustard morphometrics data base: an introduction and preliminary findings. *Bustard Studies* **4**, 125–34.

—, Collar, N. and Goriup, P. (1984). *Bustards*. Dubai Wildlife Research Centre, Dubai.

Otero Muerza, C. (1985a). The Spanish great bustard census conducted by Recursos Naturalis, S.A., in 1982. *Bustard Studies* **2**, 21–30.

— (1985b). Criteria for the establishment of recovery areas for the Spanish great bustard population: area classification and population studies. *Bustard Studies* **2**, 31–4.

— (1985c). A guide to sexing and ageing little bustards. *Bustard Studies* **2**, 172–8.

Palacios, F., Garzon, J. and Castroviejo, J. (1975). La alimentación de la avutarda (*Otis tarda*) in España, especialmente en primavera. *Ardeola* **21**, 347–406.

Parker, W.K. (1864). On the osteology of gallinaceous birds and tinamous. *Transactions of the Zoological Society of London* **5**, 149–241.

Parkes, K.C. (1949). A new button-quail from New Guinea. *Auk* **66**, 84–6.

— (1968). An undescribed subspecies of button-quail from the Philippines. *Bulletin of the British Ornithologists' Club* **88**, 24–5.

Paton, P.W.C., Ashman, P.R. and McEldowney, H. (1982). Chestnut-bellied sandgrouse in Hawaii. *Elepaio* **43** (2), 9–11.

Pendleton, R.C. (1947). Field observations on the spotted button-quail on Guadalcanal. *Auk* **64**, 417–21.

Peters, J.L. (1934). *Check-list of the birds of the world*, Vol.2. Harvard University Press, Cambridge, MA.

— (1937). *Check-list of the birds of the world*, Vol.3. Harvard University Press, Cambridge, MA.

Petretti, F. (1985). Preliminary data on the status of the little bustard in Italy. *Bustard Studies* **2**, 165–70.

— (1988). An inventory of steppe habitats in southern Italy. In Goriup (1988a), pp. 125–44.

Phillips, A.R. (1947). The button quails and tree sparrows of the Riu Kiu islands. *Auk* **64**, 126–7.

Pitman, C.R.S. (1957). Uganda's bustards. *Uganda Wildlife and Sport* **1** (2).

Pizzey, G. (1980). *A field guide to the birds of Australia*. Princeton University Press, Princeton.

Ponomareva, T. (1983). (Reproductive behavior and distribution of houbara bustard on the nesting ground.) *Zoologicheskii Zhurnal* **62**, 592–602. (In Russian, English summary.)

— (1985). The houbara bustard: present status and conservation prospects. *Bustard Studies* **3**, 93–6.

Prakash, I. (1983). Current status of the great Indian bustard (*Choriotis nigriceps*) in the Thar desert. In Goriup and Vardhan (1983), pp. 39–43.

Quinton, W.F. (1948). The Karroo korhaan (*Eupodotis vigorsii vigorsii*). *Ostrich* **19**, 235–6.

Rabor, D.S. (1977). *Philippine birds and mammals.* University of the Philippines Press, Quezon City.

Rahmani, A.R. (1985). Present status of the great Indian bustard. *Bustard Studies* **3**, 123–32.

—— (1986). Status of the great Indian bustard in Rajasthan. *Bombay Natural History Society Technical Report* **11**, 1–324.

—— (1989). *The great Indian bustard.* Final report. Study of ecology of certain endangered species and their habitats. Bombay Natural History Society, Bombay.

—— and Manakadan, R. (1986). Movement and flock composition of the great Indian bustard at Nannaj, Solapur district, Maharashtra, India. *Journal of the Bombay Natural History Society* **83**, 17–31.

—— and —— (1987). Interspecific behaviour of the great Indian bustard *Ardeotis nigriceps* (Vigors). *Journal of the Bombay Natural History Society* **84**, 317–31.

—— and Shobrak, M. (1987). *Arabian bustard survey, 1987*, Technical Report No.1. National Committee for Wildlife Conservation and Development, Saudi Arabia. 13 pp.

—— and —— (1988). Bustard sanctuaries in India. *Bombay Natural History Society Technical Report* **13**, 1–40.

—— and Yahya, H.A.S. (1985). The lesser florican in Maharashtra. In Ali *et al.* (1985), pp. 74–5.

——, Lachungpa, U.G. and Lachungpa, G. (1985). Lesser florican survey in Gujarat. In Ali *et al.* (1985), pp. 51–73.

——, Narayan, G. and Rosalind, L. (1988a). Bengal florican survey in Uttar Pradesh, Bihar, West Bengal and Assam. In Rahmani *et al.* (1988b), pp. 65–120.

——, ——, Sankaran, R. and Rosalind, L. (eds) (1988b). *The Bengal florican: status and ecology;* 148 pp. Annual Report, Bombay Natural History Society, Bombay.

Rand, A. (1936). The distribution and habits of Madagascan birds. *Bulletin of the American Museum of Natural History* **73**, 143–499.

Rand, A.L. and Gilliard, E.T. (1968). *Handbook of New Guinea birds.* Natural History Press, Garden City.

Rands, M. (1986). An autumn in Arabia, the OSME North Yemen expedition 1985. *Bulletin of the Ornithological Society of the Middle East* **16**, 1–4.

Razdan, T. and Mansoori, J. (1989). A review of the bustard situation in the Islamic Republic of Iran. *Bustard Studies* **4**, 135–45.

Rego, L.H.A. (1983). The status of bustards in Maharashtra. In Goriup and Vardhan (1983), pp. 60–75.

Ridley, M.W. (1983). A review of the ecology and behaviour of button-quails. *World Pheasant Association Journal* **8**, 51–61.

——, Magrath, R.D. and Woinarski, J.C.Z. (1985). Display leap of the lesser florican *Sypheotides indica. Journal of the Bombay Natural History Society* **82**, 271–7.

Roberts, A. (1957). *Birds of South Africa*, rev. edn (ed. G.R. McLachlan and R. Liversidge). Trustees of South African Bird Fund, Cape Town.

Rockingham-Gill, D.V. (1983). On the distribution of the kori bustard (*Otis kori*) in Zimbabwe. In Goriup and Vardhan (1983), pp. 97–103.

St Quintin, W.H. (1905). The breeding of *Pterocles exustus. Avicultural Magazine*, new series, **3**, 64–6.

Saleh, M.A. (1989). The status of the houbara bustard in Egypt. *Bustard Studies* **4**, 151–6.

Samarin, E.G. (1977). (Distribution and growth dynamics of the little bustard in the Ural Valley and adjacent clay plains. In *Rare and vanishing animals and birds of Kazakhstan* (ed. A.A. Sludski, Alma Ata.) (In Russian; translated for ICBP Bustard Group by M.G. Wilson.)

Sankaran, R. and Rahmani, A.R. (1985a). Immigration chronology of the lesser florican at Sailana: preliminary results. In Ali *et al.* (1985), pp. 20–4.

—— and —— (1985b). Intra- and inter-specific behaviour of the lesser florican. In Ali *et al.* (1985), pp. 25–50.

—— and —— (1986). *Study of ecology of certain endangered species of wildlife and their habitats. The lesser florican*, Annual Report No.2. Bombay Natural History Society, Bombay.

—— and —— (1988). Status of Bengal florican in Dudway National Park. In Rahmani *et al.* (1988b), pp. 43–64.

Saxena, V.S. and Meena, B.L. (1985). Occurrence of lesser floricans in forest plantations in Rajasthan. *Bustard Studies* **3**, 183–4.

Schaller, G.B. (1973). *Golden shadows, flying hooves.* A.A. Knopf, New York.

Schenk, H. and Aresu, M. (1985). On the distribution, number and conservation of the little bustard in Sardinia (Italy), 1971–1982. *Bustard Studies* **2**, 161–4.

Schönwetter, M. (1967). *Handbuch der Oologie*, Band 1 (*Nonpasseres*). Akademie-Verlag, Berlin.

Schulz, H. (1980). Zur Bruthabitatwahl der Zwergtrappe *Tetrax t. tetrax* in der Crau (Südfrankreich). *Braunschweiger Naturkundliche Schriften* **1**, 141–60.

—— (1985a). A review of the world status and breeding distribution of the little bustard. *Bustard Studies* **2**, 131–52.

—— (1985b). *Grundlagenforschung zur Biologie der Zwergtrappe;* 401 pp. Staatliches Naturhistorisches Museum Braunschweig, Braunschweig.

—— (1985c). On the social behaviour of the little bustard *Tetrax tetrax*: a preliminary report. *Bustard Studies* **2**, 179–82.

—— (1986a). Agonistisches Verhalten, Territorialverhalten und Balz der Zwergtrappe (*Tetrax tetrax*). *Journal für Ornithologie* **127**, 125–204.

—— (1986b). Zum Geschlechtverhältnis der Zwergtrappe (*Tetrax tetrax*). *Vogelwelt* **107**, 210–10.

—— (in preparation). Die Balz der Trappen (Otididae)—Signalcharakten und Evolution.

—— and Schulz, M. (1986). Beitrag zur Biologie der Schwarzbauchtrappe (*Eupodotis melanogaster*) in Senegal. *Vogelwelt* **107**, 140–56.

Serventy, D.L. and Whittell, H.M. (1962). *Birds of western Australia.* Paterson Brokensha Pty, Perth.

Seth-Smith, D. (1903). On the breeding in captivity of *Turnix tanki* with some notes on the habits of the species. *Avicultural Magazine* **1**, 317–24.

—— (1905). On the breeding of *Turnix varia. Avicultural Magazine* **2**, 295–301.

Shams, K.M. (1985). Occurrence and distribution of bustards in Baluchistan. *Bustard Studies* **3**, 51–4.

Sharpe, R.B. (1894). Catalogue of the Fulicariae (Rallidae and Heliornithidae) and Alectorides (Aramidae, Eurypygidae, Mesitidae, Rhinochetidae, Gruidae, Psophidae, and Otididae) in the collection of the British Museum. *Catalogue of the birds of the British Museum*, Vol.23, 1–353. British Museum (Natural History), London.

—— (1896). Catalogue of the Limicolae in the collection of the British Museum. *Catalogue of the birds of the British Museum*, Vol.24, 1–794. British Museum (Natural History), London.

Shufeldt, R.W. (1901). On the systematic position of the sand grouse (*Pterocles, Syrrhaptes*). *American Naturalist* **35**, 11–6.

Sibley, C.G. and Ahlquist, J. (1972). A comparative study of the egg white proteins of non-passerine birds. *Bulletin Peabody Museum of Natural History* **39**, 1–276.

—— and —— (1985). The relationships of some groups of African birds, based on comparisons of the genetic material. In Proceedings of the Symposium on African Vertebrates (ed. K.L. Schumann), pp. 115–61. Museum A. Koenig, Bonn.

——, —— and Monroe, B.L. Jr (1988). A classification of the living birds of the world based on DNA–DNA hybridization studies. *Auk* **105**, 409–23.

Sich, H.L. (1927). The breeding of the Lepurana bustard-quail (*Turnix lepurana*). *Avicultural Magazine*, series 4, **5**, 264–6.

Siewert, H. (1939). Die Balz der Grosstrappen. *Zeitschrift für Jagdkunde* **1**, 6–36.

Sims, R.W. (1954). A new race of button-quail (*Turnix maculosa*) from New Guinea. *Bulletin of the British Ornithologists' Club* **74**, 37–40.

Singh, Y.D. (1983). Some experiences of rearing great Indian bustards in captivity. In Goriup and Vardhan (1983), pp. 235–38.

Sinha, S.K. (1983). Status of the great Indian bustard in Gujarat. In Goriup and Vardhan (1983), pp. 51–3.

Skead, C.J. (1965). Birds of the Albany (Grahamstown) district. *South African Avifauna Series* **30**, 1–45.

Sludski, A.A. (ed.) (1977). (Rare and vanishing animals and birds of Kazakhstan. Alma Ata.) (In Russian; translated for ICBP Bustard Group by M.G. Wilson.)

Smith, M.J. (1987). Observations on Denham's bustard *Neotis denhami* at Maralal, Kenya. *Scopus* **11**, 47–51.

Snow, D.W. (ed.) (1978). *An atlas of speciation in African non-passerine birds*. British Museum (Natural History) Publications No. 787.

Someren, V.G.L. van. (1926). The birds of Kenya and Uganda. *Journal of the East Africa and Uganda National History Society* **27**, 197–212.

—— (1933). The birds of Kenya and Uganda. *Journal of the East Africa and Uganda Natural History Society* **(47–48)**, 89–127.

Spenkelink-van Schaik, J.L. (1984). Notes on keeping and breeding little button quail *Turnix sylvatica*. *Avicultural Magazine* **90**, 219–20.

Stegmann, B.K. (1968). Ueber die phyletische Beziehungen zwischen Regenfeifervögeln, Tauben und Flughühner. *Journal für Ornithologie* **109**, 141–5.

—— (1969). Uber die systematische Stellung der Tauben und Flughühnern. *Zoologischer Jahrbücher, Abteilung für Systematik, Ökologie und Geographie der Tiere* **96**, 1–51.

Sterbetz, I. (1981). Comparative investigation in the reproduction behaviour of monogamous, polygamous and unmated great bustard populations in south-east Hungary. *Aquila* **87**, 31–47.

Stresemann, E. (1927–34). Aves. In *Handbuch der Zoologie* (ed. W. Kükenthal and T. Krumbach), Vol.7, no.2. W. de Gruyter, Berlin.

Surahio, M.I. (1985). Ecology and distribution of houbara bustards in Sind. *Bustard Studies* **3**, 55–8.

Sutter, E. (1955a). Ueber die Mauser einiger Laufhühnchen und die Rassen von *Turnix maculosa* und *sylvatica* im indo-australischen Gebiet. *Verhandlungen der Naturforschenden Gesellschaft in Basel* **66**, 85–139.

—— (1955b). *Turnix maculosa obiensis* subsp. nova. *Journal für Ornithologie* **96**, 220–2.

—— (1964). Button-quail. *A new dictionary of birds*. In (ed. A.L. Thompson), pp. 116–8. British Ornithological Union, London.

—— and Cornaz, N. (1963). Ueber die Gewichtsentwicklung und das Schwingwachtum junger Spitzschwanz-Laufhühnchen, *Turnix sylvatica*. Ornithologischer Beobobachter **60**, 213–23.

Taka-Tsukasa, N. (1967). *The birds of Nippon*. Maruzen Co., Tokyo.

Tarboton, W.R. (1989). Breeding behaviour of Denham's bustard. *Bustard Studies* **4**, 160–9.

——, Kemp, M.I. and Kemp, A.C. (1987). *Birds of the Transvaal*. Transvaal Museum, Pretoria.

Taylor, N. (1985). Live-capture of the houbara bustard. *Bustard Studies* **3**, 59–80.

Thomas, D.H. (1984a). Sandgrouse as models of avian adaptations to deserts. *South African Journal of Zoology* **19**, 113–20.

—— (1984b). Adaptations of desert birds: sandgrouse (Pteroclididae) as highly successful inhabitants of Afro-Asian arid lands. *Journal of Arid Environments* **7**, 157–81.

—— and Maclean, G.L. (1981). Comparison of physiological and behavioural thermoregulation and osmoregulation in two sympatric sandgrouse species (Aves: Pteroclididae). *Journal of Arid Environments* **4**, 335–58.

—— and Robin, A.P. (1977). Comparative studies of thermoregulatory and osmoregulatory behaviour and physiology of five species of sandgrouse (Aves: Pteroclididae) in Morocco. *Journal of Zoology, London* **183**, 229–49.

—— and —— (1983). Description of the downy young of Lichtenstein's sandgrouse *Pterocles lichtensteinii* and the significance of 'unpatterned' downy young in the Pteroclididae. *Bulletin of the British Ornithologists' Club* **103**, 40–3.

——, —— and Clinning, C.F. (1981). Daily patterns of behaviour compared between two sandgrouse species (Aves: Pteroclididae) in captivity. *Madoqua* **12**, 187–98.

Timmermann, G. (1969). Zur Frage der systematische Stellung der Flughühner (Pteroclididae) im vergleichen parasitologischer Sicht. *Journal für Ornithologie* **110**, 103–4.

Trollope, J. (1967). Breeding a hemipode, *Turnix suscitator*. *Avicultural Magazine* **73**, 184–8.

—— (1970). Behaviour notes on the barred and Andalusian

hemipodes (*Turnix suscitator* and *Turnix sylvatica*). *Avicultural Magazine* **76**, 219–27.

Turner, D.A. (1982). The status and distribution of the Arabian bustard *Otis arabs* in northeastern Africa and its possible occurrence in northern Kenya. *Scopus* **6**, 20–1.

—— (1989). The status of Denham's bustard in Kenya. *Bustard Studies* **4**, 170–3.

Urban, E.K., Brown, L.H. and Newman, K.B. (1978). Kori bustard eating gum. *Bokmakierie* **30**, 105.

——, Fry, C.H. and Keith, S. (1986). *The birds of Africa*, Vol.2. Academic Press, London.

Uys, C.J. (1963). Some observations on the Stanley bustard (*Neotis denhami*) at the nest. *Bokmakierie* **15** (2), 2–4.

Valverde, J.A. (1957). *Aves de Sahara Español*. Consejo Superior de Investigaciones Científicas, Madrid.

van Praet Lucas, L. (1976). Breeding successes for 1976. *Foreign Birds* **42**, 88–9.

Vardhan, H. (1985). A report on the status of bustard in India. *Bustard Studies* **3**, 113–5.

Vaurie, C. (1961). Systematic notes on Palearctic birds, No. 50. The Pteroclidae. *American Museum Novitates* **2071**, 1–13.

—— (1965). *The birds of the Palearctic fauna. Non Passeriformes*. H.F. and G. Witherby, London.

Vernon, C. (1983). Notes from the border. 11. Blue korhaan group size. *Bee-eater* Suppl. 10, 1–2.

Verheyen, R. (1957). Analyse du potentiel morphologique et projet de classification des Columbiformes (Wetmore 1934). *Bulletin, Institut Royal des Sciences Naturelles de Belgique* **33** (3), 1–42.

—— (1958). Contribution au démembrement de l'ordo artificiel des Gruiformes. IV. Les Turniciformes. *Bulletin, Institut Royal des Sciences Naturelles de Belgique* **34** (2), 1–18.

—— (1961). A new classification for the non-passerine birds of the world. *Bulletin, Institut Royal des Sciences Naturelles de Belgique* **37** (27), 1–36.

Viljoen, P.J. (1983). Distribution, numbers and group size of the Karoo korhaan in Kaokoland, South West Africa. *Ostrich* **54**, 50–1.

Voous, K.H. (1973). List of Recent Holarctic bird species: Non-passerines. *Ibis* **115**, 612–38.

Vyas, D.K., Jacob, D. and Mathur, K.M. (1983). Ecology, behaviour and breeding biology of the great Indian bustard. In Goriup and Vardhan (1983), pp. 184–93.

Ward, P. (1972). The functional significance of mass drinking flights by sandgrouse: Pteroclididae. *Ibis* **114**, 533–636.

Waterson, J. (1928). The Mallophaga of sand-grouse. *Proceedings of the Zoological Society of London*, 333–56.

Weinitschke, H. (ed.) (1983). Verbreitung und Schutz der Grosstrappe (*Otis tarda* L.) in der D.D.R. *Naturschutzarbeit in Berlin und Brandenburg*, Beiheft 6. (Not seen, cited in Collar 1985.)

Welch, G. and Welch, H. (1986). Djibouti 2 expedition. *Bulletin of the Ornithological Society of the Middle East* (16), 4–7.

—— and —— (1989). A preliminary survey of the Arabian bustard in Djibouti. *Bustard Studies* **4**, 177–84.

Werner, K.W. (1975). Zur Haltung und Pflege des Laufhühnchens (*Turnix sylvatica*). *Gefiederte Welt* **99**, 236–7.

Wetmore, A. (1930). A systematic classification of the birds of the world. *Proceedings of the U.S. National Museum* **54**, 577–86.

—— (1960). A classification for the birds of the world. *Smithsonian Miscellaneous Collections* **117**, 1–22.

White, C.M.N. and Bruce, M.D. (1986). *The birds of Wallacea*. British Ornithological Union, London.

White, D.M. (1985). A report on the captive breeding of Australian bustards at Serendip Wildlife Research Station. *Bustard Studies* **3**, 195–212.

Wickler, W. (1961). Über die Stammesgeschichte und den taxonomischen Wert einer Verhaltenweisen der Vögel. *Zeitschrift für Tierpsychologie* **18**, 320–42.

Wilkinson, R. and Manning, N. (1986). Breeding Pallas' sandgrouse *Syrrhaptes paradoxus* at Chester zoo. *Avicultural Magazine* 121–4.

Wilson, V.J. (1972). Notes on *Otis denhami jacksoni* from the Nyika Plateau. *Bulletin of the British Ornithologists' Club* **92**, 77–81.

Winkel, W. (1969). Einige ergänzende Volierenbeobachtungen am Schwarzbrustlaufhühnchen (*Turnix suscitator*). *Gefiederte Welt* **93**, 169–70.

Winkler, H. (1973). Beitrag zur Ethologie der Zwergtrappe, *Tetrax t. tetrax*. Anzeiger der österreichischen Akademie der Wissenschaften, *Mathematische-Naturwissenschaftliche Klasse* **109**, 61–72.

Wintle, C.C. (1975). Notes on the breeding habits of the Kurricane buttonquail. *Honeyguide* **82**, 27–30.

Wolters, H.E. (1974). Aus der ornithologischer Sammlung des Museums Alexander Koenig III. *Bonner Zoologische Beiträge* **25**, 283–91.

—— (1975). *Die Vogelarten der Erde*, Pt.1. Paul Parey, Hamburg.

Index

This index includes all generic, specific, and subspecific epithets of the groups discussed in this book, and also all English vernacular names as used herein. Commonly used alternative English vernacular names for species and subspecies are provided for cross-reference, but all non-English vernacular names are excluded. Complete page citations are shown only under the English names used in the text. Principal accounts of genera and species are indicated by italics; pages mentioning subspecies as indexed separately. Plates are indexed by number; pages with drawings are indicated by asterisks, as are distribution maps when they occur outside the principal species account.

afra, *Eupodotis* 9, *185–9*
afraoides, *Eupodotis* 185, 186
African buttonquail, *see* black-rumped buttonquail
Afrotis 7, 8, 106
agaze, *Neotis* 164
albiventris, *Turnix* 5
alchata, *Pterocles* 8, 12, 13, *221–7*
alleni, *Turnix* 79
Andalusian buttonquail, *see* striped buttonquail
ansorgei, *Pterocles* 249
Arabian bustard Plate 19, 24*, 25*, 44, 106, *150–3*
arabicus, *Pterocles* 246, 253
arabs, *Ardeotis* 9, *150–3*, 157
Ardeotis 7, 19, 25, 40, 43, 105, 113, 114, *137–58*, 163
arenaria, *Turnix* 79
arenarius, *Pterocles* 12
atratus, *Pterocles* 240
atroqularis, *Turnix* 73, 101
Australian bustard Plate 17, 24*, 25*, 39, 40, 44, 55, 57, 106, 129*, *137–42*
australis, *Ardeotis* 9, *137–42*, 149

bangsi, *Pterocles* 237
barlowi, *Eupodotis* 182, 184
barred buttonquail Plate 12, 22*, 36, 38, 43, 45, 47, 48, 67, 77, 93*, *101–4*
Barrow's bustard, *see* Denham's bustard
barrowii, *Eupodotis* 177
bartelsorum, *Turnix* 78, 80
baweanus, *Turnix* 73, 101
beccarii, *Turnix* 85, 86
Bengal florican Plate 34, 22, 24*, 39, 43, 49, 57, 63, 64, *196–202*
bengalensis, *Eupodotis* 9, *196–202*, 207
bengalensis, *Turnix* 73, 101
benquetensis, *Turnix* 72, 73
bicinctus, *Pterocles* 12, 13, 26, 37, 244, *249–51*, 252, 253
black-backed quail, *see* red-backed buttonquail
black-bellied bustard Plate 25, 22, 24*, 25*, 39, 41, 44, 49, *169–73*, 175, 179, 195, 238
black-bellied sandgrouse Plate 42, 26*, 28*, 38, 44, 53, 54, *236–9*
black-breasted buttonquail Plate 13, 22*, 38, 43, 57, 68, *98–100*
black bustard, *see* little black bustard
black-faced sandgrouse Plate 45, 26*, 27, 28*, 44, 54, 231*, *242–4*

black-fronted buttonquail, *see* black-breasted buttonquail
black-rumped buttonquail Plate 5, 22*, 38, 43, 47, 67, *77–9*, 81
black-spotted quail, *see* red-backed buttonquail
black-throated bustard Plate 29, 24*, 39, 44, 106, 179, 181, *182–4*, 185, 189
black-throated buttonquail, *see* Madagascan buttonquail
blakistoni, *Turnix* 73, 101, 102
blandini, *Eupodotis* 196, 202, 203
blanfordii, *Turnix* 5, 74, 75, 76
Blanford's bustard-quail, *see* yellow-legged buttonquail
blue bustard Plate 33, 24*, 39, 44, 49, 106, 170*, 179, *194–5*
blue-necked bustard, *see* white-bellied bustard
boemeri, *Eupodotis* 185
buff-breasted buttonquail, *see* chestnut-backed buttonquail
buff-crested bustard, *see* rufous-crested bustard
Burchell's sandgrouse Plate 51, 27, 28*, 44, 210, *258–61*
burchelli, *Pterocles* 12, 13, 26, 38, *258–61*
burchellii, *Neotis* 159
Burchell's bustard, *see* Denham's bustard
bush-lark quail, *see* lark-quail
bustard-quail, *see* appropriate species of buttonquail
butleri, *Ardeotis* 150, 151
butterfly quail, *see* painted and little buttonquails

caerulescens, *Eupodotis* 9, *194–5*
Calopterocles 261
canicollis, *Eupodotis* 176, 177
castanota, *Turnix* 5, 6, 68, 94, 95, *97–8*
Celebean bustard-quail, *see* barred buttonquail
celestinoi, *Turnix* 57, 84, 85
centralis, *Eupodotis* 185
chestnut-backed buttonquail Plate 10, 22*, 38, 43, 57, 68, *97–8*
chestnut-bellied sandgrouse Plate 40, 27, 28*, 29*, 37, 44, 53, 54, 210, *227–31*, 238, 239
chestnut-breasted buttonquail, *see* spotted and red-chested buttonquails
chiversi, *Eupodotis* 185
Chlamydotis 7, 20, 105, *106–16*
chobiensis, *Pterocles* 247, 251

Choriotis, *see* *Ardeotis*
close-barred sandgrouse, *see* Lichtenstein's sandgrouse
common bustard-quail, *see* barred buttonquail
coronatus, *Pterocles* 12, 13, 26, 221, 230, 236, *239–42*, 243, 244, 245
coronetted sandgrouse, *see* crowned sandgrouse
crested bustard, *see* rufous-crested bustard
crowned sandgrouse Plate 43, 26*, 27, 28*, 44, 210, *239–42*

Damara korhaan, *see* Ruppell's bustard
damarensis, *Eupodotis* 185
davidi, *Turnix* 79, 80
decoratus, *Pterocles* 12, 13, 26, *242–4*, 249, 251
denhami, *Neotis* 9, *158–63*, 165, 167, 168
Denham's bustard Plate 22, 24*, 25*, 39, 44, 58, 106, 154, *158–63*, 168, 169
Dilophilus 12, 26
dotterel quail, *see* little buttonquail
double-banded sandgrouse Plate 47, 26, 28*, 44, 210, 244, *249–56*
dusky buttonquail, *see* barred buttonquail
dussumier, *Turnix* 79, 80
dwarf buttonquail, *see* Hottentot buttonquail
dybowskii, *Otis* 126, 127

ellenbecki, *Pterocles* 231, 242, 243
ellioti, *Pterocles* 227, 228
enigmaticus, *Pterocles* 237
Eremialector 239
erlangeri, *Eupodotis* 177
erlangeri, *Pterocles* 227, 228
etoschae, *Eupodotis* 185, 186
Eupodotis 7, 8, 19, 23, 44, 106, 119, 163, *169–208*
Everett's buttonquail, *see* red-chested buttonquail
everetti, *Turnix* 4, 57, 67, 89, 90, 91, 92*, 94
exustus, *Pterocles* 12, 13, 20, 221, 226, *227–31*, 236, 242

fasciata, *Turnix* 73, 74, 101
fitzsimmonsi, *Eupodotis* 180, 184
florensiana, *Turnix* 85
floweri, *Pterocles* 227, 228

Index

four-banded sandgrouse Plate 48, 23*, 28*, 44, 210, 244, 249*, *251–3*, 256
fuertaventurae, Chlamydotis 58, 107, 108
furva, Pterocles 231
furva, Turnix 85

geyri, Ardeotis 151
giant bustard, *see* kori bustard
giluwensis, Turnix 85, 86
gindiana, Eupodotis 189, 190, 191
great Arabian bustard, *see* Arabian bustard
great bustard Plate 16, 24*, 25*, 31*, 36, 37, 39, 40, 44, 50, 51*, 52, 56–9, 105, 124, *125–37*, 238
great Indian bustard Plate 18, 23*, 24*, 34*, 39, 40, 44, 57, 62–3, 64, 106, 139, *142–50*
gutturalis, Pterocles 12, 13, 26, 241, *255–8*

harei, Eupodotis 182
hartlaubii, Eupodotis 9, *173–6*
Hartlaub's bustard Plate 26, 22, 24*, 25*, 44, 106, *173–6*
Heterotetrax 106
heuglinii, Neotis 9, *165–7*
Heuglin's bustard Plate 23, 24*, 44, 106, *165–7*
hilgerti, Eupodotis
hindustan, Pterocles 227, 228
horsbrughi, Turnix 81
Hottentot buttonquail, *see* black-rumped buttonquail
hottentotta, Turnix 5, 6, 72, *77–9*, 81, 84
Houbara bustard Plate 14, 7, 23*, 24*, 25*, 30, 39, 57, 61–2, 64, 105, *106–16*, 124, 135, 238
Houbaropsis 106, 201
humilis, Eupodotis 9, *180–1*

imperial sandgrouse, *see* black-bellied sandgrouse
Indian buttonquail, *see* yellow-legged buttonquail
Indian sandgrouse, *see* painted and chestnut-bellied sandgrouse
indica, Sypheotides 9, *196–202*
indicus, Pterocles 12, 13, 26, 27, 37, 244, 249, 250, 251, 252, *253–5*
ingramsi, Pterocles 246, 247
insolata, Turnix 77, 78
interrumpens, Turnix 73, 101
island buttonquail, *see* barred buttonquail

jacksoni, Neotis 158, 159
Jackson's bustard, *see* Denham's bustard

kalaharica, Eupodotis 185
Karoo bustard, *see* black-throated bustard
karooensis, Eupodotis 182
kinneari, Turnix 85
korejewi, Otis 126
korhaans, *see* corresponding species of bustards
kori, Ardeotis 9, 150, *153–8*, 160, 168

Kori bustard Plate 20, 24*, 39, 40, 44, 58, 106, 139, 150*, *153–8*
koslovae, Pterocles 237
kuiperi, Turnix 73, 101
Kurrichaine buttonquail, *see* striped buttonquail

ladas, Pterocles 240
large pin-tailed sandgrouse, *see* pin-tailed sandgrouse
lark buttonquail, *see* lark-quail
lark-quail Plate 1, 22*, 38, 39, 43, *68–70*
leggei, Turnix 73, 101
lepurana, Turnix 79, 80
lesser florican Plate 35, 22, 24*, 39, 40, 43, 44, 49, 57, 63, 106, *202–8*
Lesser Sunda buttonquail, *see* barred buttonquail
leucogaster, Turnix 87
Lichtenstein's sandgrouse Plate 46, 26*, 28*, 29*, 44, 210, 241, 244, *246–9*, 252
lichtensteinii, Pterocles 12, 13, 26, 37, 229, 244, *246–9*, 251, 253, 254, 255, 256
Lissotis 7, 8, 106
little black bustard Plate 31, 22, 25*, 34*, 39, 44, 106, *185–9*, 193, 195
little brown bustard Plate 28, 24*, 44, 58, 106, *180–1*
little bustard Plate 15, 23*, 24*, 25*, 37, 39, 40, 44, 49, 50, 52, 57, *59–61*, 106, *116–25*, 130
little buttonquail, *see* striped buttonquail
Lophotis 7, 8, 106, 113
loverridgei, Pterocles 231, 242, 243
luciana Turnix 77
Ludwig's bustard 24*, 44, 58, 106, 160, 164*, *167–9*
ludwigi, Neotis 9, *167–9*
Luzon buttonquail, *see* red-crested buttonquail
Lynes' bustard, *see* red-crested bustard
lynesi, Ardeotis 150, 151

mababiensis, Eupodotis 185
machetes, Turnix 73, 101
mackenziei, Eupodotis 177
macqueenii, Chlamydotis 106, 107, 108
Macqueen's bustard, *see* houbara bustard
maculosa, Turnix 5, 6, 77, *84–7*
Madagascan buttonquail Plate 2, 22*, 38, 43, 45, 47, 68, *70–2*
Madagascan sandgrouse Plate 44, 26*, 28*, 44, 210, 231*, *244–6*
magnifica, Turnix 97
makarikari, Pterocles 259
Malagsi sandgrouse, *see* Madagascan sandgrouse
masaaki, Turnix 57, 79, 80
masked sandgrouse, *see* Madagascan sandgrouse
mayri, Turnix 85
meiffrenii, Ortyxelos 5, 6, *68–70*
melanogaster, Eupodotis 9, *169–73*, 175, 176
melanogaster, Turnix 5, 6, 97, *98–100*, 104
melanota, Turnix 85
mikado, Turnix 79
multicolor, Pterocles 249

Namapterocles 236
namaqua, Eupodotis 182, 184
namaqua, Pterocles 12, 13, 38, 221, 226, 230, *231–6*, 242
Namaqua sandgrouse Plate 41, 27, 28*, 38, 41, 44, 52, 53, 54, 209, *231–6*
nana, Turnix 77
Natal buttonquail, *see* black-rumped buttonquail
Neotis 7, 43, 106, 113, 114, *158–69*
New Caledonian buttonquail, *see* painted buttonquail
New Holland partridge, *see* painted buttonquail
ngami, Pterocles 231
Nicobar bustard-quail, *see* yellow-legged buttonquail
nigrescens, Turnix, 73, 101
nigriceps, Ardeotis 9, 138, *142–50*
nigricollis, Turnix 5, 6, *70–2*, 100
nigrorum, Turnix 79, 80
northern bustard-quail, *see* barred buttonquail
notophila, Eupodotis, 169, 170
novaecaledoniae, Turnix 94, 95, 98
nuba, Neotis 9, *163–5*
Nubian bustard Plate 21, 24*, 25*, 35*, 58, 106, *163–5*
Nyctiperdix 249, 255

obiensis, Turnix 85, 86
ocellata, Turnix 5, 6, *72–4*
okinavensis, Turnix 73, 101
olivascens, Pterocles 227, 228
olivei, Turnix 57, 68, 94, 97, 98
Olive's buttonquail, *see* chestnut-backed buttonquail
orange-breasted quail, *see* red-backed buttonquail
orangensis, Eupodotis 182
orientalis, Pterocles 12, 13, 20, 21, 26, *242–4*, 258
orientalis, Tetrax 117
Ortyxelos 1, 21, 25, 67, *68–70*
Otis 7, 19, 20, 23, 43, 105, 113, *125–37*, 163

painted buttonquail Plate 11, 22*, 38, 43, 45, 57, 68, *94–6*
painted sandgrouse Plate 37, 23*, 27, 28*, 34*, 38, 40, 44, 52, 53, 54, 209, *210–16*
Pallas' sandgrouse Plate 39, 23*, 27, 28*, 34*, 38, 40, 44, 52, 53, 54, 209, *210–16*
pallescens, Turnix 73, 101
paradoxus, Syrrhaptes 12, 13, 21, *210–16*
parva, Eupodotis 176
personatus, Pterocles 12, 13, 26, 241, 242, *244–6*
Philippine bustard-quail, *see* barred and red-chested buttonquails
picturata, Turnix 87
pin-tailed sandgrouse Plate 39, 26, 28*, 29*, 38, 40, 44, 52, 53, 209, *221–7*
plain turkey, *see* Australian bustard
plumbipes, Turnix 73, 101
powelli, Turnix 5, 73, 101
pseutes, Turnix 85, 86
Pterocles 12, 25, 26, 53, 209, *217–61*
Pteroclurus 12, 226

pygmy bustard, *see* little bustard
pyrrhothorax, Turnix 5, 6, 79, 87, 88, 89–94

quadricintus, Pterocles 12, 13, 26, 37, 244, 249, *251–3*, 255
quail-plover, *see* lark-quail

red-backed buttonquail Plate 7, 22*, 38, 43, 57, 67, *84–7*
red-chested buttonquail Plate 9, 22*, 30, 38, 43, 48, 54, 67, 68, *89–94*
red-crested bustard, *see* rufous-crested bustard
remotus, Pterocles 217
Robinson's buttonquail, *see* chestnut-backed buttonquail
rostrata, Turnix 73, 101, 102
rueppellii, Eupodotis 9, *184–5*
rufescens, Turnix 5
ruficrista, Lophotis 9, *189–94*
rufilatus, Turnix, 5, 73, 101
rufous-chested bustard-quail, *see* red-chested buttonquail
rufous-crested bustard Plate 32, 22, 24*, 39, 44, 106, *189–94*, 195
Ruppell's bustard Plate 30, 24*, 39, 44, 58, 106, 179, 180*, *184–5*

salamonis, Turnix 85
satiratior, Pterocles 255
saturata, Turnix 5, 84, 85
saturatus, Pterocles 240
savilei, Lophotis 189, 190, 191, 193
Savile's bustard, *see* rufous-crested bustard
savuensis, Turnix 85
scintillans, Turnix 94, 95
scrub quail, *see* painted buttonquail
Senegal bustard, *see* white-bellied bustard
senegalensis, Eupodotis 9, *179–80*, 195
senegalensis, Pterocles 12
senegallus, Pterocles 12, 13, *217–21*, 226, 230, 236, 241, 242

Small buttonquail, *see* striped buttonquail
Smith's buttonquail, *see* striped buttonquail
Smith's sandgrouse, *see* yellow-throated sandgrouse
Somali black-throated bustard, *see* little brown bustard
somaliensis, Eupodotis 176
Somali little bustard, *see* little brown bustard
South African buttonquail, *see* Hottentot buttonquail
speckled quail, *see* painted buttonquail
spotted buttonquail Plate 3, 22*, 43, *72–4*
spotted sandgrouse, *see* Burchell's sandgrouse
stanleyi, Neotis 158
Stanley's bustard, *see* Denham's bustard
stieberi, Ardeotis 150, 151
striped buttonquail Plate 6, 22*, 23*, 36, 38, 39, 43, 45, 46, 67, *79–84*
struthiunculus, Ardeotis 150, 153
sukensis, Pterocles 246–7
suluensis, Turnix 57, 79, 247
Sumba buttonquail, *see* red-chested buttonquail
sumbana, Turnix 85, 86
Sumbawa buttonquail, *see* barred buttonquail
Sunda buttonquail, *see* barred buttonquail
suscitator, Turnix 5, 6, 72, 73, 76, 100, *101–4*
swift bustard-quail, *see* little buttonquail
sylvatica, Turnix 4, 5. 6, 74, 76, 78, *79–84*, 87
Sypheotides 7, 23, 106, 207
Syrrhaptes 12, 21, 27, 53, 209, *210–17*

taigoor, Turnix 5, 73, 101
tanki, Turnix 5, 6, *74–7*, 86, 87
tarda, Otis 9, 124, *125–37*
tarquis, Pterocles 246, 247
Temminck's bustard-quail, *see* red-backed buttonquail

Tetrax 7, 8, 19, 20, 113, *116–25*
tetrax, Tetrax 9, *116–25*
thai, Turnix 73, 101
tibetana, Syrrhaptes 12, 13, 215, *216–17*
Tibetan sandgrouse Plate 36, 28*, 29*, 44, 53, 209, 210*, 212, *216–17*
Trachelotis 106
Turnix 1, 2, 17, 21, 36, 38, 39, 42, 67, 69, *70–100*

undulata, Chlamydotis 9, *106–16*
usheri, Pterocles 249

varia, Turnix 5, 6, 98, *94–6*
varied turnix, *see* painted buttonquail
variegated bustard-quail, *see* painted buttonquail
variegated sandgrouse, *see* Burchell's sandgrouse
vastitas, Pterocles 240
velox Turnix 5, 6, *87–9*, 94
Vigor's bustard, *see* black-throated bustard
vigorsii, Eupodotis 9, *182–4*, 195

white-bellied bustard Plate 27, 24*, 39, 44, 106, *179–80*, 181
whiteheadi, Turnix, 74, 79, 80, 89, 92*
white-quilled korhaan, *see* little black bustard
worcesteri, Turnix 4, 57, 67, 74, 89, 90, 91, 92* 94
Worcester's buttonquail, *see* red-chested buttonquail

yellow-legged buttonquail Plate 4, 22*, 38, 43, 45, 67, *74–7*
yellow quail, *see* red-chested buttonquail
yellow-throated sandgrouse Plate 5, 26*, 27, 28*, 29*, 33*, 44, 52, 213, *255–8*
yorki, Turnix 85